# Molecular Ecology

W9-BDL-839

# Molecular Ecology

**Joanna R. Freeland**
*The Open University, Milton Keynes*

John Wiley & Sons, Ltd

Copyright © 2005    John Wiley & Sons Ltd, The Atrium, Southern Gate, Chichester,
West Sussex PO19 8SQ, England

Telephone    (+44) 1243 779777

Email (for orders and customer service enquiries): cs-books@wiley.co.uk
Visit our Home Page on www.wileyeurope.com or www.wiley.com

Reprinted October 2006, April 2007

**Other Wiley Editorial Offices**

John Wiley & Sons Inc., 111 River Street, Hoboken, NJ 07030, USA

Jossey-Bass, 989 Market Street, San Francisco, CA 94103-1741, USA

Wiley-VCH Verlag GmbH, Boschstr. 12, D-69469 Weinheim, Germany

John Wiley & Sons Australia Ltd, 42 McDougall Street, Milton, Queensland 4064, Australia

John Wiley & Sons (Asia) Pte Ltd, 2 Clementi Loop #02-01, Jin Xing Distripark, Singapore 129809

John Wiley & Sons Canada Ltd, 22 Worcester Road, Etobicoke, Ontario, Canada M9W 1L1

Wiley also publishes its books in a variety of electronic formats. Some content that appears in print may not
be available in electronic books.

**Library of Congress Cataloging-in-Publication Data**

Freeland, Joanna.
    Molecular ecology / Joanna Freeland.
        p. cm.
    Includes bibliographical references.
    ISBN-13 978-0-470-09061-9 (HB)      ISBN-13 978-0-470-09062-6 (PB)
    ISBN-10 0-470-09061-8 (HB)          ISBN-10 0-470-09062-6 (PB)
        1.  Molecular ecology.  I. Title.
    QH541.15.M63F74 2005
    577–dc22                                          2005027904

**British Library Cataloguing in Publication Data**

A catalogue record for this book is available from the British Library

ISBN-13 978-0-470-09061-9 (H/B)    ISBN-13 978-0-470-09062-6 (P/B)

Typeset in 10.5/13pt Minion-Regular by Thomson Press (India) Limited, New Delhi, India.
Printed and bound in Great Britain by Antony Rowe Ltd, Chippenham, Wiltshire.
This book is printed on acid-free paper responsibly manufactured from sustainable forestry
in which at least two trees are planted for each one used for paper production.

# Contents

## 6 Molecular Approaches to Behavioural Ecology     201

# Preface

The theory and practice of molecular ecology draw on a number of subjects, particularly genetics, ecology and evolutionary biology. Although the foundations of molecular ecology are not particularly new, it did not emerge until the 1980s as the discipline that we now recognize. Since that time the growth of molecular ecology has been explosive, in part because molecular data are becoming increasingly accessible and also because it is, by its very nature, a collaborative discipline. Molecular ecology is now a broad area of research that embraces topics as varied as population genetics, conservation genetics, molecular evolution, behavioural ecology and biogeography, and has added much to our understanding of ecology. Researchers in molecular ecology now routinely publish their work in a wide range of ecological and evolutionary journals (including *Molecular Ecology*, first published in 1992), and also in more general publications such as *Science* and *Nature*.

Although somewhat varied, the areas of research within molecular ecology are united by the fact that they all use molecular genetic data to help us understand the ecology and evolution of organisms in the wild. Although there are many excellent texts that cover general ecology and evolution, there is currently a shortage of books that provide a comprehensive overview of molecular ecology. The most important goal of this book, therefore, has been to present molecular genetics, population genetics and applied molecular ecology in a logical and uncomplicated – but not oversimplified – manner, using up-to-date examples from a wide range of taxa. This text is aimed at upper-level undergraduate and postgraduate students, as well as at researchers who may be relatively new to molecular ecology or are thinking about different ways to address their research questions using molecular data.

Each chapter may be read in isolation, but there is a structure to the book that should be particularly useful to students who read the text in its entirety. The first two chapters provide a brief history of molecular ecology and a review of genetics, followed by an overview of molecular markers and the types of data they generate. Chapters 3 and 4 then build on this foundation by looking at ways in which molecular data can be used to characterize single and multiple populations. Having read Chapters 1–4, readers should have a good understanding of the relevant theory and practice behind molecular markers and population genetics.

Chapter 5 then builds on this by adding an explicit evolutionary component within the context of phylogeography. Chapters 6 and 7 then focus on two additional, specific applications of molecular ecology, namely behavioural ecology and conservation genetics. Finally, chapter 8 provides a more general overview of the practical applications of molecular ecology, paying particular attention to questions surrounding law enforcement, agriculture and fishing, which will be of interest to biologists and non-biologists alike.

As an aid to the reader, each chapter is followed by a summary, a list of useful websites and software and some recommended further reading. Suggestions for further reading also can, of course, come from the extensive reference list at the end of the book. There are review questions after each chapter that students can use to identify key points and test their knowledge. There is also a glossary at the end of the book, and glossary words are highlighted in bold when they first appear in the text. An ongoing website (www.wiley.com/go/freeland) will be maintained upon which corrections and new developments will be reported, and from which figures that may be used as teaching aids, can be downloaded.

**Joanna Freeland**

# Acknowledgements

Many thanks to James Austin, Amanda Callaghan Colin Ferris, Hélène Fréville, Trevor Hodkinson and Steve Lougheed for reading part or all of this book and providing helpful comments. The cover photo and design concept were by Kelvin Conrad. Thanks also to James Austin, Spencer Barrett, P.G. Bentz, David Bilton, Kelvin Conrad, Mike Dodd, Claude Gascon, Beth Okamura, Kate Orr and Jon Slate for providing photos. Kelvin Conrad also helped with some figures and provided essential technical advice. This book is dedicated to Eva and William.

**Joanna Freeland**

# 1
# Molecular Genetics in Ecology

## What is Molecular Ecology?

Over the past 20 years, molecular biology has revolutionized ecological research. During that time, methods for genetically characterizing individuals, populations and species have become almost routine, and have provided us with a wealth of novel data and fascinating new insights into the ecology and evolution of plants, animals, fungi, algae and bacteria. Molecular markers allow us, among other things, to quantify genetic diversity, track the movements of individuals, measure inbreeding, identify the remains of individuals, characterize new species and retrace historical patterns of dispersal. These applications are of great academic interest and are used frequently to address practical ecological questions such as which endangered populations are most at risk, from inbreeding, and how much hybridization has occurred between genetically modified crops and their wild relatives. Every year it becomes easier and more cost-effective to acquire molecular genetic data and, as a result, laboratories around the world can now regularly accomplish previously unthinkable tasks such as identifying the geographic source of invasive species from only a few samples, or monitoring populations of elusive species such as jaguar or bears based on little more than hair or scat samples.

In later chapters we will take a detailed look at many of the applications of molecular ecology, but before reaching that stage we must first understand just why molecular markers are such a tremendous source of information. The simplest answer to this is that they generate data from the infinitely variable **deoxyribonucleic acid (DNA)** molecules that can be found in almost all living things. The extraordinarily high levels of genetic variation that can be found in most species, together with some of the methods that allow us to tap into the goldmine of information that is stored within DNA, will therefore provide the focus of this chapter. We will start, however, with a retrospective look at how

*Molecular Ecology*   Joanna Freeland
© 2005 John Wiley & Sons, Ltd.

the characterization of proteins from fruitfly populations changed forever our understanding of ecology and evolution.

## The Emergence of Molecular Ecology

Ecology is a branch of biology that is primarily interested in how organisms in the wild interact with one another and with their physical environment. Historically, these interactions were studied through field observations and experimental manipulations. These provided phenotypic data, which are based on one or more aspects of an organism's morphology, physiology, biochemistry or behaviour. What we may think of as traditional ecological studies have greatly enhanced our knowledge of many different species, and have made invaluable contributions to our understanding of the processes that maintain ecosystems.

At the same time, when used on their own, phenotypic data have some limitations. We may suspect that a dwindling butterfly population, for example, is suffering from low genetic diversity, which in turn may leave it particularly susceptible to pests and pathogens. If we have only phenotypic data then we may try to infer genetic diversity from a variable morphological character such as wing pattern, the idea being that morphologically diverse populations will also be genetically diverse. We may also use what appear to be population-specific wing patterns to track the movements of individuals, which can be important because immigrants will bring in new genes and therefore could increase the genetic diversity of a population. There is, however, a potential problem with using phenotypic data to infer the genetic variation of populations and the origins of individuals: although some physical characteristics are under strict genetic control, the influence of environmental conditions means that there is usually no overall one-to-one relationship between an organism's **genotype** (set of genes) and its **phenotype**. The wing patterns of African butterflies in the genus *Bicyclus*, for example, will vary depending on the amount of rainfall during their larval development period; as a result, the same genotype can give rise to either a wet season form or a dry season form (Roskam and Brakefield, 1999).

The potential for a single genotype to develop into multiple alternative phenotypes under different environmental conditions is known as **phenotypic plasticity**. A spectacular example of phenotypic plasticity is found in the oak caterpillar *Nemoria arizonaria* that lives in the southwest USA and feeds on a few species of oaks in the genus *Quercus*. The morphology of the caterpillars varies, depending on which part of the tree it feeds on. Caterpillars that eat catkins (inflorescences) camouflage themselves by developing into catkin-mimics, whereas those feeding on leaves will develop into twig mimics. Experiments have shown that it is diet alone that triggers this developmental response (Greene, 1996). The difference in morphology between twig-mimics and catkin-mimics is so pronounced that for many years they were believed to be two different species. There

**Table 1.1** Some examples of how environmental factors can influence phenotypic traits, leading to phenotypic plasticity

| Characteristic | Environmental influence | Example |
|---|---|---|
| Gender | Temperature during embryonic development | Eggs of the American snapping turtle *Chelydra serpentina* develop primarily into females at cool temperatures, primarily into males at moderate temperatures, and exclusively into females at warm temperatures (Ewert, Lang and Nelson, 2005) |
| Growth patterns in plants | Soil nutrients and water availability | Southern coastal violet (*Viola septemloba*) allocated a greater proportion of biomass to roots and rhizomes in poor-quality environments (Moriuchi and Winn, 2005) |
| Leaf size | Light intensity | Dandelions (*Taraxacum officinale*) produce larger leaves under conditions of relatively strong light intensity (Brock, Weinig and Galen, 2005) |
| Migration between host plants | Age and nutritional quality of host plants | Diamond-back moths (*Plutella xylostella*) are most likely to migrate as adults if the juvenile stage feed on mature plants (Campos, Schoereder and Sperber, 2004). |
| Feeding-related morphology | Food availability | Sea-urchin larvae (*Strongylocentrotus purpuratus* and *S. franciscanus*) produce longer food-gathering arms and smaller stomachs when food is scarce (Miner, 2005) |
| Plumage colouration | Carotenoids in diet | The plumage of male house finches (*Carpodacus mexicanus*) shows varying degrees of red, orange and yellow depending on the carotenoids in each bird's diet (Hill, Inouye and Montgomerie, 2002) |

is also a behavioural component to these phenotypes, because if either is placed on a part of the tree that it does not normally frequent, the catkin-mimics will seek out catkins against which to disguise themselves, and the twig-mimics will seek out leaves or twigs. Some other examples of phenotypic plasticity are given in Table 1.1.

Phenotypic plasticity can lead to overestimates of genetic variation when these are based on morphological variation. In addition, phenotypic plasticity may obscure the movements of individuals and their genes between populations if it causes the offspring of immigrants to bear a closer resemblance to individuals in their natal population than to their parents. Complex interactions between genotype, phenotype and environment provided an important reason why biologists sought long and hard to find a reliable way to genotype wild organisms; genetic data would, at the very least, allow them to directly quantify genetic variation, and to track the movements of genes – and therefore individuals or **gametes** – between populations. The first milestone in this quest occurred around 40 years ago, when researchers discovered how to quantify individual genetic variation by identifying structural differences in proteins (Harris, 1966; Lewontin and Hubby, 1966). This discovery is considered by many to mark the birth of molecular ecology.

## Protein allozymes

In the 1960s a method known as starch gel **electrophoresis** of allozymic proteins was an extremely important breakthrough that allowed biologists to obtain direct information on some of the genetic properties of individuals, populations, species and higher taxa. Note that we are not yet talking about DNA markers but about proteins that are encoded by DNA. This distinction is extremely important, and to eliminate any confusion we will take a minute to review the relationship between DNA, genes and proteins. **Prokaryotes**, which lack cell nuclei, have their DNA arranged in a closed double-stranded loop that lies free within the cell's cytoplasm. Most of the DNA within the cells of **eukaryotes**, on the other hand, is organized into **chromosomes** that can be found within the nucleus of each cell; these constitute the nuclear genome (also referred to as **nuclear DNA** or **nrDNA**). Each chromosome is made up of a single DNA molecule that is functionally divided into units called genes. The site that each gene occupies on a particular chromosome is referred to as its **locus** (plural **loci**). At each locus, different forms of the same gene may occur, and these are known as **alleles**.

Each allele is made up of a specific sequence of DNA. The DNA sequences are determined by the arrangement of four nucleotides, each of which has a different chemical constituent known as a base. The four DNA bases are adenine (A), thymine (T), guanine (G) and cytosine (C), and these are linked together by a sugar–phosphate backbone to form a strand of DNA. In its native state, DNA is arranged as two strands of complementary sequences that are held together by hydrogen bonds in a double-helix formation (Figure 1.1). No two alleles have exactly the same DNA sequence, although the similarity between two alleles from the same locus can be very high.

The function of many genes is to encode a particular protein, and the process in which genetic information is transferred from DNA into protein is known as **gene expression**. The sequence of a protein-coding gene will determine the structure of the protein that is synthesized. The first step of protein synthesis occurs when the coding region of DNA is transcribed into **ribonucleic acid** (**RNA**) through a process known as **transcription**. The RNA sequences, which are single stranded, are complementary to DNA sequences and have the same bases with the exception of uracil (U), which replaces thymine (T). After transcription, the **introns** (non-coding segments of DNA) are excised and the RNA sequences are translated into protein sequences following a process known as **translation**.

Translation is possible because each RNA molecule can be divided into triplets of bases (known as **codons**), most of which encode one of 20 different **amino acids**, which are the constituents of proteins (Table 1.2). Transcription and

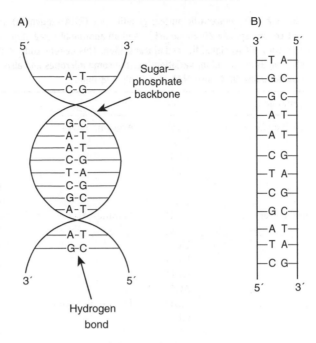

**Figure 1.1** (A) A DNA double helix. Each sequence is linked together by a sugar–phosphate backbone, and complementary sequences are held together by hydrogen bonds; 3′ and 5′ refer to the orientation of the DNA: one end of a sequence has an unreacted 5′ phosphate group and the other end has an unreacted 3′ hydroxyl group. (B) Denatured (single-stranded) DNA showing the two complementary sequences. The DNA becomes denatured following the application of heat or certain chemicals

translation involve three types of RNA: **ribosomal RNA (rRNA)**, **messenger RNA (mRNA)** and **transfer RNA (tRNA)**. Ribosomal RNA is a major component of ribosomes, which are the **organelles** on which mRNA codons are translated into proteins, i.e. it is here that protein synthesis takes place. Messenger RNA molecules act as templates for protein synthesis by carrying the protein-coding information that was encoded in the relevant DNA sequence, and tRNA molecules incorporate particular amino acids into a growing protein by matching amino acids to mRNA codons (Figure 1.2).

Specific combinations of amino acids give rise to **polypeptides**, which may form either part or all of a particular protein or, in combination with other molecules, a protein complex. If the DNA sequences from two or more alleles at the same locus are sufficiently divergent, the corresponding RNA triplets will encode different amino acids and this will lead to multiple variants of the same protein. These variants are known as **allozymes**. However, not all changes in DNA sequences will result in different proteins. Table 1.2 shows that there is some redundancy in the **genetic code**, e.g. leucine is specified by six different codons. This redundancy means that it is possible for two different DNA sequences to produce the same polypetide product.

**Table 1.2**   The eukaryotic nuclear genetic code (RNA sequences): a total of 61 codons specify 20 amino acids, and an additional three stop-codons (UAA, UAG, UGA) signal the end of translation. This genetic code is almost universal, although minor variations exist in some microbes and also in the **mitochondrial DNA (mtDNA)** of animals and fungi

| Amino acid | Codon | Amino acid | Codon |
|---|---|---|---|
| Leucine (Leu) | UUA | Arginine (Arg) | CGU |
|  | UUG |  | CGC |
|  | CUU |  | CGA |
|  | CUC |  | CGG |
|  | CUA |  | AGA |
|  | CUG |  | AGG |
| Serine (Ser) | UCU | Alanine (Ala) | GCU |
|  | UCC |  | GCC |
|  | UCA |  | GCA |
|  | UCG |  | GCG |
|  | AGU |  |  |
|  | AGC |  |  |
| Valine (Val) | GUU | Threonine (Thr) | ACU |
|  | GUC |  | ACC |
|  | GUA |  | ACA |
|  | GUG |  | ACG |
| Proline (Pro) | CCU | Glycine (Gly) | GGU |
|  | CCC |  | GGC |
|  | CCA |  | GGA |
|  | CCG |  | GGG |
| Glutamine (Gln) | CAA | Aspartic acid (Asp) | GAU |
|  | CAG |  | GAC |
| Asparagine (Asn) | AAU | Glutamic acid (Glu) | GAA |
|  | AAC |  | GAG |
| Lysine (Lys) | AAA | Cysteine (Cys) | UGU |
|  | AAG |  | UGC |
| Tyrosine (Tyr) | UAU | Histidine (His) | CAU |
|  | UAC |  | CAC |
| Isoleucine (Ile) | AUU | Phenylalanine (Phe) | UUU |
|  | AUC |  | UUC |
|  | AUA |  |  |
| Methionine (Met) | AUG[a] | Tryptophan (Trp) | UGG |

[a]Codes for Met when within the gene and signals the start of translation when at the beginning of the gene.

### Allozymes as genetic markers

The first step in allozyme genotyping is to collect tissue samples or, in the case of smaller species, entire organisms. These samples are then ground up with appropriate buffer solutions to release the proteins into solution, and the allozymes then can be visualized following a two-step process of gel electrophoresis

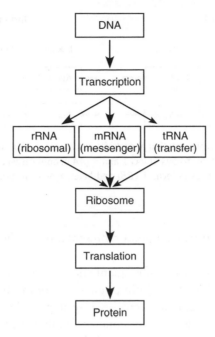

**Figure 1.2** DNA codes for RNA via transcription, and RNA codes for proteins via translation

and staining. Electrophoresis refers here to the process in which allozymes are separated in a solid medium such as starch, using an electric field. Once an electric charge is applied, molecules will migrate through the medium at different rates depending on the size, shape and, most importantly, electrical charge of the molecules, characteristics that are determined by the amino acid composition of the allozymes in question. Allozymes then can be visualized by staining the gel with a reagent that will acquire colour in the presence of a particular, active **enzyme**. A coloured band will then appear on the gel wherever the enzyme is located. In this way, allozymes can be differentiated on the basis of their structures, which affect the rate at which they migrate through the gel during electrophoresis.

Genotypes that are inferred from allozyme data provide some information about the amount of genetic variation within individuals; if an individual has only one allele at a particular locus then it is **homozygous,** but if it has more than one allele at the same locus then it is **heterozygous** (Figure 1.3). Furthermore, if enough individuals are characterized then the genetic variation of populations can be quantified and the genetic profiles of different populations can be compared. This distinction between individuals and populations will be made repeatedly throughout this book because it is fundamental to many applications of molecular ecology. Keep in mind that data are usually collected from individuals, but if the sample size from any given population is big enough then we often assume that the

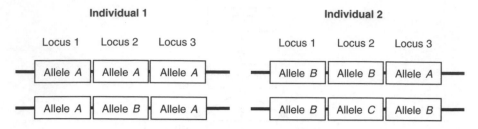

**Figure 1.3** Diagrammatic representation of part of a chromosome, showing which alleles are present at three loci. Individual 1 is homozygous at loci 1 and 3 (*AA* in both cases) and heterozygous at locus 2 (*AB*). Individual 2 is homozygous at locus 1 (*BB*) and heterozygous at locus 2 (*BC*) and locus 3 (*AB*)

individuals collectively provide a good representation of the genetic properties of that population.

We will return to allozymes in subsequent chapters, but at this point it is enough to realize that the identification within populations of multiple allozymes (alleles) at individual loci was a seminal event because it provided the first snapshot of genetic variation in the wild. In 1966, one of the first studies based on allozyme data was conducted on five populations of the fruitfly *Drosophila pseudoobscura*. This revealed substantially higher levels of genetic variation within populations than were previously believed (Lewontin and Hubby, 1966). In this study 18 loci were characterized from multiple individuals, and in each population up to six of these loci were found to be **polymorphic** (having multiple alleles). There was also evidence of genetic variation within individuals, as revealed by the **observed heterozygosity** ($H_o$) values, which are calculated by averaging the heterozygosity values across all characterized loci (Table 1.3).

Although unarguably a major breakthrough in population genetics, and still an important source of information in molecular ecology, allozyme markers do have some drawbacks. One limitation is that, as we saw in Table 1.2, not all variation in

**Table 1.3** Levels of polymorphism and observed heterozygosity ($H_o$) at 18 enzyme loci calculated for five populations of *Drosophila pseudoobscura* (data from Lewontin and Hubby, 1966). This was one of the first studies to show that genetic variation in the wild is much higher than was previously believed

| Population | Number of polymorphic loci | Proportion of polymorphic loci | Observed heterozygosity |
|---|---|---|---|
| Strawberry Canyon | 6 | 6/18 = 0.33 | 0.148 |
| Wildrose | 5 | 5/18 = 0.28 | 0.106 |
| Cimarron | 5 | 5/18 = 0.28 | 0.099 |
| Mather | 6 | 6/18 = 0.33 | 0.143 |
| Flagstaff | 5 | 5/18 = 0.28 | 0.081 |

DNA sequences will translate into variable protein products, because some DNA base changes will produce the same amino acid following translation. A wealth of information is contained within every organism's **genome**, and allozyme studies capture only a small portion of this. Less than 2 per cent of the human genome, for example, codes for proteins (Li, 1997). The acquisition of allozyme data is also a cumbersome technique because organisms often have to be killed before adequate tissue can be collected, and this tissue then must be stored at very cold temperatures (up to $-70°C$), which is a logistical challenge in most field studies. These drawbacks can be overcome by using appropriate DNA markers, which are now the most common source of data in molecular ecology because they can potentially provide an endless source of information, and they also allow a more humane approach to sampling study organisms. In the following sections, therefore, we shall switch our focus from proteins to DNA.

## An Unlimited Source of Data

Even very small organisms have extremely complex genomes. The unicellular yeast *Saccharomyces cerevisiae*, despite being so small that around four billion of them can fit in a teaspoon, has a genome size of around 12 megabases (Mb; 1 Mb = 1 million base pairs) (Goffeau *et al.*, 1996). The genome of the considerably larger nematode worm *Caenorhabditis elegans*, which is 1 mm long, is approximately 97 Mb (*Caenorhabditis elegans* Sequencing Consortium, 1998), and that of the flowering plant *Arabidopsis thaliana* is around 157 Mb (*Arabidopsis* Genome Initiative, 2000). The relatively enormous mouse *Mus musculus* contains somewhere in the region of 2600 Mb (Waterston *et al.*, 2002), which is not too far off the human genome size of around 3200 Mb (International Human Genome Mapping Consortium, 2001). Within each genome there is a tremendous diversity of DNA. This diversity is partly attributable to the incredible range of functional products that are encoded by different genes. Furthermore, not all DNA codes for a functional product; in fact, the International Human Genome Sequencing Consortium has suggested that the human genome contains only around 20 000–25 000 genes, which is not much more than the ~19 500 found in the substantially smaller *C. elegans* genome (International Human Genome Sequencing Consortium, 2004). Non-coding DNA includes **introns** (intervening sequences) and **pseudogenes** (derived from functional genes but having undergone mutations that prevent transcription).

Many stretches of nucleotide sequences are repeated anywhere from several times to several million times throughout the genome. Short, highly repetitive sequences include **minisatellites** (motifs of 10–100 bp repeated many times in succession) and **microsatellites** (repeated motifs of 1–6 bp). Another class of repetitive gene regions that has been used sometimes in molecular ecology is middle-repetitive DNA. These are sequences of hundreds or thousands of base

**Figure 1.4** Diagram showing the arrangement of the nuclear ribosomal DNA gene family as it occurs in animals. The regions coding for the 5.8S, 18S and 28S subunits of rRNA are shown by bars; NTS = non-transcribed spacer, ETS = external transcribed spacer and ITS = internal transcribed spacer. The entire array is repeated many times

pairs that occur anywhere from dozens to hundreds of times in the genome. Examples of these include the composite region that codes for nuclear ribosomal DNA (Figure 1.4). In contrast, **single-copy nuclear DNA (scnDNA)** occurs only once in a genome, and it is within scnDNA that most transcribed genes are located. The proportion of scnDNA varies greatly between species, e.g. it comprises approximately 95 per cent of the genome in the midge *Chironomus tentans* but only 12 per cent of the genome in the mudpuppy salamander *Necturus maculosus* (John and Miklos, 1988).

Although the structure and function of genes vary between species, they are typically conserved among members of the same species. This does not, however, mean that all members of the same species are genetically alike. Variations in both coding and non-coding DNA sequences mean that, with the possible exception of clones, no two individuals have exactly the same genome. This is because DNA is altered by events during replication that include recombination, duplication and mutation. It is worth examining in some detail how these occur, because if we remain ignorant about the mechanisms that generate DNA variation then our understanding of genetic diversity will be incomplete.

## Mutation and recombination

Genetic variation is created by two processes: **mutation** and **recombination**. Most mutations occur during **DNA replication**, when the sequence of a DNA molecule is used as a template to create new DNA or RNA sequences. Neither reproduction nor gene expression could occur without replication, and therefore its importance cannot be overstated. During replication, the hydrogen bonds that join the two strands in the parent DNA duplex are broken, thereby creating two separate strands that act as templates along which new DNA strands can be synthesized. The mechanics of replication are complicated by the fact that the synthesis of new strands can occur only in the 5′–3′ direction (Figure 1.5). Synthesis requires an enzyme known as **DNA polymerase**, which adds single nucleotides along the template strand in the order necessary to create a complementary sequence in which G is paired with C, and A is paired with T (or U in RNA). Successive nucleotides are added until the process is complete, by which time a single parent

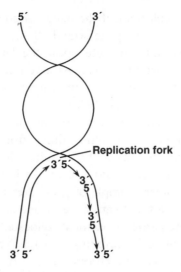

**Figure 1.5** During DNA replication, nucleotides are added one at a time to the strand that grows in a 5′ to 3′ direction. In eukaryotes, replication is bi-directional and can be initiated at multiple sites by a primer (a short segment of DNA)

DNA duplex (double-stranded segment) has been replaced by two newly synthesized daughter duplexes.

Errors in DNA replication can lead to **nucleotide substitutions** if one nucleotide is replaced with another. These can be of two types: **transitions**, which involve changes between either **purines** (A and G) or **pyrimidines** (C and T); and **transversions**, in which a purine is replaced by a pyrimidine or vice versa. Generally speaking, transitions are much more common than transversions. When a substitution does not change the amino acid that is coded for, it is known as a **synonymous substitution**, i.e. the DNA sequence has been altered but the encoded product remains the same. Alternatively, **non-synonymous substitution** occurs when a nucleotide substitution creates a codon that specifies a different amino acid, in which case the function of that stretch of DNA may be altered. Although single nucleotide changes often will have no phenotypic outcome, they can at times be highly significant. Sickle-cell anaemia in humans is the result of a single base-pair change that replaces a glutamic acid with a valine, a mutation that is generally fatal in homozygous individuals.

Errors in DNA replication also include nucleotide insertions or deletions (collectively known as **indels**), which occur when one or more nucleotides are added to, or removed from, a sequence. If an indel occurs in a coding region it will often shift the reading frame of all subsequent codons, in which case it is known as a **frameshift mutation**. When this happens, the gene sequence is usually rendered dysfunctional. Mutations can also involve **slipped-strand mis-pairing**, which sometimes occurs during replication if the daughter strand of DNA temporarily becomes dissociated from the template strand. If this occurs in a region of a repetitive sequence such as a microsatellite repeat, the daughter strand may lose its place and re-anneal to the 'wrong' repeat. As a result, the completed daughter strand will be either longer or shorter than the parent strand because it contains a different number of repeats (Hancock, 1999).

Mutations are by no means restricted to one or a few nucleotides. **Gene conversion** occurs when genotypic ratios differ from those expected under Mendelian inheritance, an aberration that results when one allele at a locus apparently converts the other allele into a form like itself. In the 1940s, Barbara McClintock discovered another example of gene alterations called **transposable elements**, which are sequences that can move to one of several places within the genome. Not only are these particular elements relocated, but they may take with them one or more adjacent genes, resulting in a relatively large-scale rearrangement of genes within or between chromosomes. Transposable elements can interrupt function when they are inserted into the middles of other genes and can also replicate so that their transposition may include an increase in their copy number throughout the genome. Many also are capable of moving from one species to another following a process called **horizontal transfer**, a possibility that is being investigated by some researchers interested in the potential hazards associated with genetically modified foods.

The other key process that frequently alters DNA sequences is recombination. Most individuals start life as a single cell, and this cell and its derivatives must replicate many times during the growth and development of an organism. This type of replication is known as **mitosis**, and involves the duplication of an individual's entire complement of chromosomes – in other words the daughter cells contain exactly the same number and type of chromosomes as the parental cells. Mitosis occurs regularly within **somatic** (non-reproductive) cells.

Although necessary for normal body growth, mitosis would cause difficulties if it were used to generate reproductive cells. Sexual reproduction typically involves the fusion of an egg and a sperm to create an embryo. If the egg and the sperm were produced by mitosis then they would each have the full complement of chromosomes that were present in each parent, and the fused embryo would have twice as many chromosomes. This number would double in each generation, rapidly leading to an unsustainable amount of DNA in each individual. This is circumvented by **meiosis**, a means of cellular replication that is found only in **germ** cells (cells that give rise to eggs, sperm, ovules, pollen and spores). In **diploid** species (Box 1.1), meiosis leads to **gametes** that have only one set of chromosomes ($n$), and when these fuse they create a diploid ($2n$) embryo. During meiosis, recombination occurs when **homologous chromosomes** exchange genetic material. This leads to novel combinations of genes along a single chromosome (Figure 1.6) and is an important contributor to genetic diversity in sexually reproducing **taxa**.

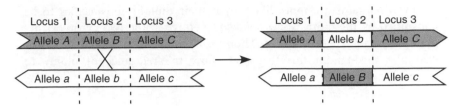

**Figure 1.6** Recombination at the gene level, after which the gene sequence at chromosome 1 changes from *ABC* to *AbC*. Recombination often involves only part of a gene, which typically leads to the generation of unique alleles

---

**Box 1.1  Chromosomes and polyploidy**

The **karyotype** (the complement of chromosomes in a somatic cell) of many species includes both **autosomes**, which usually have the same complement and arrangement of genes in both sexes, and **sex chromosomes**. The number of copies of the full set of chromosomes determines an individual's ploidy. Diploid species have two sets of chromosomes (2*n*), and if they reproduce sexually then one complete set of chromosomes will be inherited from each parent. Humans are diploid and have 22 pairs of autosomes and two sex chromosomes (either two X chromosomes in a female or one X and one Y chromosome in a male), which means that their karyotype is 2*n* = 46. **Polyploid** organisms have more than two complete sets of chromosomes. The creation of new polyploids sometimes results in the formation of new species, although a single species can comprise multiple races, or **cytotypes**. In **autopolyploid** individuals, all chromosomes originated from a single ancestral species after chromosomes failed to separate during meiosis. In this way, a diploid individual (2*n*) can give rise to a tetraploid individual (4*n*), which would have four copies of the original set of chromosomes. This contrasts with **allopolyploid** individuals, which have chromosomes that originated from multiple species following hybridization.

Polyploidy is very common in flowering plants and also occurs to a lesser degree in fungi, vertebrates (primarily fishes, reptiles and amphibians) and invertebrates (including insects and crustaceans). Polyploidy is of ecological interest for a number of reasons, for example newly formed polyploids may either outcompete their diploid parents or co-exist with them by exploiting an alternative habitat. Habitat differentiation among cytotypes of the same species has been documented in a number of plant species. Ecological differences between cytotypes also may depend on unrelated species, for example tetraploid individuals of the alumroot plant *Heuchera grossulariifolia* living in the

Rocky Mountains are more likely than their diploid conspecifics to be consumed by the moth *Greya politella*, even when the two cytotypes are living together (Nuismer and Thompson, 2001). There will be other examples throughout this text that show the relevance of ploidy to molecular ecology.

## Neutralists and selectionists

Prior to the 1960s, most biologists believed that a genetic mutation would either increase or decrease an individual's **fitness** and therefore mutations were maintained within a population as a result of natural selection (the selectionist point of view). However, many people felt that this theory became less plausible following the discovery of the high levels of genetic diversity in natural populations that were revealed by allozyme data in the 1960s, since there was no obvious reason why natural selection should maintain so many different genotypes within a population. At this time the Neutral Theory of Molecular Evolution began to take shape (Kimura, 1968). This proposed that although some mutations confer a selective advantage or disadvantage, most are neutral or nearly so, that is to say they have no or little effect on an organism's fitness. The majority of genetic polymorphisms therefore arise by chance and are maintained or lost as a result of random processes (the neutralist point of view). For a while, reconciliation between selectionists and neutralists seemed unlikely, but the copious amount of genetic data that we now have access to suggests that molecular change can be attributed to both random and selective processes. As a result, many well-supported theories of molecular evolution and population genetics now embrace elements of both neutralist and selectionist theories (Li and Graur, 1991).

There are a number of predictions that can be made about mutation rates under the neutral theory. For example, synonymous substitutions should accumulate much more rapidly than non-synonymous substitutions because they are far less likely to cause phenotypic changes. In general, this prediction has been borne out. Data from 32 *Drosophila* genes revealed an average synonymous substitution rate of 15.6 substitutions per site per $10^9$ years compared with an average non-synonymous substitution rate of 1.91; similarly, the synonymous substitution rate averaged across various mammalian protein-coding genes was 3.51 compared with a rate of 0.74 in non-synonymous substitutions (Li, 1997, and references therein). As we may expect under the neutral theory, mutations tend to accumulate more rapidly in introns (non-coding regions) compared with exons (non-coding regions), and pseudogenes appear to have higher substitution rates compared with functional genes, although this conclusion is based on limited data (Li, 1997).

A combination of chance and natural selection means that a proportion of mutations will inevitably be maintained within a species and this accumulation of mutations, along with recombination, means that even members of the same species often have fairly divergent genomes. Overall, around 0.1 per cent of the human genome (approximately three million nucleotides) is variable (Li and Sadler, 1991), compared with around 0.67 per cent of the rice (*Oryza sativa*) genome (Yu *et al.*, 2002). In molecular ecology, studies are typically based on multiple individuals from one or more populations of the species in question, and overall levels of sequence variability are usually expected to be around 0.2 – 0.5 per cent (Fu and Li, 1999), although this may be considerably higher depending on the gene regions that are compared. Sequence divergence also tends to be higher between more distantly related groups, and therefore comparisons of populations, species, genera and familes will often show increasingly disparate genomes, although there are exceptions to this rule (Figure 1.7). Part of the challenge to finding suitable genetic markers for ecological research involves

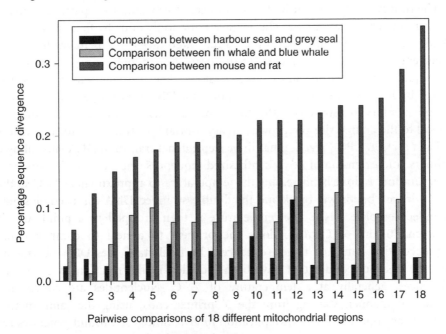

Pairwise comparisons of 18 different mitochondrial regions

**Figure 1.7** Sequence divergence based on pairwise comparisons of 18 different randomly numbered regions of mtDNA for members of two different genera from the same family (harbour seal *Phoca vitulina* and grey seal *Halichoerus grypus*); members of the same genus (fin whale *Balaenoptera physalus* and blue whale *B. musculus*); and members of two different families (mouse *Mus musculus* and rat *Rattus norvegicus*). As we might expect, sequence divergences are highest in the comparison between families (mouse and rat). However, contrary to what we might expect, the congeneric whale species are genetically less similar to one another than are the two seal genera. This is an example of how taxonomic relationships do not always provide a useful guide to overall genetic similarities. Data from Lopez *et al.* (1997) and references therein

identifying which regions of the genome have levels of variability that are appropriate to the questions being asked.

## Polymerase chain reaction

A wealth of information in the genome is of no use to molecular ecologists if it cannot be accessed and quantified, and after 1985 this became possible thanks to Dr Kary Mullis, who invented a method known as the **polymerase chain reaction** (**PCR**) (Mullis and Faloona, 1987). This was a phenomenal breakthrough that allowed researchers to isolate and amplify specific regions of DNA from the background of large and complex genomes. The importance of PCR to many biological disciplines, including molecular ecology, cannot be overstated, and its contributions were recognized in 1993 when Mullis was one of the recipients of the Nobel Prize for Chemistry.

The beauty of PCR is that it allows us to selectively amplify a particular area of the genome with relative ease. This is most commonly done by first isolating total DNA from a sample and then using paired **oligonucleotide primers** to amplify repeatedly a target DNA region until there are enough copies to allow its subsequent manipulation and characterization. The primers, which are usually 15–35 bp long, are a necessary starting point for DNA synthesis and they must be complementary to a stretch of DNA that flanks the target sequence so that they will anneal to the desired site and provide an appropriate starting point for replication.

Each cycle in a PCR reaction has three steps: denaturation of DNA, annealing of primers, and extension of newly synthesized sequences (Figure 1.8). The first step, denaturation, is done by increasing the temperature to approximately 94°C so that the hydrogen bonds will break and the double-stranded DNA will become single-stranded template DNA. The temperature is then dropped to a point, usually between 40 and 65°C, that allows the primers to anneal to complementary sequences that flank the target sequence. The final stage uses DNA polymerase and the free nucleotides that have been included in the reaction to extend the sequences, generally at a temperature of 72°C. Nucleotides are added in a sequential manner, starting from the 3′ primer ends, using the same method that is used routinely for DNA replication *in vivo*. Since each round generates two daughter strands for every parent strand, the number of sequences increases exponentially throughout the PCR reaction. A typical PCR reaction follows 35 cycles, which is enough to amplify a single template sequence into 68 billion copies!

These days, PCR reactions use a heat-stable polymerase, most commonly *Taq* polymerase, so called because it was isolated originally from a bacterium called *Thermus aquaticus* that lives in hot springs. Since *Taq* is not deactivated at high temperatures, it need be added only once at the beginning of the reaction, which runs in computerized thermal cyclers (PCR machines) that repeatedly cycle through different temperatures. Some optimization is generally required when

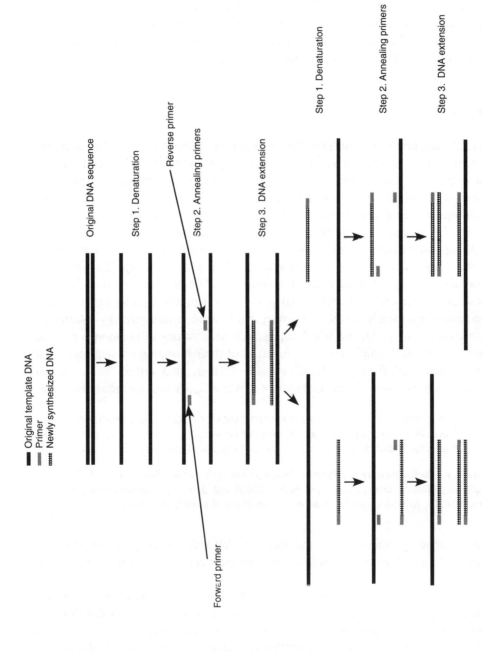

**Figure 1.8** The first two cycles in a PCR reaction. Solid black lines represent the original DNA template, short grey lines represent the primers and hatched lines represent DNA fragments that have been synthesized in the PCR reaction

starting with new primers or targeting the DNA of multiple species, such as altering the annealing temperature or using different salt concentrations to sustain polymerase activity. However, once this has been done, all the researcher has to do is set up the reactions, program the machine and come back when all the cycles have been completed, which usually takes 2–3 h. By this time copies of the target region will vastly outnumber any background non-amplified DNA and the final product can be characterized in one of several different ways, some of which will be outlined later in this chapter, and also in Chapter 2.

## *Primers*

An extremely important consideration in PCR is the sequence of the DNA primers. Primers can be classified as either universal or species-specific. Universal primers will amplify the same region of DNA in a variety of species (although, despite their name, no universal primers will work on all species). This is possible because **homologous sequences** in different species often show a degree of similarity to one another because they are descended from the same ancestral gene. Examples of homology can be obtained easily by searching a databank such as those maintained by the National Centre of Biotechnology Information (NCBI), the European Molecular Biology Laboratory (EMBL), and the DNA Data Bank of Japan (DDBJ). These are extremely large public databanks that contain, among other things, hundreds of thousands of DNA sequences that have been submitted by researchers from around the world, and which represent a wide variety of taxonomic groups. Figure 1.9 shows the high sequence similarity of three homologous sequences that were downloaded from the NCBI website. Primers that anneal to conserved regions such as those shown in Figure 1.9 will amplify specific gene sequences

mouse   **CGGTCGA**ACTTGACTAT**C**TAGAGGAAGTAAAAGTCGTAA**C**AAGGTTTCCGTAGGTGAACCTGCGGAAGGA
frog      **CGATCAA**ACTTGACTAT**C**TAGAGGAAGTAAAAGTCGTAA**C**AAGGTTTCCGTAGGTGAACCTGCGGAAGGA
chicken  **GAGTGTGA**ATTGAGTAT**G**TAGAGGAAGTAAAAGTCGTAA**G**AAGGTTTCCGTAGGTGAACCTGCGGAAGGA

**Figure 1.9**   Partial sequence of 28S rDNA showing homology in mouse (*Mus musculus*; Michot *et al.*, 1982), frog (*Xenopus laevis*; Furlong and Maden, 1983) and chicken (*Gallus domesticus*; Azad and Deacon, 1980). All sites except for those in bold are identical in all three species

from a range of different taxa and therefore fall into the category of universal primers. Table 1.4 shows a sample of universal primers and the range of taxa from which they will amplify the target sequence.

Universal primers are popular because often they can be used on species for which no previous sequence data exist. They can also be used to discriminate among individual species in a composite sample. For example, samples taken from soil, sediment or water columns generally will harbour a microbial community. We may wish to identify the species within this community so that we can address

**Table 1.4** Some examples of universal primers and the range of taxa in which they have been used successfully. Forward and reverse primers anneal to either side of the target DNA sequence (see Figure 1.8)

| Sequences of primer pairs | Region amplified | Taxonomic applications | Reference |
|---|---|---|---|
| Forward primer[a]: <br> ACATCKARTACKGGACCAATAA <br> Reverse primer[a]: <br> AACACCAGCTTTRAATCCAA | A non-coding spacer in **chloroplast DNA** | Mosses, ferns, coniferous plants, flowering plants | Chiang, Schaal and Peng (1998) |
| Forward primer: <br> TTTCATGTTTCCTTGCGGTAC <br> Reverse primer: <br> AAAGCACGGCACTGAAGATGC | Portion of mitochondrial 12S rRNA | Fishes, amphibians, reptiles, birds, mammals | Wang *et al.* (2000) |
| Forward primer: <br> TACACACCGCCCGTCGCTACTA <br> Reverse primer: <br> ATGTGCGTTCA/GAAATGTCGATGTTCA | Internal transcribed spacer (ITS) of nuclear rRNA | Insects, arthropods, fishes | Ji, Zhang and He (2003) |
| Forward primer: <br> TTCCCCGGTCTTGTAAACC <br> Reverse primer: <br> ATTTTCAGTGTCTTGCTTT | Mitochondrial control region | Mammals, fishes | Hoelzel, Hancock and Dover (1991) |
| Forward primer: <br> CCATCCAACATCTCAGCATGAAA <br> Reverse primer: <br> CCCCTCAGAATGATATTTGTCCTCA | Portion of mitochondrial cytochrome *b* | Mammals, birds, amphibians, reptiles, fishes | Kocher *et al.* (1989) |

[a]K = G or T; R = A or G.

ecological questions such as bacterial nutrient cycling, or identify any bacteria that may pose a health risk. Composite extractions of microbial DNA can be characterized using universal primers that bind to the 16S rRNA gene of many prokaryotic microorganisms or the 18S rRNA gene of numerous eukaryotic microorganisms (Velazquez *et al.*, 2004). In one study, the generation of species-specific rRNA PCR products enabled researchers to identify the individual species that make up different soil and rhizosphere microbial communities (Kent and Triplett, 2002).

Unlike universal primers, species-specific primers amplify target sequences from only one species (or possibly a few species if they are closely related). They can be designed only if relevant sequences are already available for the species in question. One way to generate species-specific primers is to use universal primers to initially amplify the product of interest and then to sequence this product (see below). By aligning this sequence with homologous sequences obtained from public databases, it is possible to identify regions that are unique to the species of interest. Primers can then be designed that will anneal to these unique regions. Although their initial development requires more work than universal primers, species-specific primers will decrease the likelihood of amplifying undesirable DNA (contamination).

There are two ways in which contamination can occur during PCR. The first is through improper laboratory technique. When setting up PCR reactions, steps should be taken at all times to ensure that no foreign DNA is introduced. For example, disposable gloves should be worn to decrease the likelihood of researchers inadvertently adding their own DNA to the samples, and equipment and solutions should be sterilized whenever possible. The other possible source of contamination is in the samples themselves. Leaf material should be examined carefully for the presence of invertebrates, fungi or other possible contaminants. If the entire bodies of small invertebrates are to be used as a source of DNA then they should be viewed under a dissecting scope to check for visible parasites and, if possible, left overnight to expel the stomach contents. Fortunately most parasites, predators and prey are not closely related to the species of interest and therefore their sequences should be divergent enough to sound alarm bells if they are amplified in error.

Despite potential problems with contamination, PCR is generally a robust technique and it is difficult to overstate its importance in molecular ecology. The ability to amplify particular regions of the genome has greatly contributed to the growth of this discipline. Futhermore, because only a very small amount of template DNA is required for PCR, we can genetically characterize individuals from an amazingly wide range of samples, many of which can be collected without causing lasting harm to the organism from which they originated.

## Sources of DNA

There are many different methods for extracting DNA from tissue, blood, hair, feathers, leaves, roots and other sources. In recent years kits have become widely available and reasonably priced, and as a result the extraction of DNA from many different sample types is often a fairly routine procedure. The amount of starting material can be very small; because a successful PCR reaction can be accomplished with only a tiny amount of DNA, samples as small as a single hair follicle may be adequate (Box 1.2) and therefore lethal sampling of animals is no longer necessary before individuals can be characterized genetically. Examples of non-lethal samples that have been used successfully for DNA analysis include wing tips from butterflies (Rose, Brookes and Mallet, 1994), faecal DNA from elusive species such as red wolves and coyotes (Adams, Kelly and Waits, 2003), single feathers from birds (Morin and Woodruff, 1996) and single scales from fish (Yue and Orban, 2001). Apart from the obvious humane considerations, this has been incredibly useful for conservation studies that require genetic data without reducing the size of an endangered population. When working with small samples, however, particular care must be taken to avoid contamination, because very small amounts of target DNA can be overwhelmed easily by 'foreign' DNA.

From a practical perspective, the storage of samples destined for PCR is relatively easy during field trips; whereas material suitable for allozyme analysis will often require surgery or even sacrifice of the entire animals, and then must be stored in liquid nitrogen or on dry ice until it is brought back to the laboratory, samples for PCR analysis can be removed without dissection and can be stored either as dried specimens or in small vials of 70–95 per cent ethanol or buffer that can be kept at room temperature. One thing to watch for, however, is the problem of degraded DNA that can reduce the efficacy of PCR. The DNA in freshly harvested blood or tissue will remain in good condition provided that it is placed quickly into suitable buffer or ethanol, but improperly stored DNA will degrade rapidly as the DNA molecules become fragmented. The DNA extracted from a non-living sample, such as faecal material or museum specimens, will already be at least partially degraded. If the DNA fragments in a degraded sample are smaller than the size of the desired region, then amplification will be impossible and therefore relatively short DNA sequences should be targeted.

---

### Box 1.2 Novel sources of DNA

One source of material for PCR, which would have been assigned to the realm of science fiction before the 1980s, is ancient DNA from samples that are thousands of years old. Most fossils do not contain any biological material and therefore do not yield any DNA, but organisms that have been preserved in arid conditions or in sealed environments such as ice or amber may retain DNA fragments that are large enough to amplify using PCR (Landweber, 1999). However, even if some genetic material has been preserved, characterizing ancient DNA is never straightforward because there is typically very little material to work with. This makes amplification problematic, particularly if the degraded DNA fragments are very short. Chemical modifications also may interfere with the PCR reaction (Landweber, 1999).

Even if amplification is possible, the risk of contamination is quite high because foreign DNA such as fungi or bacteria that invaded the organism after death may be more abundant than the target DNA. Furthermore, as is always the case with very small samples of DNA, contamination from modern sources, including humans, can be a problem. In 1994 Woodward and colleagues claimed to have sequenced the DNA from dinosaur bones that were 80 million years old (Woodward, Weyond and Bunnell, 1994), but further investigation showed that the most likely source of this DNA was human contamination (Zischler *et al.*, 1995). Nevertheless, characterization of ancient DNA has been successful on numerous occasions, for example DNA sequences from 3000-year-old moas provided novel insight into the evolution of flightless birds (Cooper *et al.*, 1992), and ancient DNA from Neanderthals lends support to the theory that modern humans originated in Africa relatively recently (Krings *et al.*, 1997).

Other relatively unusual sources of DNA for PCR amplification that are particularly useful in molecular ecology include: faeces, hair and urine, all of which have been used to genotype elusive species such as wolves, lynxes and wombats (Sloane *et al.*, 2000; Valiere and Taberlet, 2000; Pires and Fernandes, 2003); sperm collected from the membranes of birds' eggs, which provides a novel view of some aspects of mating behaviour (Carter, Robertson and Kempenaers, 2000); gut contents from ladybirds and lacewings that were analysed to determine which species of aphids they had consumed (Chen et al., 2000); museum specimens, which may be particularly useful when characterizing widespread species or locally extinct populations (Lodge and Freeland, 2003; Murata *et al.*, 2004); and ancient fungi from glacial ice cores up to 140 000 years old, which can provide data on fungal ecology and evolution (Ma *et al.*, 2000).

## Getting data from PCR

Once a particular gene region has been amplified from the requisite number of samples, a genetic identity must be assigned to each individual. The simplest way to do this is from the size of the amplified product, which can be quantified by running out the completed PCR reaction on an agarose or acrylamide gel following the same principle of gel electrophoresis that is used for allozymes. The gel solutions are made up in liquid form, and combs are left in place while the gel is hardening, after which time the combs are removed to leave a solid gel with a number of wells at one end. The DNA samples are loaded into the wells and an electrical field is applied. The DNA molecules are negatively charged and therefore will migrate towards the positive electrode.

The speed at which fragments of DNA migrate through electrophoresis gels depends primarily on their size, with the largest fragments moving most slowly. Once DNA fragments have segregated across a gel, they can be visualized using a dye called ethidium bromide, which binds to DNA molecules and can be seen with the human eye when illuminated with short-wavelength ultraviolet light. The sizes of DNA bands then can be extrapolated from ladders that consist of DNA fragments of known sizes (Figure 1.10). If the amplified products are of variable

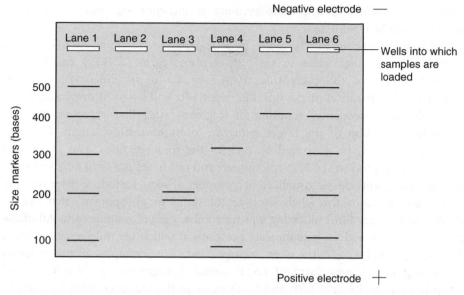

**Figure 1.10** Representation of an agarose gel after DNA fragments have been stained with ethidium bromide. Lanes 1 and 6 are size markers; note that the smaller fragments migrate through the gel more rapidly, and therefore further, than the larger fragments, which is why fragments of different sizes will separate at a predictable rate. The samples in lanes 2 and 5 have a single band of just over 400 bases long. The sample in lane 3 has two bands that are both close to 200 bases long, and in lane 4 the two bands are close to 100 and 300 bases long

sizes then at this stage we may be able to assign individual genetic identities. However, it is often the case that sequences with different compositions will be of the same length, in which case sequencing the PCR products will allow us to identify different genotypes.

*DNA Sequencing*

The most common method of DNA sequencing, known as dideoxy or sanger sequencing, was invented by Frederick Sanger in the mid-1970s (work that helped him to win a shared Nobel prize in 1980). His protocol was designed to synthesize a strand of DNA using a DNA polymerase plus single nucleotides in a manner analogous to PCR, but with two important differences. First, only a single primer is used as the starting point for synthesis so that sequences are built along the template in only one direction. Second, some of the nucleotides contain the sugar **dideoxyribose** instead of deoxyribose, the sugar normally found in DNA. Dideoxyribose lacks the 3'-hydroxyl group found in deoxyribose, and without this the next nucleotide cannot be added to the growing DNA strand; therefore, whenever a nucleotide with dideoxyribose is incorporated into the reaction, synthesis will be terminated.

Dideoxy sequencing can be done in four separate reactions, each of which include all four nucleotides in their deoxyribose form (dNTPs) and a small amount of one of the nucleotides (G,A,T or C) in its dideoxyribose form (ddNTP). Incorporation of the ddNTPs eventually will occur at every single site along the DNA sequence, resulting in fragment sizes that represent the full spectrum from 1 bp of the target sequence to its maximum length. Different fragment sizes will be generated by each of the four reactions, and in manual sequencing the products of each reaction are run out in separate but adjacent lanes on a gel. Fragments can be visualized in a number of ways, including silver staining or the use of radioactive labels (isotopes of sulfur or phosphorus) that can be developed on x-ray films following a process known as autoradiography. All of the fragment sizes in a given lane indicate positions at which the dideoxyribose bases for that particular reaction were incorporated. For example, if the reaction containing the dideoxy form of dATP contains fragments that are 1, 5, and 10 bp long, then the first, fifth and tenth bases in the sequence must be adenine (A). The fragments from each of the four reactions can be pieced together to recreate the entire sequence (Figure 1.11).

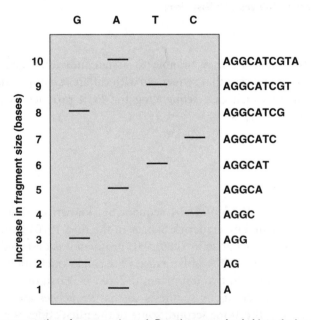

**Figure 1.11**   Representation of a sequencing gel. Reactions were loaded into the lanes labeled G, A, T and C, depending on which nucleotide was present in the dideoxyribose form. Because the smallest fragments migrate most rapidly, we can work from the bottom to the top of the gel to generate the cumulative sequences that are shown on the right-hand side

Although manual dideoxy sequencing was the norm for a number of years, it is now being replaced in many universities and research institutions by automated sequencing. Many brands and models of automated sequencers are currently available but the principle remains the same in all. The different fragments that make up a sequence are generated as before, but the nucleotides that contain dideoxyribose are labelled with different-coloured fluorescent dyes. The reactions do not need to be kept separate in the same way as they do with manual sequencing, because different colours represent the size fragments that were terminated by each type of ddNTP. When these reactions are run out on automated sequencers, lasers activate the colour of the fluorescent label of each band (typically black for G, green for A, red for T and blue for C). Each colour is then read by a photocell and stored on a computer file that records the fragments as a series of different-coloured peaks. By substituting the appropriate base for each coloured peak, the entire sequence can be read from a single image.

### Real-time PCR

So far we have looked at how PCR can provide valuable genotypic information from the sizes or sequences of amplified products. One thing that conventional PCR cannot do, however, is to supply us with accurate estimates of the amount of DNA that is present in a particular sample. This is because there is no correspondence between the amount of template at the start of the reaction and the amount of DNA that has been amplified by the end of the reaction. In the 1990s, however, a technique known as **real-time PCR** (RT-PCR, also known as **quantitative PCR**) was developed, and this does allow researchers to quantify the amount of DNA in a particular sample.

Real-time PCR allows users to monitor a PCR reaction in 'real time', i.e. as it occurs, instead of waiting until all of the cycles in a PCR reaction have finished. The fragments produced in RT-PCR are labelled by either fluorescent probes or DNA binding dyes and can be quantified after each cycle. There is a correlation between the first significant increase in the amount of PCR product and the total amount of the original template. Real-time PCR can quantify DNA or RNA in either an absolute or a relative manner. Absolute quantification determines the number of copies that have been made of a particular template, usually by comparing the amount of DNA generated in each cycle to a standard curve based on a sample of known quantity. Relative quantification allows the user to determine which samples have more or less of a particular gene product.

There are several ways in which quantitative PCR can benefit ecological studies. For one thing, it can provide important insight into the relationship between gene expression and the development of particular phenotypic attributes. Since RNA is transcribed only during gene expression, the amount of RNA in a sample is indicative of the amount of gene expression that is taking place. Researchers can synthesize DNA from an RNA template using the enzyme reverse transcriptase (such 'reverse-engineered' DNA is called **complementary DNA, or cDNA**). Real-time PCR then can use cDNA as a basis for quantifying gene expression because the amount of cDNA in a sample will be directly proportional to the amount of gene expression that has occurred. Quantitative PCR has been used to identify overexpression of 19 different genes in oysters (*Crassostrea virginica* and *C. gigas*) that had been infected with the protozoan pathogen *Perkinsus marinus*, compared with those that were uninfected (Tanguy, Guo and Ford, 2004), and also to identify variable levels of expression in several key genes that promote salt tolerance in the Euphrates poplar tree *Populus euphratica* (Gu *et al.*, 2004).

Another application for Real-Time PCR in molecular ecology involves estimation of the numbers (as opposed to simply the identities) of different species within a composite DNA sample. This was done in a study of two species of branching corals, *Acropora tenuis* and *A. valida*, living in the Great Barrier Reef. Researchers wished to identify which species of symbiotic algae (zooanthellae) were living within the coral colonies. The identity of these zooanthellae is of interest because they are apparently essential for the maintenance of healthy shallow tropical coral reefs. Bleaching, which is a major threat to coral reefs, occurs when the symbiotic algae living in coral die or lose their pigment because of stresses such as elevated sea temperatures or pollutants, and the resistance of coral reefs to bleaching may be influenced by which species of algae live within coral colonies.

In this particular study the researchers extracted DNA from the coral colonies and used primers specific for different species of the zooanthellae genus *Symbiodinium* to identify which zooanthellae were living within different colonies. By using quantitative PCR they were able to determine not only which *Symbiodinium* species were present but also the extent to which various *Symbiodinium* species occurred in different coral colonies. These data revealed a relationship between the abundance of *Symbiodinium* species and the availability of light, which suggests that local adaptation plays a role in the distribution of genetically distinct zooanthellae (Ulstrup and Van Oppen, 2003).

## Overview

In this chapter we summarized why the application of molecular data to ecological studies has been so important. We have also considered why DNA is so variable, both within and among species. Now that we know how techniques such as PCR and sequencing allow us to tap into some of the information that is stored within genomes, we will build on this information in the next chapter by taking a more detailed look at the properties of the different genomes and genetic markers that are used in molecular ecology.

# Chapter Summary

- Before the emergence of molecular ecology it was very difficult to obtain genetic data from wild populations, and biologists often had to rely on visible polymorphisms. Phenotypic data are useful for many things, although phenotypic plasticity often obscures the relationships between phenotypes and genotypes.

- The first studies to link molecular genetics and ecology were based on allozyme data. Proteins are encoded by DNA, and therefore reflect some of the variation in DNA sequences. Allozymes provided some of the first population genetics data, and in early studies these data revealed significantly higher levels of genetic variation than had been anticipated.

- Although undoubtedly a major breakthrough, redundancy in the genetic code and relevance to only a small portion of the genome mean that estimates of genetic diversity based on allozyme data will be conservative. Furthermore, collecting samples for allozyme analysis is often problematic.

- Many different types of gene regions exist within a genome, both coding (repetitive and single copy) and non-coding (including introns and pseudogenes) regions.

- Genetic diversity is generated continually through recombination and mutations, which include slipped-strand mis-pairings, nucleotide insertions/deletions and nucleotide substitutions.

- These days, laboratories routinely use PCR to selectively amplify specific regions of DNA, a technique that allows researchers to genetically characterize individuals by generating enough copies of a particular segment of DNA to allow subsequent manipulation and characterization.

- Universal and species-specific primers enable researchers to amplify sequences from very small amounts of DNA that can be isolated from a range of samples, including faeces, feathers, hair, leaves and scales; this permits the humane sampling of wild organisms.

- Gel electrophoresis allows us to separate and identify the fragments that are amplified by PCR on the basis of their sizes. DNA sequencing reactions will generate the precise sequences of amplified DNA products.

- The quantities of DNA and RNA in a given sample can be estimated using RT-PCR, a technique that has been used to study gene expression and to estimate both the numbers and identities of species in composite samples such as microbial communities.

## Useful Websites and Software

- National Center for Biotechnology Information (NCBI): http://www.ncbi.nlm. nih.gov/. Includes public databases of DNA sequences

- European Molecular Biology Laboratory (EMBL): http://www.embl.org/. Includes the EMBL Nucleotide Sequence Database (also known as EMBL-Bank)

- DNA Data Bank of Japan (DDBJ): www.ddbj.nig.ac.jp

- The Web Guide of Polymerase Chain Reaction: www.pcrlinks.com

- List of primer-design software provided by the UK Human Genome Mapping Project Resource Centre: http://www.hgmp.mrc.ac.uk/GenomeWeb/ nuc-primer.html

- ClustalX (Thompson *et al.*, 1997), software for aligning sequences: ftp://ftp-igbmc.u-strasbg.fr/pub/ClustalX/

- Molecular Evolutionary Genetics Analysis (MEGA) software, which includes automatic codon translation, the calculation of transition/transversion ratios and identification of synonymous sites: www.megasoftware.net

## Further Reading

### Books

Hartl, D.L. and Jones, E.W. 1998. *Genetics: Principles and Analysis*, (4th edn). Jones and Bartlett Publishers, Boston, MA.
Li, W.-H. 1997. *Molecular Evolution*. Sinauer Associates, Sunderland, Massachusetts.

### Review articles

Baker, G.C., Smith, J.J. and Cowan, D.A. 2003. Review and re-analysis of domain-specific 16S primers. *Journal of Microbial Methods* **55**: 541–555.
Benard, M.F. 2004. Predator-induced phenotypic plasticity in organisms with complex life histories. *Annual Review of Ecology, Evolution, and Systematics* **35**: 651–673.
Gachon, C., Mingam, A. and Charrier, B. 2004. Real-time PCR: what relevance to plant studies? *Journal of Experimental Botany* **55**: 1445–1454.

Gugerli, F., Parducci, L. and Petit, R.J. 2005. Ancient plant DNA: review and prospects. *New Phytologist* **166**: 409–418.

Pääbo, S., Poinar, H., Serre, D., Jaenicke-Després, V., Hebler, J., Rohland, N., Kuch, M., Krause, J., Vigilan, L., Hofreiter, M. 2004. Genetic analyses from ancient DNA. *Annual Review of Genetics* **38**: 645–679.

# Review Questions

**1.1.** A particular DNA sequence is transcribed into the following RNA sequence in which the bases are grouped as codons:

GCG CCC CAA UGU UGG ACC

Within a population, two mutations occur that result in slight modifications of the same sequence:

GCG **G**CC CCA AUG UUG GAC C
GCG CC**A** CAA UGU UGG ACC

In each case, what type of mutation has occurred, and is it likely to be selectively neutral?

**1.2.** Below are two partial cytochrome *b* mitochondrial sequences (180 bp each) from a pair of neotropical warbling-finches in the genus *Poospiza* (after Lougheed *et al.*, 2000). Sequence differences are in bold type. What is the ratio of transitions to transversions in these two sequences?

```
P. alticola  CTCACTTTCCTCCACGAAACAGGCTCAAACAACCCAACGGGCATCCCCTCAGATTGCGAC
P. baeri     CTGACTTTCCTACACGAAACAGGCTCAAACAATCCAATAGGAATCCCCTCAGACTGCGAC

P. alticola  AAAATTCCCTTCCACCCATACTACACCATCAAAGACATCTTAGGATTCGTACTAATACTC
P. baeri     AAAATCCCCTTCCACCCCTACTACACTATCAAAGACATCCTAGGCTTCGTAATCATACTC

P. alticola  ACCCTATTAGTCTCACTAGCTTTATTCTCCCCCAATCTCCTAGGCGACCCAGAAAATTTC
P. baeri     TCCCTACTCGCATCACTAGCCCTATTCGCCCCCAACCTACTAGGAGACCCAGAAAACTTC
```

**1.3.** Provide sequences for a pair of primers, each 15 bases long, that could be used to amplify the following sequence in a PCR reaction:

5′—CTCACTTTCCTCCACGAAACAGGCTCAAACAACCCAACGGGCATCCCCTCAGATTGCGAC — 3′
3′—GAGTGAAAGGAGGTGCTTTGTCCGAGTTTGTTGGGTTGCCCGTAGGGGAGTCTAACGCTG — 5′

**1.4.** A dideoxy sequencing reaction gives target DNA sequence fragments of the following sizes:

Reaction with ddG: 2, 4, 5 and 8 bases
Reaction with ddA: 6, 7, 9, 13 and 14 bases
Reaction with ddT: 1, 3, 12 and 15 bases
Reaction with ddC: 10 and 11 bases

What is the sequence of this region of DNA?

# 2

# Molecular Markers in Ecology

## Understanding Molecular Markers

In Chapter 1 we started to look at the extraordinary wealth of genetic information that is present in every individual, and to explore how some of this information can be accessed and used in ecological studies. We will build on this foundation by looking in more detail at some of the properties of the genetic markers that are used in molecular ecology. We will start with an overview of the different genomes, because the use and interpretation of all markers will be influenced by the way in which they are inherited. The second half of the chapter will be an overview of those molecular markers that are most commonly used in ecology. After reading this chapter you should understand enough about different molecular markers to suggest which would be applied most appropriately to general research questions.

## Modes of Inheritance

Genetic material is transmitted from parents to offspring in a predictable manner, and this is why molecular markers allow us to infer the genetic relationships of individuals. This does not simply mean that we can use genetic data to determine whether two individuals are siblings or first-cousins. In molecular ecology, the calculation of genetic relationships often takes into account the transmission of particular alleles through hundreds, thousands or even millions of generations. In later chapters we will look at some of the ways in which both recent and historical genealogical relationships can be unravelled, but first we must understand how different genomic regions are passed down from one generation to the next. Not all DNA is inherited in the same way, and understanding different modes of inheritance is crucial before we can predict how different regions of DNA might behave under various ecological and evolutionary scenarios.

*Molecular Ecology*   Joanna Freeland
© 2005 John Wiley & Sons, Ltd.

## Nuclear versus organelle

The offspring of sexually reproducing organisms inherit approximately half of their DNA from each parent. In a diploid, sexually reproducing organism for example, this means that within the nuclear genome one allele at each locus came from the mother and the other allele came from the father. This is known as **biparental inheritance**. However, even in sexually reproducing species, not all DNA is inherited from both parents. Two important exceptions are the **uniparentally inherited** organelle genomes of mitochondria (mtDNA) and **plastids**, with the latter including chloroplasts (cpDNA). These are both located outside the cell nucleus. Mitochondria are found in both plants and animals, whereas plastids are found only in plants. Organelle DNA typically occurs in the form of supercoiled circles of double-stranded DNA, and these genomes are much smaller than the nuclear genome. For example, at between 15 000 and 17 000 bp the mammalian mitochondrial genome is approximately 1/10 000 the size of the smallest animal nuclear genome, but what they lack in size they partially make up for in number – a single human cell normally contains anywhere from 1 000 to 10 000 mitochondria. Molecular markers from organelle genomes, particularly animal mtDNA, have been exceedingly popular in ecological studies because, as we shall see below, they have a number of useful attributes that are not found in nuclear genomes.

### Animal mitochondrial DNA

Mitochondrial DNA is involved primarily in cellular respiration, the process by which energy is extracted from food. Animal mtDNA contains 13 protein-coding genes, 22 transfer RNAs and two ribosomal RNAs. There is also a control region that contains sites for replication and transcription initiation. Most of the sequences are unique, i.e. they are non-repetitive, and there is little evidence of either spacer sequences between genes or intervening sequences within transcribed genes. Although some rearrangement of mitochondrial genes has been found in different animal species, the overall structure, size and arrangement of genes are relatively conserved (Figure 2.1). In most animals, mitochondrial DNA is inherited maternally, meaning that it is passed down from mothers to their offspring (although there are exceptions; see Box 2.1).

There are several reasons why mtDNA markers have been used extensively in studies of animal population genetics. First of all, mtDNA is relatively easy to work with. Its small size, coupled with the conserved arrangement of genes, means that many pairs of universal primers will amplify regions of the mitochondria in a wide variety of vertebrates and invertebrates. This means that data often can be obtained without any *a priori* knowledge about a particular species' mitochondrial DNA sequence. Second, although the arrangement of genes is conserved, the overall mutation rate is high. The rate of synonymous substitutions in mammalian mtDNA

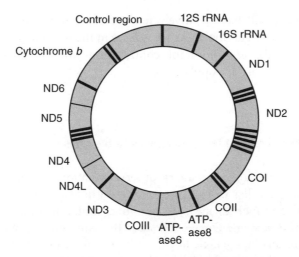

**Figure 2.1** Typical gene organization of vertebrate mtDNA. Unlabelled dark bands represent 22 transfer RNAs (tRNAs). Gene abbreviations starting with ND are subunits of NADH dehydrogenase, and those starting with CO are subunits of cytochrome *c* oxidase

has been estimated at $5.7 \times 10^{-8}$ substitutions per site per year (Brown *et al.*, 1982), which is around ten times the average rate of synonymous substitutions in protein-coding nuclear genes. The non-coding control region, which includes the **displacement (D) loop**, evolves particularly rapidly in many taxa. The high mutation rate in mtDNA may be due partly to the by-products of metabolic respiration and also to less-stringent repair mechanisms compared with those acting on nuclear DNA (Wilson *et al.*, 1985). Regardless of the cause, these high mutation rates mean that mtDNA generally shows relatively high levels of polymorphism and therefore will often reveal multiple genetic lineages both within and among populations.

The third relevant property of mtDNA is its general lack of recombination, which means that offspring usually will have (barring mutation) exactly the same mitochondrial genome as the mother. As a result, mtDNA is effectively a single **haplotype** that is transmitted from mothers to their offspring. This means that mitochondrial lineages can be identified in a much more straightforward manner than nuclear lineages, which, in sexually reproducing species, are continuously pooling genes from two individuals and undergoing recombination. The effectively clonal inheritance of mtDNA means that individual lineages can be tracked over time and space with relative ease, and this is why, as we will see in Chapter 5, mtDNA sequences are commonly used in studies of phylogeography.

Finally, because mtDNA is haploid and uniparentally inherited, it is effectively a quarter of the population size of diploid nuclear DNA. Because there are fewer copies of mtDNA to start with, it is relatively sensitive to demographic events such as bottlenecks. These occur when the size of a population is temporarily reduced, e.g. following a disease outbreak or a catastrophic event. Even if the population recovers quickly, it will have relatively few surviving mitochondrial haplotypes compared with nuclear genotypes. As we will see in the next chapter, inferring past bottlenecks can make an important contribution towards understanding the current genetic make-up of populations.

---

**Box 2.1   Mitochondrial DNA: exceptions to the rules**

Uniparental inheritance and a lack of recombination have made mtDNA the molecular marker of choice in many studies of animal populations because these properties mean that, until a mutation occurs, all of the descendants of a single female will share the same mitochondrial haplotype. This means that genetic lineages can be retraced through time using relatively straightforward models. However, as with so many things in biology, there are exceptions to the rules of mitochondrial inheritance. For one thing, not all mitochondria in animals are inherited maternally. Instances of paternal leakage (transmission of mitochondria from father to offspring) have been found in a number of species, including mice (Gyllensten *et al.*, 1991), birds (Kvist *et al.*, 2003) and humans (Schwartz and Vissing, 2002). Nevertheless, the extent of paternal leakage is believed to be low in most animals with the exception of certain mussel species within the families Mytilidae, Veneridae and Unionidae, which follow double biparental inheritance. This means that females generally inherit their mitochondria from their mothers, but males inherit both maternal and paternal mtDNA. The males therefore represent a classic case of **heteroplasmy** (more than one type of mitochondria within a single individual).

   Bivalves by no means represent the only group in which heteroplasmy has been documented, but its prevalence in some mussel species makes them an ideal taxonomic group in which to investigate another question that has been posed recently: do animal mitochondria sometimes undergo recombination? The lack of identifiable recombinant mtDNA haplotypes in natural populations has led to the assumption that no recombination was taking place, but there was always the possibility that recombination was occurring but was remaining undetected because it involved two identical haplotypes. The presence of more than one mtDNA haplotype in male mussels meant that recombination could, at least in theory, produce a novel haplotype, and this indeed has proved to be the case (Ladoukakis and Zouros, 2001; Burzynski *et al.*, 2003). These findings suggest that mtDNA recombination may be more common than was previously believed in the animal kingdom, although at the moment there is little evidence that

recombination, heteroplasmy or paternal leakage are compromising those
ecological studies that are based on mtDNA sequences.

### Plant mitochondrial DNA

As with animals, mtDNA in most higher plants is maternally inherited. There are a
few exceptions to these rules, for example mtDNA is transmitted paternally in the
redwood tree *Sequoia sempervirens* and biparentally inherited in some plants in the
genus *Pelargonium* (Metzlaff, Borner and Hagemann, 1981). The overall function
of plant and animal mitochondria is similar but their structures differ markedly.
Unlike animal mtDNA, plant mitochondrial genomes regularly undergo recombination and therefore evolve rapidly with respect to gene rearrangements and
duplications. As a result, their sizes vary considerably (40 000 – 2 500 000 bp). This
variability makes it difficult to generalize; to take one example, the mitochondrial
genome of the liverwort *Marchantia polymorpha* is around 186 608 bp long and
appears to include three ribosomal RNA genes, 29 transfer RNA genes, 30 protein-
coding genes with known functions and around 32 genes of unknown function
(Palmer, 1991). Although the organization of plant mitochondria regularly
changes, evolution is slow with respect to nucleotide substitutions. In fact, in
most plant species the mitochondria are the slowest evolving genomes (Wolfe, Li
and Shorg, 1987). The overall rates of nucleotide substitutions in plant mtDNA are
up to 100 times slower than those found in animal mitochondria (Palmer and
Herbon, 1988), and this low mutation rate combined with an elevated recombination rate means that plant mitochondrial genomes have not featured prominently
in studies of molecular ecology.

That is not to say that there are no useful applications of mtDNA data in plant
studies. For many plant species, dispersal is possible through either seeds or pollen,
which often vary markedly in the distances over which they can travel. For
example, if seeds are eaten by small mammals they may travel relatively short
distances before being deposited. In contrast, pollen may be dispersed by the wind,
in which case it could travel a long way from its natal site. Even if seeds are wind
dispersed they are heavier than pollen and therefore still likely to travel shorter
distances. Nevertheless, it is also possible that the opposite scenario could occur,
for example seeds that are ingested by migratory birds may travel much further
than wind-blown pollen. Tracking seeds and pollen is extremely difficult, but the
different dispersal abilities of the two sometimes can be inferred by comparing the
distributions of mitochondrial and nuclear genes. Because mtDNA is usually
inherited maternally, its distribution will reflect the patterns of seed dispersal but
will not be influenced by the spread of pollen, which contains only the paternal
genotype.

Canadian populations of the black spruce (*Picea mariana*) grow in areas that were covered in ice until approximately 6000 years ago, so we know that populations must have been established since that time. Researchers found that current populations share the same mtDNA haplotypes but not the same nuclear alleles (Gamache *et al.*, 2003). This difference was attributed to the widespread dispersal of nuclear genes that were carried by wind-blown pollen, coupled with a much more restricted dispersal of mitochondrial genes that can be transported only in seeds. Because seeds usually do not travel very far, it is likely that only a few were involved in establishing populations once the ice had retreated, hence the lack of variability in mtDNA. On the other hand, the pollen that blew to these sites most likely originated in multiple populations, and this has led to a much higher diversity in nuclear genotypes. If this study had been based solely on data from either nuclear or mitochondrial DNA we would have an incomplete, and possibly misleading, picture denoting the dispersal of this coniferous species.

### Plastids, including chloroplast DNA

The relatively low variability of plant mtDNA means that when haploid markers are desirable in plant studies, researchers more commonly turn to plastid genomes, including chloroplast DNA (cpDNA). Like mtDNA, cpDNA is inherited maternally in most **angiosperms** (flowering plants), although in most **gymnosperms** (conifers and cycads) it is usually inherited paternally. Chloroplast genomes, which in most plants are key to the process of photosynthesis, typically range from 120 000 to 220 000 bp (the average size is around 150 000 bp). Although recombination sometimes occurs, chloroplasts are for the most part structurally stable, and most of the size variation can be attributed to differences in the lengths of sequence repeats, as opposed to the gene rearrangement and duplication found in plant mtDNA.

In tobacco (*Nicotiana tabacum*), the cpDNA genome contains approximately 113 genes, which include 21 ribosomal proteins, 4 ribosomal RNAs, 30 transfer RNAs, 29 genes that are necessary for functions associated with photosynthesis and 11 genes that are involved with chlororespiration (Sugiura, 1992). A partial arrangement of chloroplast genes in the liverwort (*Marchantia polymorpha*) is shown in Figure 2.2. The average rate of synonymous substitutions in the chloroplast genome, at least in higher plants, is estimated as nearly three times higher than that in plant mtDNA (Wolfe, Li and Sharp, 1987), although this is still four to five times slower than the estimated overall rate of synonymous substitutions in plant nuclear genomes (Wolfe, Sharp and Li, 1989). However, this is an average mutation rate, and the use of cpDNA markers in plant population genetic studies has escalated in recent years following the discovery of highly variable microsatellite regions within the chloroplast genome (see below).

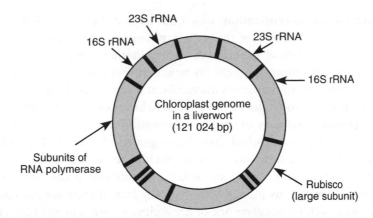

**Figure 2.2** The genome of the chloroplasts found in the liverwort *Marchantia polymorpha* contains 121 024 base pairs (Ohyama *et al.*, 1986). These make up an estimated 128 genes, and the approximate locations of some of these are shown on this figure. The dark lines mark the locations of 12 of the 37 tRNAs

Even when variability is low, genetic data from chloroplasts continue to play an important role in ecological studies, in part because of their uniparental mode of inheritance. In the previous section, we saw how a comparison of data from mtDNA (dispersed only in seeds) and nuclear DNA (dispersed in both seeds and pollen) provided insight into the relative contributions that seeds and pollen make to the dispersal patterns of black spruce. Another way to approach this question in conifers, in which chloroplasts are inherited paternally, is to compare the distribution of chloroplast genes (dispersed in both seeds and pollen) with the distribution of mitochondrial genes (dispersed only in seeds). Latta and Mitton (1997) followed this approach in a study of Limber pine (*Pinus flexilis* James) in Colorado. The seeds of this species are dispersed by Clark's nutcracker (*Nucifraga columbiana*), which caches seeds within a limited radius of the natal tree. If these caches are subsequently abandoned, they may grow into seedlings. Pollen, on the other hand, is dispersed by the wind, and therefore should travel further than the seeds. As expected, the mitochondrial haplotypes of Limber pine were distributed over much smaller areas than the chloroplast haplotypes, once again supporting the notion of relatively widespread pollen dispersal (Latta and Mitton, 1997).

## Haploid chromosomes

When discussing the inheritance of nuclear and organelle markers we usually refer to nuclear genes as being inherited biparentally following sexual reproduction. For the most part this is true, but sex chromosomes (chromosomes that have a role in the determination of sex) provide an exception to this rule. Not all species have sex chromosomes, for example crocodiles and many turtles and lizards follow

**environmental sex determination**, which means that the sex of an individual is determined by the temperature that it is exposed to during early development. Many other species follow **genetic sex determination**, which occurs when an individual's sex is determined genetically by sex chromosomes. This can happen in a number of different ways. In most mammals, and some **dioecious** plants, females are **homogametic** ( two copies of the same sex chromosome: XX), whereas males are **heterogametic** (one copy of each sex chromosome: XY). The opposite is true in birds and lepidopterans, which have heterogametic females (ZW) and homo-gametic males (ZZ). In some other species such as the nematode *Caenorhabditis elegans*, the heterogametic (male) sex is XO, meaning that it has only a single X chromosome. **Monoecious** plant species typically lack discrete sex chromosomes.

In mammals, each female gives one of her X chromosomes to all of her children, male and female alike. It is the male parent's contribution that determines the sex of the offspring; if he donates an X chromosome it will be female, and if he donates a Y chromosome then the offspring will be male. The Y chromosome therefore follows a pattern of **patrilineal descent** because it is passed down only through the male lineage, from father to son (Table 2.1). Because there is never more than

**Table 2.1** Usual mode of inheritance of different genomic regions in sexually reproducing taxa

| Genomic region | Typical mode of inheritance |
| --- | --- |
| **Animals** | |
| Autosomal chromosomes | Biparental |
| Mitochondrial DNA | Maternal in most animals |
| | Biparental in some bivalves |
| Y chromosome | Paternal |
| **Higher plant** | |
| Autosomal chromosomes | Biparental |
| Mitochondrial DNA | Usually maternal |
| Plastid DNA (including chloroplast DNA) | Maternal in most angiosperms |
| | Paternal in most gymnosperms |
| | Biparental in some plants |
| Y chromosome | Paternal in some dioecious plants |

one copy of a Y chromosome in the same individual (barring genetic abnormalities), Y chromosomes are the only mammalian chromosomes that are effectively haploid. In addition, like mtDNA, Y chromosomes for the most part do not undergo recombination. There are two small pseudo-autosomal regions at the tips of the chromosome that recombine with the X chromosome, but in between these are approximately 60 Mb of non-recombining sequence (Figure 2.3).

The mutation rate of Y chromosomes is relatively low. One study found that the variability of three genes on the Y chromosome was approximately five times lower

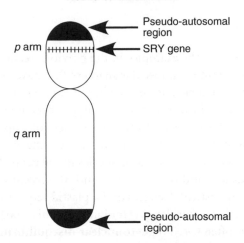

**Figure 2.3**  Mammalian Y chromosome. The SRY gene (sex-determining region Y) effectively converts an embryo into a male

than that of the corresponding regions of autosomal genes (Shen *et al.*, 2000). Reasons for this remain unclear, although it may be due in part to its smaller population size: the total number of Y chromosomes in any given species is a quarter of that for autosomes and a third of that for X chromosomes. This may seem initially confusing because the population size of mtDNA and Y chromosomes is essentially the same and yet mtDNA is relatively variable; however, Y chromosomes have the same repair processes that are found in other regions of nuclear DNA but are lacking in mtDNA. Despite relatively low levels of variability, the Y chromosome still has the potential to be a significant source of information because it is much larger than mtDNA and, unlike mitochondria, contains substantial amounts of non-coding DNA.

One important property of Y chromosomes is that they allow biologists to follow the transmission of paternal genotypes in animals in much the same way that chloroplast markers can be used for gymnosperms. A recent study (Zerjal *et al.*, 2003) found that a particular Y chromosome haplotype is abundant in human populations in a large area of Asia, from the Pacific to the Caspian Sea. Approximately 8 per cent of the men in this region carry it and, because of the high population density in this part of the world, this translates into approximately 0.5 per cent of the world's total population. The researchers who conducted this study suggested that this prevalent Y chromosome haplotype can be traced back to the infamous warrior Genghis Khan. Born in the 12th century, Khan created the biggest land-based empire that the world has seen (Genghis Khan means supreme ruler). He had a vast number of descendants, many of whom were fathered following his conquests, and apparently his sons were also extremely prolific. His policy of slaughtering millions of people and then reproducing en masse is one possible explanation for the widespread occurrence of a single Y chromosome haplotype in Asia today.

*Identifying hybrids*

It should be apparent from the examples in the previous section that there is often an advantage to using multiple molecular markers that have contrasting modes of inheritance. These potential benefits are further illustrated by studies of hybridization. **Hybrids** can be identified from their genotypes because hybridization results in **introgression**, the flow of alleles from one species (or population) to another. As a result, hybrids typically contain a mixture of alleles from both parental species, for example a comparison of grass species in the genus *Miscanthus* revealed that *M. giganteus* was a hybrid of *M. sinensis* and *M. sacchariflorus* because it had one ITS allele from each parental species and a plastid sequence that identified the maternal lineage as *M. sacchariflorus* (Hodkinson *et al.*, 2002). Identification of hybrids in the wild is often based on **cytonuclear disequilibrium**, which occurs in hybrids that have cytoplasmic markers (another name for mitochondria and chloroplasts) from one species or population and nuclear markers from another. Some examples of cytonuclear disequilibrium are given in Table 2.2. By using a combination of markers that have maternal, paternal or biparental inheritance, we may be able to identify which species – or even which population – the hybrid's mother and father came from.

**Table 2.2**    Some examples of cytonuclear disequilibrium in hybrids

| Hybrid | Nuclear DNA | mtDNA or cpDNA | Reference |
|---|---|---|---|
| Freshwater crustaceans *Daphnia pulex* × *D. pulicaria* | *D. pulicaria* | *D. pulex* | Crease *et al.* (1997) |
| Grey wolf (*Canis lupus*) × coyote (*C. latrans*) | Wolf | Coyote | Lehman *et al.* (1991) |
| House mice *Mus musculus* × *M. domesticus* | *M. musculus* | *M. domesticus* | Gyllensten and Wilson (1987) |
| Northern red-backed vole (*Clethrionomys rutilus*) × bank vole (*C. glareolus*) | Bank vole | Northern red-backed vole | Tegelström (1987) |
| White poplar (*Populus alba*) × black poplar (*P. nigra*) | Black poplar | White poplar | Smith and Sytsma (1990) |

Researchers used multiple markers to determine whether or not members of the declining wolf (*Canis lupus*) populations in Europe have been hybridizing with domestic dogs (*C. familiaris*), a question that has been open to debate for some time. One study that used mitochondrial markers to investigate this possibility found only a few instances of haplotype sharing between wolves and dogs and therefore concluded that hybridization between dogs and declining wolf populations was not a cause for concern (Randi *et al.* 2000). However, because mtDNA is inherited maternally, dog haplotypes would appear in wolf populations only if hybridization occurred between female dogs and male wolves. In a more recent study, researchers used a combination of markers from mtDNA, Y chromosome and autosomal nuclear DNA to conduct a more detailed assessment of a possible wolf–dog hybrid from Norway. The mitochondrial haplotype in this suspected hybrid came from a Scandinavian wolf. The Y chromosome data suggested that the father of the hybrid was not a Scandinavian wolf, but did not provide enough information for researchers to discriminate between dogs or migrant wolves from Finland or Russia. However, data from autosomal chromosomes suggested that the most likely father was a dog. It was therefore the combination of uniparental and biparental markers that identified this specimen as a hybrid that had resulted from a cross between a female Scandinavian wolf and a male dog (Vilà *et al.*, 2003b).

## Uniparental markers: a cautionary note

Although uniparental markers have many useful applications in molecular ecology, it is important to keep in mind that they also have some drawbacks. We have referred already to the high recombination rates of plant mtDNA and the low mutation rates of cpDNA, and whereas neither of these considerations is particularly relevant to animal mtDNA, there are other limitations that are common to all mtDNA and cpDNA markers. For one thing, organelles behave as single inherited units and are therefore effectively single locus markers. As we will see throughout this text, data from a single locus allow us to retrace the history of only a single genetic unit (gene or genome), which may or may not be concordant with the history of the species in question. This is particularly true of mtDNA, cpDNA and Y chromosomes because their reduced effective population sizes, relative to autosomal DNA, mean that their haplotypes have a greater probability of going extinct. This loss of haplotypes can cause researchers to infer an oversimplified population history or to underestimate levels of genetic diversity.

An additional drawback to uniparentally inherited markers is that they may not be representative of populations as a whole. We have seen already how the inheritance of markers can influence our understanding of the dispersal patterns of plants depending on whether we target mtDNA, cpDNA or nuclear genomes. The same can be true in animals if dispersal is undertaken by only one gender, e.g. if males disperse and females do not, the mtDNA haplotypes would be population-specific and mtDNA data may lead us to conclude erroneously that individuals never move between populations.

Another risk associated with mtDNA markers involves copies of mtDNA that have been translocated to the nuclear genome; these are known as **mitochondrial pseudogenes**, or **numts** (*nu*clear copies of *mt*DNA sequences). Once they have been transposed into the nucleus, these non-functional pseudogenes continue to evolve independently of mtDNA. Problems will arise from this during PCR if the primer-binding sites have been conserved in the pseudogene, which then may be amplified in addition to, or instead of, the desired mitochondrial region. Although not evenly distributed across taxa, nuclear copies of mitochondrial DNA have been found in more than 80 eukaryotic species, including fungi, plants, invertebrates and vertebrates (reviewed in Bensasson *et al.*, 2001). Steps can be taken that often greatly reduce the likelihood of amplifying these mitochondrial pseudogenes, although one study that specifically set out to investigate this problem concluded that the high frequency of numts in gorillas, combined with their overall similarity to true mtDNA sequences, meant that the application of mtDNA analysis in this species should be undertaken only with extreme caution (Thalman *et al.*, 2004).

A final note about uniparentally inherited markers is that their applications are somewhat different in asexually reproducing organisms. Up to this point our discussion has centred, either implicitly or explicitly, on sexually reproducing organisms, but of course not all organisms reproduce in this way. A large proportion of prokaryotes, plus numerous eukaryotes (plants, invertebrates and some vertebrates), can reproduce asexually (Table 2.3). In the absence of sex, there is no distinction between genomes that are uniparentally and biparentally inherited. Note, however, that the picture is often complicated by the fact that many species are capable of both sexual and asexual reproduction. For example, reproduction of the grain aphid (*Sitobion avenae*) appears, for the most part, to be predominantly asexual in the north and sexual in the south of Britain, although

**Table 2.3** Some methods of asexual reproduction

| Method of reproduction | Examples |
| --- | --- |
| Vegetative reproduction (asexual reproduction from somatic cells). Includes: | |
| • Budding. An offspring grows out of the body of the parent | Hydra, planarians |
| • Fragmentation. Body of parent breaks into distinct pieces, each of which can form offspring | *Opuntia* cactus |
| • Rhizomes and stolons. Runners that give rise to new individuals | Strawberries |
| • Regeneration. If a piece of a parent is detached, it can grow and develop into a new individual | Echinoderms |
| **Parthenogenesis** (asexual reproduction via eggs). Includes: | |
| • Apomixis (mitotic parthenogenesis). The development of an individual from an egg that has not been fertilized, and which has a full complement of the mother's chromosomes. Because there is no involvement of a male gamete, it leads to the production of offspring that are genetically identical to the mother | Aphids, dandelions, flatworms, water fleas, rotifers, whiptail lizards |
| • Amphimixis (meiotic parthenogenesis). Female parent produces eggs by meiosis, which develop without uniting with a male gamete. The diploid state is restored either by a cell division that doubles the number of chromosomes or by fusion of the egg nucleus with another maternal nucleus | Bagworm moth *Solenobia*; nematodes (genus *Heterodera*) |

this seems to depend partly on the climate (Llewellyn *et al.*, 2003). Research into the population genetics of species that have multiple reproductive modes often requires more data than can be obtained from a uniparentally inherited marker.

# Molecular Markers

So far in this chapter we have learned that molecular markers will be influenced by the manner in which they are inherited. We will now take a more detailed look at the properties of some of the most common markers that are used to generate data from one or more genomes. An important point to remember is that molecular markers are simply tools that can be used for generating data; like anything else, the job can be done properly only if the correct tools are chosen. Since we cannot make an informed choice until we understand how the tools operate, we need to learn about some of the key characteristics of different markers. The other point to bear in mind while learning about molecular markers is that we are using them to answer ecological questions. In molecular ecology, genetic data are often combined with ecological data that have been obtained either in the wild or in controlled laboratory experiments, following, for example, observations of mating behaviour, census counts, capture–mark–recapture studies, comparisons of growth patterns under different environmental conditions, and descriptions of morphological characters. In this chapter we will limit ourselves to a general understanding of what markers are, and how they can be used in molecular ecology; methods for analysing data from different markers and more detailed discussions of how genetic data can be applied to ecological questions can be found in later chapters.

Several factors need to be taken into account when choosing a marker. First, it is important to consider the expected level of variability. Some genetic regions are expected to evolve more rapidly than others, and the desired variability will depend on the question that is being asked. Generally speaking, markers that allow us to differentiate between closely related organisms will need to be highly variable, whereas the relationships among more distantly related taxa may be resolved by less variable markers. This ties back to an earlier part of this chapter and also to Chapter 1, when we noted that mutation rates vary both within and between genomes. As we shall see, many types of markers can be applied to either nuclear or organelle genomes; however, different mutation rates combined with alternative modes of inheritance means that the decision about which DNA region or genome to target should not be taken lightly.

There are also practical concerns surrounding the choice of marker. Time, money and expertise are all relevant, and there is often a trade-off between precision and convenience. This brings us to the two main categories of markers that will be described in this chapter: **co-dominant** and **dominant**. Co-dominant markers allow us to identify all of the alleles that are present at a particular locus, whereas dominant markers will reveal only a single dominant allele. As a result, co-dominant data are generally more precise than dominant data, although dominant markers usually require less development time and may therefore be a more convenient way to obtain data. In the rest of this section we will highlight some of the main features of markers in each of the two categories. Note that laboratory practice is, for the most part, beyond the scope of this text; readers looking for specific protocols are referred to the Further Reading section at the end of the chapter.

## Co-dominant markers

In a diploid species, each dominant marker will identify one allele in a homozygous individual and two alleles in a heterozygous individual (Figure 2.4). This ability to distinguish between homozygotes and heterozygotes is one of the most important features of co-dominant markers because it means that we can calculate easily the **allele frequencies** for pooled samples (such as populations). Allele frequency simply refers to the frequency of any given allele within a population, i.e. it tells us how common a particular allele is. If we had a diploid population with 30 individuals then there will be a total of $30 \times 2 = 60$ alleles at any autosomal locus. If 12 individuals had the homozygous genotype $AA$ and 18 individuals had the heterozygous genotype $Aa$ at a particular locus, then the frequency of allele $A$

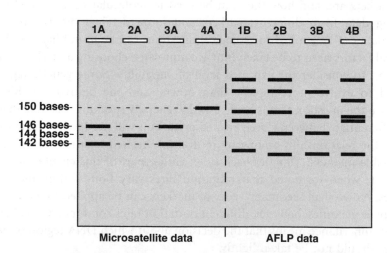

**Figure 2.4** Gel showing the genotypes of four individuals based on one microsatellite (co-dominant) locus (1A–4A) and several AFLP (dominant) loci (1B–4B). According to the microsatellite locus, individuals 1 and 3 are heterozygous for alleles that are 142 and 146 bases long, whereas individuals 2 and 4 are homozygous for alleles that are 144 and 150 bases, respectively. Since there are two of each allele in this sample of eight alleles, the frequency of each microsatellite allele is 0.25. According to the AFLP marker, which screens multiple loci, all four individuals are genetically distinct but we cannot identify homozygotes and heterozygotes, nor can we readily calculate allele frequencies

is $[2(12) + 18]/60 = 42/60 = 0.7$ *or* 70 per cent and the frequency of allele *a* is $18/60 = 0.3$ or 30 per cent. As we will see in later chapters, numerous analytical methods in population genetics are based at least partially on allele frequencies.

It is important to note that although each co-dominant marker characterizes a single locus, most projects will use multiple co-dominant markers to generate data from a number of different loci so that conclusions are not based on a single, possibly atypical, locus. The main drawback to using these types of markers is that they tend to be a relatively time-consuming and expensive way to generate data, and in practice this can limit the number of loci that are genotyped.

## *Allozymes*

In Chapter 1 we learned that allozymes were among the first markers to unite molecular genetics and ecology when they were used to quantify the levels of genetic variation within populations. Since their inception in the 1960s, allozymes have played an ongoing role in studies of animal and plant populations, although in recent years they have featured less prominently than DNA markers. Allozymes benefit from their co-dominant nature and may be more time- and cost-effective than some other markers because they do not require any DNA sequence inform-ation. However, as we noted in Chapter 1, they provide conservative estimates of genetic variation because their variability depends entirely on non-synonymous substitutions in protein-coding genes. In addition, allozymes are of limited utility when we are interested in the evolutionary relationships between different alleles; if an individual has allele B, there is no reason to believe that its ancestor had allele A, in other words it is not always possible to identify an ancestor and its descendant.

Another property of allozymes is that they are functional proteins and therefore are not always selectively neutral. This can be both an advantage and a disadvan-tage. A lack of neutrality can be a disadvantage if a marker is being used to test whether or not populations are genetically distinct from one another. The free-swimming larvae of the American oyster (*Crassostrea virginica*) can travel relatively long distances if swept along on ocean currents. Populations that are not connected by currents may therefore be genetically distinct from one other, a hypothesis that was tested in a genetic survey of populations located along the Atlantic and Gulf coasts of the USA (Karl and Avise, 1992). A comparison of mtDNA and six anonymous nuclear sequences clearly showed that populations around the Gulf of Mexico were, in fact, genetically distinct from those located along the Atlantic coast, a finding that is consistent with the expectation of very low dispersal between coastlines that are not connected by currents. Variation at six allozyme loci, on the other hand, revealed no genetic differences between the two geographical areas, presumably because natural selection has been maintaining the same alleles in different populations. If the researchers had looked at only allozyme data they probably would have concluded that larvae regularly travelled between the Atlantic and Gulf coasts, a finding that would have been difficult to reconcile with the ocean currents in that region.

On the other hand, a non-neutral marker can be useful if we are looking for evidence of adaptation. Mead's sulphur butterfly *Colias meadii* showed some interesting patterns of variation in the glycolytic enzyme phosphoglucose isomerase (PGI), an enzyme involved in glycolysis, which provides fuel for insect flight (Watt *et al.*, 2003). Because flight ability is related to fitness, the allele that confers the best flight ability should be selected for, and therefore a level of genetic uniformity may be expected at the locus coding for PGI. This prediction was supported only partially by a comparison of PGI alleles from *C. meadii* that were sampled from lowland (below the tree line) and alpine (above the tree line) habitats in central USA. Populations showed a high level of genetic uniformity over several hundred kilometres *within* habitats but a marked and abrupt shift in allele frequencies *between* habitats.

Both of these trends are apparently driven by natural selection. *Colias* butterflies spend their adult life within a neighbourhood radius that seldom exceeds 1.5 km. These low levels of dispersal mean that genetic uniformity over hundreds of kilometres must be maintained by a selective force, in this case the relationship between PGI alleles and fitness. Selection also explains the contrasting allele frequencies between alpine and lowland habitats. Because the two habitats are delineated by the tree line, they may be expected to have different thermal (and other abiotic) properties. The activity of PGI varies with temperature, and the authors of this study suggest that alternative PGI alleles may be selected for under different thermal regimes (Watt *et al.*, 2003).

The markers that we will be discussing in the rest of this chapter all target variation in DNA as opposed to proteins. Although allozymes are often subject to selection pressures, DNA markers are more likely to be neutral because they often target relatively variable sequences that, in turn, are less likely to be selectively constrained. However, it is important to bear in mind that not all DNA markers are selectively neutral. In some cases we will specifically discuss non-neutral DNA markers. In other cases neutrality may be assumed, although it is always possible that an apparently non-functional region of DNA is subject to selective pressures that are acting on a genetic region to which it is linked – the so-called **hitch-hiking effect** (Maynard Smith and Haigh, 1974). A more detailed discussion of genetic markers and natural selection is included in Chapter 4.

### Restriction fragment length polymorphisms

The first widespread markers that quantified variation in DNA sequences (as opposed to proteins) were **restriction fragment length polymorphism (RFLPs)**. RFLP data are generated using **restriction enzymes**, which cut DNA at short (usually four to six base pairs), specific sequences. Examples of restriction enzymes include *Alu*I, which cuts DNA when it encounters the sequence AGCT, and *Eco*RV, which cuts in the middle of the sequence GATATC. Digesting purified DNA with one or more restriction enzymes can turn a single piece of DNA into multiple fragments. If two individuals have different distances between two restriction sites, the resulting fragments will be of different lengths. The RFLPs therefore do not survey the entire DNA sequence, but any mutations that add or remove a recognition site for a particular enzyme, or that change the length of sequence between two restriction sites, will be reflected in the sizes and numbers of the fragments that are run out on a gel (Figure 2.5).

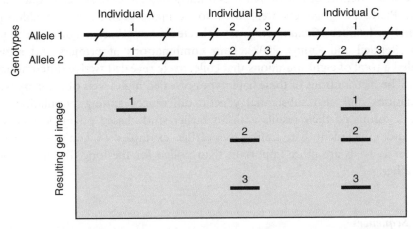

**Figure 2.5** Three different RFLP genotypes result from sequence differences that affect the restriction enzyme recognition sites (designated as /). At this locus, individuals A and B are homozygous for alleles that have two and three restriction sites, respectively. Individual C is heterozygous, with two restriction sites at one allele and three restriction sites at the other allele. The numbers of bands that would be generated by the RFLP profiles are shown in the resulting gel image

Analysis of RFLPs can be done on either an entire genome (nuclear or organelle) or a specific fragment of DNA. The traditional method involves digesting DNA with one or more enzymes and then running out the fragments on a gel. These are then transferred onto a membrane that is placed in a solution containing multiple single-stranded copies of a particular sequence, all of which have been labelled radioactively or fluorescently (this is known as a **probe**). The single-stranded probe will hybridize to the bands that contain its complementary sequence, and these bands then can be identified from the radioactive or fluorescent label. The number of bands produced will depend on the region surveyed and the enzyme used. For example, a digestion with *Alu*I produces around 341 bands in tobacco chloroplast DNA, whereas *Eco*RV produces only 36 bands (Shinozaki *et al.*, 1986). The same enzymes generate approximately 64 bands and three bands, respectively, in human mitochondrial DNA (Anderson *et al.*, 1981). A comparison of the number and sizes of labelled bands among individuals provides an estimate of overall genetic similarity. This method is useful for screening relatively large amounts of DNA but is fairly cumbersome and time-consuming.

A more straightforward method of generating RFLP data is to first amplify a specific fragment of DNA using PCR and then digest the amplified product with enzymes. The fragments then can be visualized after they are run out on a high-resolution gel. This technique is known as PCR-RFLP. Development of PCR-RFLP markers inevitably involves a period of trial and error during which different combinations of primers and enzymes must be screened before enough variable sites can be identified, but overall it is a fairly straightforward technique. In one study, PCR-RFLP markers were used to compare regions of the chloroplast genome of heather (*Calluna vulgaris*) collected from Western European populations (Rendell and Ennos, 2002). Four combinations of primers and enzymes revealed a total of eight mutations that collectively revealed twelve different haplotypes. The distributions of these haplotypes revealed high levels of diversity within populations and also substantial genetic differences among populations. The authors compared their results with an earlier study based on nuclear allozyme data and concluded that, unlike the earlier examples of coniferous trees in Chapter 1, seeds are more important than pollen for the long-distance dispersal of heather.

### DNA Sequences

In Chapter 1 we saw how DNA sequences can be obtained from fragments of DNA that have been amplified by PCR. Although all genetic markers quantify variations in DNA, sequencing is the only method that identifies the exact base pair differences between individuals. This is an important feature of DNA sequencing because it leaves little room for ambiguity: by comparing two sequences we can identify exactly where and how they are different. As a result, sequencing allows us to infer the evolutionary relationships of alternate alleles. This is possible because, barring back-mutations, each mutation acquired by a specific lineage remains there, even after additional mutations occur. In other words, if an allele with a sequence of GGGATATACGATACG mutates to a new allele with a sequence of CGGATATACGATACG, then all descendants of the individual with the new allele will have a C instead of a G at the first base position, even if subsequent mutations occur at other sites along the sequence. Generally speaking, the more mutations that a pair of individuals has in common, the more closely related they are to one another, a concept that will be developed further in Chapter 5.

Sequence data were used to unravel the evolutionary history of the Hawaiian silversword alliance, a group of 28 endemic Hawaiian plant species in the sunflower family (Baldwin and Robichaux, 1995). These plants are of interest because collectively they demonstrate substantial morphological and ecological variation.

Throughout the Hawaiian archipelago they inhabit a range of wet (including sedgeland and forest) and dry (including grassland and shrubland) habitats, in contrast to their less ecologically diverse continental relatives. By comparing sequences from coding and non-coding regions of the nuclear ribosomal DNA genes, the authors of this study were able to identify both shared and unique mutations, which in turn allowed them to conclude which species are most closely related to one another. They were then able to reconstruct the events that led to the evolution of such a diverse group. The sequence data suggest that this group of species arose from a single ancestor that was dispersed, presumably by birds, to the Hawaiian archipelago some time in the past. By combining ecological and genetic data, the authors then could go one step further and conclude that shifts between wet and dry habitats occurred on multiple occasions. This would suggest that ecological diversification played an important role in the speciation of this alliance.

The variable rates of sequence evolution (Table 2.4) mean that we can use relatively rapidly evolving sequences for comparing closely related taxa, and more slowly evolving sequences for comparing distantly related taxa. In recent years, the choice of appropriate gene regions has been facilitated by the growing availability of sequence data. Nevertheless, although sequencing theoretically can be applied to any genomic region, our knowledge of chromosomal sequences is still inadequate

**Table 2.4** Evolutionary rates of some DNA sequences. All estimates are from Li (1997), with the exception of the value given for mitochondrial protein-coding regions in mammals, which is from Brown, George and Wilson (1979). To put the low values in perspective, recall that the diversity of 0.1 % in the human nuclear genome translates into a roughly three million base pair difference between individuals

| Type of sequence | Organism | Average divergence (% per million years)[a] |
|---|---|---|
| **Nuclear DNA** | | |
| Non-synonymous sites | Mammals | 0.15 |
| | *Drosophila* | 0.38 |
| | Plant (monocot) | 0.014 |
| Synonymous sites | Mammals | 0.7 |
| | *Drosophila* | 3.12 |
| | Plant (monocot) | 0.114 |
| Introns | Mammals | 0.7 |
| **Chloroplast DNA** | | |
| Non-synonymous sites | Plant (angiosperm) | 0.004–0.01 |
| Synonymous sites | Plant (angiosperm) | 0.024–0.116 |
| **Mitochondrial DNA** | | |
| Non-synonymous sites | Plant (angiosperm) | 0.004–0.008 |
| Synonymous sites | Plant (angiosperm) | 0.01–0.042 |
| Protein-coding regions | Mammals | 2.0 |

[a]These are estimates averaged over multiple loci.

in most taxa and there is a shortage of universal nuclear primers. Universal primers are more abundant for plant and particularly animal organelle genomes; in fact, animal mtDNA has been the source of data in most sequence-based ecological studies.

Although sequence data can be extremely informative, obtaining these data is quite expensive (although decreasingly so) and time-consuming. Development time will be longer if a number of sequences need to be screened before appropriately variable regions are identified. Furthermore, many studies benefit by having data from more than one genetic region, which further adds to the time and expense. Recently, however, a relatively new method known as **single nucleotide polymorphisms** (**SNPs**) has been gaining in popularity because it is specifically designed to target variable DNA bases in multiple loci.

### Single nucleotide polymorphisms

Single nucleotide polymorphisms (SNPs) refer to single base pair positions along a DNA sequence that vary between individuals. Most SNPs (pronounced snips) have only two alternative states (i.e. each individual has one of two possible nucleotides at a given SNP locus) and are therefore referred to as biallelic markers. Although technically just another way of looking at sequence variation, SNPs are given their own classification because they provide a new approach for finding informative sequence data; DNA sequencing generally entails a comparison of sequences between individuals to see how much variation exists, whereas individual polymorphic sites must be identified before they can be classified as SNPs. Once we know that particular sites are variable, we can use these SNPs to genetically characterize both individuals and populations. Although still in its infancy, the use of SNPs as molecular markers seems to hold great potential.

There is no doubt that SNPs are widespread. In the human genome they account for approximately 90 per cent of genetic variation (Collins, Brooks and Chakravarti, 1998). A survey of multiple taxa including plants, mammals, birds, insects and fungi suggested that a SNP will be revealed for every 200–500 bp of non-coding DNA and every 500–1000 bp of coding DNA that are sequenced (Brumfield *et al.*, 2003, and references therein). This proved to be a conservative estimate in pied and collared flycatchers, *Ficedula hypoleuca* and *F. albicollis*, because when researchers screened around 9000 bp from each species, they discovered 52 SNPs in pied flycatchers and 61 SNPs in collared flycatchers (Primmer *et al.*, 2002). This translates into an average frequency of approximately one SNP per 175 bp and 150 bp for pied and collared flycatchers, respectively. This is encouraging, as the search for SNPs in the nuclear genome will initially be random in most non-model species.

SNPs can be identified in a relatively straightforward manner by sequencing PCR products that have been amplified using universal or species-specific primers, or by sequencing anonymous loci such as those amplified by the multi-locus methods outlined below. By targeting multiple loci, researchers should be able to identify a number of SNPs distributed across multiple unlinked sites throughout the nuclear or organelle genomes. The mutation rates of SNPs appear to be in the order of $10^{-8}-10^{-9}$ (Brumfield *et al.*, 2003). This range is lower than the mutation rates of some other markers such as microsatellites (see below) and therefore the most promising application of SNPs in molecular ecology currently appears to be the elucidation of processes that occurred some time in the past. However, SNPs have been developed only recently, and as increasing numbers are characterized, SNPs are likely to prove suitable for a range of other applications, such as using SNP genotypes to identify individuals and to assess levels of genetic variation within populations (Morin, Luikart and Wayne, 2004).

## Microsatellites

Microsatellites, also known as simple sequence repeats (SSRs) or short tandem repeats (STRs), are stretches of DNA that consist of tandem repeats of 1–6 bp. An example of a microsatellite sequence is the dinucleotide repeat $(CA)_{12}$, which consists of 12 repeats of the sequence CA (CACACACACACACACACACACACA). In this case the complementary DNA sequence would have the microsatellite $(TG)_{12}$. Microsatellites are located throughout nuclear and chloroplast genomes and have also been found in the mitochondrial genomes of some species (Figure 2.6). The initial development of microsatellite markers can take considerable time and money. The usual approach is to clone random fragments of DNA

|  | Locus 1 | Locus 2 | Locus 3 |
| --- | --- | --- | --- |
|  | ▬▬TATATATA | ▬▬TAATAATAATAATAATAATAA | ▬ GCGCGCGCGCGCGC ▬▬ |
|  | ▬▬TATATATA | ▬▬TAATAATAATAATAATAA | ▬▬ GCGCGCGCGCGCGCGC ▬ |

**Figure 2.6** Diagrammatic representation showing part of a chromosome across which three microsatellite loci are distributed (note that sequences are provided for only one strand of DNA from each chromosome). This particular individual is homozygous at locus 1 because both alleles are $(TA)_4$, heterozygous at locus 2 because one allele is $(TAA)_8$ and the other is $(TAA)_7$, and heterozygous at locus 3 because one allele is $(GC)_7$ and the other allele is $(GC)_8$

into a library and then screen this library with a microsatellite probe (in much the same way as RFLPs are identified with probes). Clones that contain microsatellites are then isolated and sequenced, and primers that will amplify the repeat region are designed from flanking non-repetitive sequences (Figure 2.7). Once primers have been designed, data can be acquired rapidly by using these primers to amplify microsatellite alleles in PCR reactions. The PCR products then can be run out on a high-resolution gel that will reveal the size of each allele. The number of species from which microsatellite loci have been characterized is growing almost daily, and the sequences that flank microsatellite loci are often conserved between closely

**TTGTCAAAGAGTTCAGCCGAATA** CAATTTATTAAGTGAGCTTAATACAGTT
AGCGACGACAAAGGAAGAAGTACAAC<u>AGAGAGAGAGAGAGAGAGAGAGAGAG</u>
<u>AGAGAGAGAGAGAGAGAGAGAGAGAGAGAGAGAGAGAGAGAGAGAGAGAGA</u>
<u>GAGAGAGAGAGAGAGAGAGAGAGAGAGAGAGAG</u>TGTAAAGATATAGGGAG
ACTAGCTAGAG**CCAAGCACTAAGATACAACACGC**

**Figure 2.7** A DNA sequence that includes a microsatellite region that was isolated from the freshwater bryozoan *Cristatella mucedo* (Freeland *et al.*, 1999). The microsatellite, which is $(AG)_{53}$, is underlined. The flanking sequence regions in bold show the locations of the primers that were used to amplify this microsatellite in a PCR reaction

related species, which means that microsatellite primers sometimes can be used to generate data from multiple species (e.g. Lippe, Dumont and Berhatchez, 2004; Provan *et al.*, 2004).

Microsatellites mutate much more rapidly than most other types of sequences, with estimated mutation rates of around $10^{-4}$–$10^{-5}$ events per locus per replication in yeast (Strand *et al.*, 1993) and around $10^{-3}$–$10^{-4}$ in mice (Dallas, 1992). This is substantially higher than the estimated overall point mutation rate of around $10^{-9}$–$10^{-10}$ (Li, 1997). These high rates of mutation in microsatellites are ascribed most commonly to slipped-strand mis-pairing during DNA replication (Chapter 1), which, because it can result in either the gain or loss of a single repeat unit, has given rise to the **stepwise mutation model** (SMM; Kimura and Ohta, 1978). Alternatively, the **infinite alleles model** (IAM; Kimura and Crow, 1964) allows for mutations in which multiple repeats are simultaneously gained or lost, but it also assumes that any new allele size has not been encountered previously within a population. Mutation models are important for the analysis of genetic data, but reconciling particular models with the evolution of microsatellites is complicated by the fact that, although mutations often involve single repeats, multiple repeats are periodically gained or lost following a single mutation. At other times, insertions or deletions in the flanking sequences will alter the size of the amplified region. There is also considerable evidence suggesting that the mutation of microsatellites is influenced by the number and size of the repeat motif and also by the complexity of the microsatellite, e.g. whether it is composed of one or multiple repeat motifs (Estoup and Cornuet, 1999).

Microsatellite data are not particularly useful for inferring evolutionary events that occurred in the relatively distant past. Their rapid rate of mutation and their tendency to either increase or decrease in size means that **size homoplasy** may often occur. This can be illustrated by an example of two ancestral alleles at the same locus, one with 20 dinucleotide repeats and the other with 16 dinucleotide repeats. If the larger allele loses one repeat and the smaller allele gains one repeat, then both mutations will have led to alleles with 18 repeats. These two new alleles, each with 18 repeats, may appear to be two copies of the same ancestral sequence, but the evolutionary histories of the two alleles are in fact quite different. Size homoplasy means that ancestor-descendant relationships may be difficult to untangle from microsatellite data.

On the other hand, the high mutation rates of microsatellites mean that there are often multiple alleles at each locus, and this high level of polymorphism makes them suitable for inferring relatively recent population genetic events. East African cichlid fishes were therefore prime candidates for microsatellite analysis, because thousands of endemic species evolved in Lakes Malawi and Victoria within the last 700 000 years, and some species are believed to be only around 200 years old (Kornfield and Smith, 2000). Initial explorations of these species using mtDNA or allozymes revealed little information. The problem was that, even when polymorphic genetic regions were identified, the recency of speciation events meant that most alleles were still shared among taxa because there had not been enough time for species-specific alleles to evolve. In the 1990s, however, microsatellite markers identified much higher levels of variation within and among cichlid species (reviewed in Markert, Danley and Avnegard, 2001). As a result, researchers have been able to use microsatellite data to resolve some aspects of the evolutionary history of cichlid groups (Kornfield and Parker, 1997; Sültmann and Mayer, 1997), although their analyses were somewhat hampered by size homoplasy.

The variability of microsatellites means that, unlike some of the more slowly evolving gene regions, they can also be used to discriminate genetically between individuals and populations. This application has provided some interesting insights into cichlid mating systems. In one study, a combination of behavioural and microsatellite data was used to investigate the role of assortative mating in speciation. Although hybrids from the same lake are often fully fertile, females were found to consistently select males based on their highly divergent colour patterns, ignoring the overall shape similarity that might otherwise blur the division between species. Conclusions from the behavioural data were supported by microsatellite data, which identified the different morphs as genetically distinct taxonomic groups (van Oppen et al., 1998). The high variability, co-dominant nature and increasing availability of microsatellites have made them one of the most popular types of markers in population genetics; however, their extensive development time means that there is still considerable support for dominant markers, to which we shall now turn.

## Dominant markers

Dominant markers are also known as multi-locus markers because they simultaneously generate data from multiple loci (Figure 2.4). They typically work by using random primers to amplify anonymous regions of the genome, producing a pattern of multiple bands from each individual. Because they use random primers to amplify fragments of DNA, no prior sequence knowledge is required and therefore the development time may be relatively short. Furthermore, because dominant markers each characterize multiple regions of the genome, they often show reasonably high levels of polymorphism that can be useful for inferring close genetic relationships; however, dominant markers are generally unable to resolve more distant evolutionary relationships. Perhaps the biggest drawback to these markers is that their dominant nature means that only one allele can be identified at each locus, and therefore heterozygotes cannot be differentiated from homozygotes. The presence of a band means that an individual is either homozygous (*AA*, with both alleles producing fragments) or heterozygous (*Aa*, with only the dominant allele producing a fragment) at that particular locus. Individuals that are homozygous recessive will not produce a band.

The inability to differentiate between homozygotes and heterozygotes makes it difficult to calculate allele frequencies from dominant markers. Methods for doing this do exist but several assumptions must be made, such as the population being in Hardy–Weinberg equilibrium, and as we shall see in the next chapter this is not always the case. The anonymity of dominant markers also can make it difficult to detect contamination and to compare data between studies. Nevertheless, as we will see below, dominant markers have been employed successfully in many studies of molecular ecology.

### Random amplified polymorphic DNA

In 1990 a PCR-based technique known as **random amplified polymorphic DNA** (RAPDs) was introduced as a method for genotyping individuals at multiple loci (Welsh and McClelland, 1990; Williams *et al.*, 1990). RAPDs are generated using short (usually 10 bp) random primers in a PCR reaction. As there are about one million possible primers that can be made from ten bases, there is plenty of scope for detecting polymorphism by using a different primer in each reaction. A single primer is added to each PCR reaction, and multiple bands arise by chance when the primer happens to anneal to two reasonably proximate sites. The banding pattern for each individual will depend on where suitable primer binding sites are located throughout that individual's genome (Figure 2.8). Although RAPDs can provide a relatively quick and straightforward method for quantifying the genetic

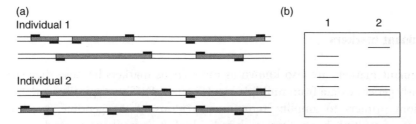

**Figure 2.8** (a) RAPD priming sites (indicated by black boxes) are distributed throughout the genome, although here only two partial chromosomes are represented. The sizes of the products (shaded in grey) that will be amplified during PCR will depend on the locations of these priming sites. (b) Diagrammatic representation of the gel that would follow PCR-RAPD screening of these two individuals. Recall that the rate at which a band migrates through the gel is inversely proportional to its size

similarity of individuals, its reproducibility depends on stringent laboratory conditions. The amplification of bands will often vary, depending on a variety of factors including the starting concentration of DNA, the parameters that are used in the PCR cycle, the type of PCR machine used (results often vary between laboratories) and the particular brand of the reagents used. Furthermore, the random nature of RAPDs means that it can be difficult to identify bands that have been amplified from non-target DNA.

If the problems of reproducibility and contamination are overcome, RAPDs can be used to estimate the genetic similarity between two individuals based on the number of bands they share. However, despite their ease of use, a general lack of reproducibility, combined with their dominant nature, has decreased the popularity of RAPDs in recent years. The journal *Molecular Ecology* actively discourages researchers from submitting manuscripts that report population genetic studies that are based primarily on RAPD data, in part because other more reliable markers now are widely available. That is not to say that there is no role for RAPDs at all. As the *Molecular Ecology* editorial board points out, RAPDs can be useful in genetic mapping studies. They can also provide markers that are diagnostic of a given species or trait if RAPD screening identifies a band that consistently differs between two groups. This was the case in the toxic dinoflagellate *Gymnodinium catenatum*, which apparently was introduced to Tasmanian waters in the ballast water of cargo ships (Bolch *et al.*, 1999). Unique RAPD banding patterns from different populations allowed the authors of this study to eliminate Spain, Portugal and Japan as source populations, and led them to conclude that the populations causing algal blooms around Tasmania were restricted largely to local estuaries. They did not, however, have enough data to pinpoint the source of the original Tasmanian introductions.

### Amplified fragment length polymorphisms

A more labour intensive, but also more reliable, method than RAPDs for generating PCR-based multi-band profiles is known as **amplified fragment length polymorphism** (**AFLPs**) (Vos *et al.*, 1995). AFLP markers are generated by first digesting DNA with two different restriction enzymes that cut the DNA so that one strand overhangs the other strand by one or a few bases, thereby producing overlapping ('sticky') ends. Meanwhile, short DNA linkers are synthesized so that one end of the linker is compatible with the overhanging sequence of DNA. The linkers are ligated to the original DNA fragments, leaving a collection of fragments that have the same DNA sequence at the end. These fragments can then be amplified using primers that anneal to the linker DNA sequence (Figure 2.9). Specificity of primers is usually increased by adding one to three nucleotides at one end of the sequence, because PCR requires a perfect match between the target sequence and the 3' end of the primer. This results in the amplification of multiple

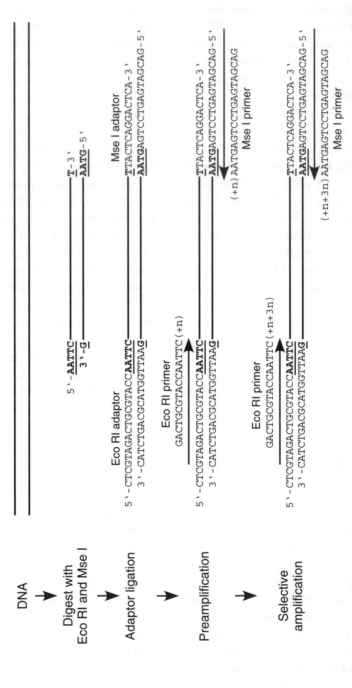

**Figure 2.9** Schematic diagram showing how AFLP genotypes are generated. Digestion with two restriction enzymes produces sticky ends to which linkers can be ligated. During preamplification, the addition of a single base to the 3′ end of each primer will reduce the number of amplified fragments to 1/16 of the number of fragments that otherwise would be amplified. The addition of three more bases to the 3′ primer ends during selective amplification further reduces the chance of a perfect match between primers and target sequences, and as a result only 1/65 536 of the original set of fragments will be amplified

fragments that appear as a series of different-sized bands when run out on a gel. The pattern of bands will depend on the sequences that are immediately adjacent to the linkers, and also on the distances between the restriction sites. As with RAPDs, the generation of bands is essentially random; in contrast to RAPDs, however, this method has a much higher level of reproducibility and therefore has become a more popular method of dominant genotyping.

The genetic similarity of individuals and populations can be inferred from the numbers of AFLP bands that they have in common. Additional information can be obtained by modifying the standard AFLP method to study gene expression. By ligating linkers to digests of cDNA, researchers can compare the banding patterns of genes that have been expressed, as opposed to the entire genome. This method was used to compare two genetically distinct lines of the endoparasitic wasp *Venturia canescens* that differ in a number of ways, including oviposition behaviour, numbers of eggs laid and growth rates in the early stages of embryonic development. Researchers used the cDNA-AFLP method to compare gene expression in ovarian tissue between the two groups and found differences in a number of expressed genes, including some that apparently are involved in the regulation of protein degradation during stress responses (Reineke, Schmidt and Zebitz, 2003). This study provided some interesting suggestions about the importance of gene expression during early development.

## Overview

A range of molecular markers are available for studying populations in the wild (see Table 2.5). Different types of markers will provide different sorts of information, depending, for example, on whether they are inherited biparentally or uniparentally, or are dominant or co-dominant. In the next chapter we will start to look at how various markers can be used to characterize genetically a single population, and in so doing we will start to discuss some of the different ways in which molecular data can be analysed.

## Chapter Summary

- Different genomes are inherited in different ways. In sexually reproducing taxa, most of the nuclear genome is inherited biparentally. In most animals and higher plants, mitochondrial DNA is maternally inherited. Chloroplast DNA usually is inherited maternally in flowering plants and paternally in conifers.

- Animal mitochondrial markers are popular in molecular ecology because of their lack of recombination, high mutation rate, small effective population size and readily available universal primers.

**Table 2.5  Summary of some of the properties of genetic markers**

| Genetic marker | Inheritance | Target genome | Development time[a] | Cost[b] | Comparison of data between studies | Suitability for inferring evolutionary relationships | Overall variability |
|---|---|---|---|---|---|---|---|
| Allozymes | Co-dominant | Nuclear | Low | Low | Limited | Limited | Low-moderate |
| PCR-RFLPs | Co-dominant | Nuclear Organelle | Moderate | Low | Limited | Limited | Low-moderate |
| DNA sequences | Co-dominant | Nuclear Organelle | Low-high | Moderate | Yes | High | Low-moderate |
| SNPs | Co-dominant | Nuclear Organelle | High | Moderate-high | Yes | High | Moderate |
| Microsatellites | Co-dominant | Nuclear Organelle | High | Moderate-high | Yes | Limited | High |
| RAPDs | Dominant | Nuclear | Low | Low | Limited | Limited | High |
| AFLPs | Dominant | Nuclear | Moderate | Moderate | Limited | Limited | High |

[a]Assuming relevant markers have not been developed already for the species in question (or close relative). Note that some development is required for all markers.
[b]Cost here is relative, because all molecular work is expensive. Cost will be reduced if relevant markers have been developed already for the species in question.

- Plant mtDNA generally lacks recombination and has a relatively small effective population size, but mutation rates are much lower than in animals. Chloroplast DNA regularly undergoes recombination and therefore gene rearrangements are common.

- In mammals, the Y chromosome is the paternally inherited sex chromosome. Most of it does not undergo recombination, and the recent identification of variable regions means that this can be a useful marker for retracing male lineages.

- Obtaining data from markers with contrasting patterns of inheritance can be extremely useful for detecting past hybridization events and for differentiating between dispersal of pollen versus seeds in plants, and of males versus females in animals.

- Co-dominant markers provide locus-specific information that allows us to discriminate between homozygotes and heterozygotes and to calculate allele frequencies.

- Dominant data neither discriminate between homozygotes and heterozygotes nor provide accurate estimates of allele frequencies; however, choice of marker is affected by time, money and expertise, and initial development is often easier for dominant than for co-dominant markers.

- DNA sequence data are most suitable for inferring evolutionary histories, and SNPs are a newly emerging source of sequence data from multiple loci. Microsatellite loci are highly polymorphic and are appropriate for inferring recent events such as dispersal or mate choice, whereas multi-locus markers permit the rapid and simultaneous screening of several loci.

## Useful Websites and Software

- Molecular Ecology Notes Primer database: http://tomato.bio.trinity.edu/home.html

- The Alaska Biological Science Centre website summarizing the cross-species utilization of microsatellite primers across a wide range of taxa: http://www.absc.usgs.gov/research/genetics/heterologous_primers.htm

- Microsatellite Analysis Server (MICAS) – an interactive web-based server to find non-redundant microsatellites in a given nucleotide sequence/genome sequence: http://210.212.212.7/MIC/index.html

- National Centre for Biotechnology Information science primer -- SNPs: variations on a theme: http://www.ncbi.nlm.nih.gov/About/primer/snps.html

- BioEdit software for the alignment and manipulation of sequence data: http://www.mbio.ncsu.edu/BioEdit/bioedit.html

# Further Reading

## Books

Avise, J.C. 2004. *Molecular Markers, Natural History and Evolution* (2nd edn). Sinauer Associates Sunderland, Massachusetts.

Goldstein, D.B. and Schlötterer, C. 1999. *Microsatellites: Evolution and Applications*. Oxford University Press, Oxford.

Hoelzel, R. 1998. *Molecular Genetic Analysis of Populations* (2nd edn). Oxford University Press, Oxford.

## Review articles

Bonin, A., Bellemain, E., Eidesen, P.B., Pompanon, F., Brochmann, C. and Taberlet, P. 2004. How to track and assess genotyping errors in population genetics studies. *Molecular Ecology* **13**: 3261–3273.

Jobling, M.A. and Tyler-Smith, C. 2003. The human Y chromosome: an evolutionary marker comes of age. *Nature Reviews Genetics* **4**: 598–612.

Mueller, U.G. and Wolfenbarger, L.L. 1999. AFLP genotyping and fingerprinting. *Trends in Ecology and Evolution* **14**: 389–394.

Rokas, A., Ladoukakis, E. and Zouros, E. 2003. Animal mitochondrial DNA recombination revisited. *Trends in Ecology and Evolution* **18**: 411–417.

Sunnucks, P. 2000. Efficient genetic markers for population biology. *Trends in Ecology and Evolution* **15**: 199–203.

# Review Questions

**2.1.** List some of the advantages and disadvantages of using mitochondrial markers in studies of animal population genetics.

**2.2.** Which genomes would you target, and why, if you wanted to:

    (i)   Compare pollen and seed flow in an angiosperm species.

    (ii)  Compare pollen and seed flow in a gymnosperm species.

    (iii) Compare male and female dispersal in a mammal species.

**2.3.** Twenty-eight diploid individuals from the same population were genotyped at a single locus, and only two alleles ($A_1$ and $A_2$) were found. Ten individuals were homozygous for $A_1$ and eight were homozygous for $A_2$. What are the frequencies of the two alleles in this sample?

**2.4.** Individuals 1 and 2 from the same population have the following sequences at a particular locus:

1. GATTATACATAGCTACTAGATACAGATACTATTTTTAGGGGCGTATGCTCGG
   ATCTATAGACCTAGTACTAGATACTAGGAAAACCCGTTGTGTCGCGTGCTGA

2. GATTATACATAGTTACTAGATACAGATACTATTTTTAGGGGCGTATGCTCGG
   ATCTATAGACCTAGTACTAGATACTAGGAAAACCCGTTGTGTCGCGTGCTGA

If these sequences are digested with two restriction enzymes – *Alu*I, which cuts DNA sequences of AGCT and *Rsa*I, which cuts DNA sequences of GTAC – how many bands will each sequence produce?

**2.5.** Two mtDNA lineages in a mouse population diverged around 500 000 years ago. If you are comparing a 500 bp stretch of DNA from a protein-coding region from two mtDNA haplotypes, approximately how many base pair differences would you expect to find, based on the data in Table 2.4?

**2.6.** What are some of the factors that need to be taken into account when deciding which molecular markers you would use in an ecological study?

# 3
# Genetic Analysis of Single Populations

## Why Study Single Populations?

Now that we know how molecular markers can provide us with an almost endless supply of genetic data, we need to know how these data can be used to address specific ecological questions. A logical starting point for this is an exploration of the genetic analyses of single populations, which will be the subject of this chapter. We will then build on this in Chapter 4 when we start to look at ways to analyse the genetic relationships among multiple populations. This division between single and multiple populations is somewhat artificial, as there are very few populations that exist in isolation. Nevertheless, in this chapter we shall be treating populations as if they are indeed isolated entities, an approach that can be justified in two ways. First, research programmes are often concerned with single populations, for example conservation biologists may be interested in the long-term viability of a particular population, or forestry workers may be concerned with the genetic diversity of an introduced pest population. Second, we have to be able to characterize single populations before we can start to compare multiple populations. But before we start investigating the genetics of populations, we need to review what exactly we mean by a population.

## What is a population?

A **population** is generally defined as a potentially interbreeding group of individuals that belong to the same species and live within a restricted geographical area. In theory this definition may seem fairly straightforward (at least for sexually reproducing species), but in practice there are a number of reasons why

*Molecular Ecology*   Joanna Freeland
© 2005 John Wiley & Sons, Ltd.

**Figure 3.1** A pair of copulating common green darner dragonflies (*Anax junius*). Juvenile development in this species is phenotypically plastic, depending on the temperature and photoperiod during the egg and larval stages. Photograph provided by Kelvin Conrad and reproduced with permission

populations are seldom delimited by obvious boundaries. One confounding factor may be that species live in different groups at different times of the year. This is true of many bird species that breed in northerly temperate regions and then migrate further south for the winter, because any one of these overwintering 'populations' may comprise birds from several distinct breeding populations.

The situation is even more complex in the migratory common green darner dragonfly, *Anax junius* (Figure 3.1). Throughout part of its range, *A. junius* has two alternative developmental pathways in which larvae take either 3 or 11 months to develop into adults (Trottier, 1966). Individuals that develop at different rates will not be reproductively active at the same time and therefore cannot interbreed. If developmental times are fixed there would be two distinct *A. junius* populations within a single lake or pond, but preliminary genetic data suggest that development in this species is an example of phenotypic plasticity (Freeland *et al.*, 2003). This means that, although some individuals are unable to interbreed within a particular mating season, their offspring may be able to interbreed in the following year; therefore, individuals that follow different developmental pathways can still be part of the same population.

Prolonged **diapause** (delayed development) also may cause researchers to underestimate the size or boundaries of a population, because seeds or other propagules that are in diapause will often be excluded from a census count. Many plants fall into this category, such as the flowering plant *Linanthus parryae* that thrives in the Mojave desert when conditions are favourable. When the environment becomes unfavourable, seeds can lay dormant for up to 6 years in a seed bank, waiting for conditions to improve before they germinate (Epling, Lewis and Ball, 1960). Similarly, the sediment-bound propagules of many species of freshwater zooplankton can survive for decades (Hairston, Van Brunt and Kearns, 1995).

Another complication that arises when we are defining populations is that their geographical boundaries are seldom fixed. Boundaries may be particularly unpredictable if reproduction within a population depends on an intermediate species. The population limits of a flowering plant, for example, may depend on the movements of pollinators, which can vary from one year to the next. Populations of the post-fire wood decay fungus *Daldinia loculata*, which grows in the wood of deciduous trees that have been killed by fire, are also influenced by vectors. Pyrophilous insect species moving between trees can disperse fungal conidia (clonal propagules that act as male gametes) across varying distances. Genetic data from a forest site in Sweden suggested that insects sometimes transfer conidia between trees, thereby increasing the range of potentially interbreeding individuals beyond a single tree (Guidot *et al.*, 2003).

It should be apparent from the preceding examples that population boundaries are seldom precise, although in a reasonably high proportion of cases they should correspond more or less to the distribution of potential mates. Biologists often identify discrete populations at the start of their research programme, if only as a framework for their sampling design, which often will specify the minimum number of individuals required from each presumptive population. Nevertheless, populations should not be treated as clear-cut units, and the boundaries are sometimes revised after additional ecological or genetic data have been acquired. Bearing in mind that molecular ecology is primarily concerned with wild populations, which by their very nature are variable (Box 3.1) and often unpredictable, we shall start to look at ways in which molecular genetics can help us to understand the dynamics of single populations.

### Box 3.1 Summarizing data

Ecological studies, molecular and otherwise, are often based on measurements of a trait or characteristic that have been taken from multiple individuals. These data may quantify phenotypic traits, such as wing lengths in birds, or genotypic traits, such as allele frequencies in different populations. Consider the following data set on wing lengths:

| Sample 1 | Sample 2 |
|:---:|:---:|
| 23 | 20 |
| 21 | 26 |
| 23 | 23 |
| 24 | 19 |
| 24 | 27 |

There are a number of ways in which we can summarize these wing measurements, including the **arithmetic mean**, or average, which is calculated as:

$$\bar{X} = \Sigma x_i / n \tag{3.1}$$

where $x_i$ is the value of the variable in the $i$th specimen, so

$$\bar{X} = (23 + 21 + 23 + 24 + 24)/5 = 23 \quad \text{for population 1, and}$$
$$\bar{X} = (20 + 26 + 23 + 19 + 27)/5 = 23 \quad \text{for population 2}$$

In this case both populations have the same average wing length, but this is telling us nothing about the variation within each population. The range of measurements (the minimum value subtracted from the maximum value, which equals 3 and 8 in samples 1 and 2, respectively), can give us some idea about the variability of the sample, although a single unusually large or unusually small measurement can strongly influence the range without improving our understanding of the variability. An alternative measure is **variance**, which reflects the distribution of the data around the mean. Variance is calculated as:

$$V = \sum_{i=1}^{n} (X_i - \bar{X})^2 / (n - 1) \tag{3.2}$$
$$= [(23 - 23)^2 + (21 - 23)^2 + (23 - 23)^2 + (24 - 23)^2$$
$$+ (24 - 23)^2]/(5 - 1)$$
$$= 1.5 \quad \text{for population 1, and}$$
$$= [(20 - 23)^2 + (26 - 23)^2 + (23 - 23)^2 + (19 - 23)^2$$
$$+ (27 - 23)^2]/(5 - 1)$$
$$= 12.5 \quad \text{for population 2}$$

This shows that, although the mean is the same in both samples, the variation in sample 2 is an order of magnitude higher than that in sample 1. Variance is described in square units and therefore can be quite difficult to visualize so it is sometimes replaced by its square root, which is known as the **standard deviation** (S), calculated as:

$$S = \sqrt{V} \qquad (3.3)$$
$$= \sqrt{1.5} = 1.225 \quad \text{for population 1, and}$$
$$= \sqrt{12.5} = 3.536 \quad \text{for population 2}$$

## Quantifying Genetic Diversity

Genetic diversity is one of the most important attributes of any population. Environments are constantly changing, and genetic diversity is necessary if populations are to evolve continuously and adapt to new situations. Furthermore, low genetic diversity typically leads to increased levels of inbreeding, which can reduce the fitness of individuals and populations. An assessment of genetic diversity is therefore central to population genetics and has extremely important applications in conservation biology. Many estimates of genetic diversity are based on either allele frequencies or genotype frequencies, and it is important that we understand the difference between these two measures. We shall therefore start this section with a detailed look at the expected relationship between allele and genotype frequencies when a population is in Hardy–Weinberg equilibrium.

### Hardy–Weinberg equilibrium

Under certain conditions, the genotype frequencies within a given population will follow a predictable pattern. To illustrate this point, we will use the example of the scarlet tiger moth *Panaxia dominula*. In this species a one locus/two allele system generates three alternative wing patterns that vary in the amount of white spotting on the black forewings and in the amount of black marking on the predominantly red hindwings. Since these patterns correspond to homozygous dominant, heterozygous and homozygous recessive genotypes, the allele frequencies at this locus can be calculated from phenotypic data. We will refer to the two relevant alleles as *A* and *a*. Because this is a diploid species, each individual has two alleles at this locus. The two homozygote genotypes are therefore *AA* and *aa* and the heterozygote genotype is *Aa*. Recall from Chapter 2 that allele frequencies are calculations that tell us how common an allele is within a population. In a two-allele system such as that which determines the scarlet tiger moth wing genotypes, the frequency of the dominant allele $(A)$ is conventionally referred to as $p$, and the frequency of the recessive allele $(a)$ is conventionally referred to as $q$. Because there are only two alleles at this locus, $p+q=1$.

Genotype frequencies, which refer to the proportions of different genotypes within a population (in this case $AA$, $Aa$ and $aa$), must also add up to 1.0. If we know the frequencies of the relevant alleles, we can predict the frequency of each genotype within a population provided that a number of assumptions about that population are met. These include:

- There is random mating within the population (**panmixia**). This occurs if mating is equally likely between all possible male–female combinations.

- No particular genotype is being selected for.

- The effects of migration or mutation on allele frequencies are negligible.

- The size of the population is effectively infinite.

- The alleles segregate following normal Mendelian inheritance.

If these conditions are more or less met, then a population is expected to be in **Hardy–Weinberg equilibrium** (**HWE**). The genotype frequencies of such a population can be calculated from the allele frequencies because the probability of an individual having an $AA$ genotype depends on how likely it is that one $A$ allele will unite with another $A$ allele, and under HWE this probability is the square of the frequency of that allele $(p^2)$. Similarly, the probability of an individual having an $aa$ genotype will depend on how likely it is that an $a$ allele will unite with another $a$ allele, and under HWE this probability is the square of the frequency of that allele $(q^2)$. Finally, the probability of two gametes yielding an $Aa$ individual will depend on how likely it is that either an $A$ allele from the male parent will unite with an $a$ allele from the female parent (creating an $Aa$ individual), or that an $a$ allele from the male parent will unite with an $A$ allele from the female parent (creating an $aA$ individual). Since there are two possible ways that a heterozygote individual can be created, the probability of this occurring under HWE is $2pq$.

The genotype frequencies in a population that is in HWE can therefore be expressed as:

$$p^2 + 2pq + q^2 = 1 \tag{3.4}$$

The various frequencies of heterozygotes and homozygotes under HWE are shown in Figure 3.2, and examples are calculated in Box 3.2.

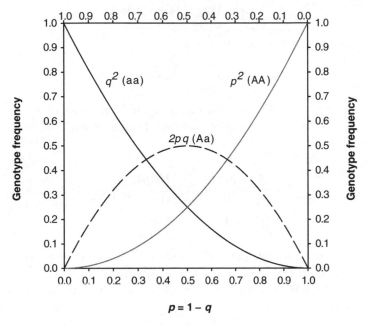

$$q = 1 - p$$

$$p = 1 - q$$

**Figure 3.2** The combinations of homozygote and heterozygote frequencies that can be found in populations that are in HWE. Note that the frequency of heterozygotes is at its maximum when $p = q = 0.5$. When the allele frequencies are between 1/3 and 2/3, the genotype with the highest frequency will be the heterozygote. Adapted from Hartl and Clark (1989)

**Table 3.1** Data from a collection of 1612 scarlet tiger moths

| Phenotype | No. of individuals | Assumed genotype | No. of A alleles | No. of a alleles |
|---|---|---|---|---|
| White spotting | 1469 | AA | 1469 × 2 = 2938 | -- |
| Intermediate | 138 | Aa | 138 | 138 |
| Little spotting | 5 | aa | -- | 5 × 2 = 10 |

**Box 3.2  Calculating Hardy–Weinberg equilibrium**

Table 3.1 is an actual data set on scarlet tiger moths that was collected by the geneticist E.B. Ford. The data in Table 3.1 tell us that in this sample there is a total of $2(1612) = 3224$ alleles at this particular locus. Of these, 3076 are A alleles (2938+138) and 148 are a alleles (138+10), therefore the frequency $p$ of the A allele in this population is:

$$p = 3076/3224 = 0.954$$

and the frequency $q$ of the $a$ allele can be calculated as either:

$$q = 148/3224 = 0.046$$

or, because $p + q = 1$, as:

$$q = 1 - p = 1 - 0.954 = 0.046$$

If we know $p$ and $q$, then we can calculate the frequencies of $AA$ ($p^2$), $Aa$ ($2pq$) and $aa$ ($q^2$) that would be expected if the population is in HWE as follows:

$$p^2 = (0.954)^2 = 0.9101$$
$$2pq = 2(0.954)(0.046) = 0.0878$$
$$q^2 = (0.046)^2 = 0.002$$

We now need to calculate the number of moths in this population that would have each genotype if this population is in HWE. We can do this by multiplying the total number of moths (1612) by each genotype frequency:

$$AA = (0.9101)(1612) = 1467$$
$$Aa = (0.0878)(1612) = 142$$
$$aa = (0.002)(1612) = 3$$

Therefore the Hardy–Weinberg ratio expressed as the number of individuals with each genotype is 1467:142:3. This is very close to the actual ratio of genotypes within the population (from Table 3.1) of 1469:138:5.

We can check whether or not there is a significant difference between the observed and expected genotype frequencies by using a chi-squared ($\chi^2$) test. This is based on the difference between the observed ($O$) number of genotypes and the number that would be expected ($E$) under the HWE, and is calculated as:

$$\chi^2 = \Sigma(O - E)^2/E \qquad (3.5)$$

The $\chi^2$ value of the scarlet tiger moth example is:

$$\chi^2 = (1469 - 1467)^2/1467 + (138 - 142)^2/142 + (5 - 3)^2/3$$
$$= 0.0027 + 0.11 + 1.33$$
$$= 1.44$$

The number of degrees of freedom (d.f.) is determined as 3 (the number of genotypes) minus 1 (because the total number was used) minus 1 (the number of alleles), which leaves d.f. $= 1$. By using a statistical table, we learn that a $\chi^2$ value of 1.44, in conjunction with 1 d.f., leaves us with a probability of $P = 0.230$. This means that there is no significant difference between the observed genotype frequencies in the scarlet tiger moth population and the genotype frequencies that are expected under the HWE. We would conclude, therefore, that this population is in HWE.

Despite the fairly rigorous set of criteria that are associated with HWE, many large, naturally outbreeding populations are in HWE because in these populations the effects of mutation and selection will be small. There are also many populations that are not expected to be in HWE, including those that reproduce asexually. A deviation from HWE may also be an unexpected result, and when this happens researchers will try to understand *why*, because this may tell us something quite interesting about either the locus in question (e.g. natural selection) or the population in question (e.g. inbreeding). First, however, we must ensure that an unexpected result is not attributable to human error. Deviations from HWE may result from improper sampling. The ideal population sample size is often at least 30–40, although this will depend to some extent on the variability of the loci that are being characterized. Inadequate sampling will lead to flawed estimates of allele frequencies and is therefore one reason why conclusions about HWE may be unreliable.

Another possible source of error is to inadvertently sample from more than one population. We noted earlier that identifying population boundaries is often problematic. If genetic data from two or more populations that have different allele frequencies are combined then a **Wahlund effect** will be evident, which means that the proportion of homozygotes will be higher in the aggregate sample than it would be if the populations were analysed separately. This could lead us to conclude erroneously that a population was *not* in HWE, whereas if the data had been analysed separately then we may have found two or more populations that *were* in HWE. An example of this was found in a study of a diving water beetle (*Hydroporus glabriusculus*) that lives in fenland habitats in eastern England. An allozyme study of apparent populations revealed significant heterozygosity deficits (Bilton, 1992), but it was only after conducting a detailed study of the beetle's ecology that the author of this study realized that each body of water actually harbours multiple populations that seldom interbreed. This population subdivision meant that samples pooled from a single water body represented multiple populations, and therefore the heterozygosity deficits could be explained by the Wahlund effect.

## Estimates of genetic diversity

Now that we have a better understanding of allele and genotype frequencies, we will look at some ways to quantify genetic diversity within populations. One of the simplest estimates is **allelic diversity** (often designated $A$), which is simply the average number of alleles per locus. In a population that has four alleles at one locus and six alleles at another locus, $A = (4+6)/2 = 5$. Although straightforward, this method is very sensitive to sample size, meaning that the number of alleles identified will depend in part on how many individuals are screened. A second measure of genetic diversity is the **proportion of polymorphic loci** (often designated $P$). If a population is screened at ten loci and six of these are variable, then $P = 6/10 = 0.60$. This can be of some utility in studies based on relatively invariant loci such as allozymes, although it also is sensitive to sample size. Furthermore, it is often a completely uninformative measure of genetic diversity in studies based on variable markers such as microsatellites which tend to be chosen for analysis only if they are polymorphic and therefore will often have $P$ values of 1.00 in all populations. A third measure of genetic diversity that is also influenced by the number of individuals that are sampled is **observed heterozygosity** ($H_o$), which is obtained by dividing the number of heterozygotes at a particular locus by the total number of individuals sampled. The observed heterozygosity of the scarlet tiger moth based on the data in Table 3.1 is $138/1612 = 0.085$.

Although one or more of the estimates outlined in the preceding paragraph are often included in studies of genetic diversity, they are generally supplemented with an alternative measure known as **gene diversity** ($h$; Nei, 1973). The advantage of gene diversity is that it is much less sensitive than the other methods to sampling effects. Gene diversity is calculated as:

$$h = 1 - \sum_{i=1}^{m} x_i^2 \tag{3.6}$$

where $x_i$ is the frequency of allele $i$, and $m$ is the number of alleles that have been found at that locus. Note that the only data required for calculating gene diversity are the allele frequencies within a population. For any given locus, $h$ represents the probability that two alleles randomly chosen from the population will be different from one another. In a randomly mating population, $h$ is equivalent to the **expected heterozygosity** ($H_e$), and represents the frequency of heterozygotes that would be expected if a population is in HWE; for this reason, $h$ is often presented as $H_e$. Most calculations of $H_e$ will be based on multiple loci, in which case $H_e$ is calculated for each locus and then averaged over all loci to present a single estimate of diversity for each population (see Box 3.3).

**Box 3.3 Calculating $H_e$**

In the following example, we will use Equation 3.6 to calculate $H_e$ from some data that were generated by a study of the southern house mosquito (*Culex quinquefasciatus*) in the Hawaiian Islands (Fonseca, LaPointe and Fleischer, 2000). This is an introduced species that has caused considerable devastation on the Hawaiian archipelago because it is the vector for avian malaria. Table 3.2 shows the allele frequencies at one locus calculated from two populations.

Following Equation 3.6 and using the data from Table 3.2, $H_e$ from the Midway population can be calculated as:

$$H_e = 1 - (0.250^2 + 0.200^2 + 0.550^2)$$
$$= 1 - (0.0625 + 0.04 + 0.3025)$$
$$= 0.595$$

Similarly, $H_e$ from the Kauai population can be calculated as:

$$H_e = 1 - (0.022^2 + 0.333^2 + 0.333^2 + 0.311^2)$$
$$= 1 - (0.000484 + 0.111 + 0.111 + 0.0967)$$
$$= 0.68$$

In this case, $H_e$ is higher in Kauai than Midway, which is not surprising since the former population has a greater number of alleles and also a more even distribution of allele frequencies than the latter.

**Table 3.2** Allele frequency data for one microsatellite locus characterized in two Hawaiian populations of *C. quinquefasciatus*. Data are from Fonseca, LaPointe and fleischer. (2000)

| Microsatellite alleles (bp) | Allele frequencies | |
| --- | --- | --- |
| | Midway population | Kauai population |
| 212 | 0 | 0.022 |
| 216 | 0.250 | 0.333 |
| 218 | 0.200 | 0.333 |
| 224 | 0.550 | 0.311 |

Research papers typically report several different calculations of a population's genetic diversity, and these often include both $H_o$ and $H_e$. By comparing these two values, we can determine whether or not the heterozygosity within a population is significantly different from that expected under HWE. If $H_o$ is lower than $H_e$ then we may have to rule out the possibility of **null alleles**. Although potentially applicable to a range of markers, this term is used most commonly to describe microsatellite alleles that do not amplify during PCR. The most common cause of this is a mutation in one or both of the primer-binding sequences. If only one allele from a heterozygote is amplified then it will be genotyped erroneously as a homozygote. When $H_o$ is significantly less than $H_e$ we should also be open to the possibility of a Wahlund effect, which, as noted earlier, will decrease $H_o$. If neither null alleles nor a Wahlund effect have caused an observed heterozygosity deficit then we may conclude that the population is not in HWE. As noted earlier, this deviation could result from one or more of a number of factors, including non-random mating (e.g. inbreeding), natural selection or a small population size.

It can be difficult to determine just what is responsible for disparities between $H_o$ and $H_e$. In one study, estimates of $H_e$ and $H_o$ were obtained for twelve European populations of the common ash (*Fraxinus excelsior*) based on microsatellite data from five loci. Deviations from HWE were apparent in ten of these populations, which is an unusual finding in forest tree populations (Morand *et al.*, 2002). These deviations were caused by $H_o$ deficits at all five loci (Table 3.3), a

**Table 3.3**  Number of alleles, expected heterozygosity ($H_e$) and observed heterozygosity ($H_o$) for three populations of the common ash, based on data from five microsatellite loci. In most cases, $H_e$ is significantly larger than $H_o$. Data from Morand *et al.* (2002)

|  | Locus 1 | Locus 2 | Locus 3 | Locus 4 | Locus 5 |
|---|---|---|---|---|---|
| Population 1 | | | | | |
| No. of alleles | 12 | 13 | 12 | 9 | 16 |
| $H_e$ | 0.938 | 0.888 | 0.905 | 0.833 | 0.937 |
| $H_o$ | 0.385 | 0.895 | 0.571 | 0.750 | 0.737 |
| Population 2 | | | | | |
| No. of alleles | 12 | 12 | 12 | 11 | 9 |
| $H_e$ | 0.938 | 0.825 | 0.936 | 0.892 | 0.917 |
| $H_o$ | 0.462 | 0.647 | 0.333 | 0.526 | 0.500 |
| Population 3 | | | | | |
| No. of alleles | 16 | 12 | 13 | 12 | 12 |
| $H_e$ | 0.932 | 0.905 | 0.859 | 0.862 | 0.918 |
| $H_o$ | 0.667 | 0.875 | 0.750 | 0.556 | 0.882 |

consistent result that was unlikely to be attributable to natural selection acting on all five putatively neutral loci. Inbreeding also seemed unlikely in this wind-pollinated species, because long-distance dispersal of pollen should minimize mating between relatives. A comparison of microsatellite genotypes between parents and offspring suggested that null alleles were unlikely to be the cause but, because no plausible explanation for the observed heterozygote deficit has been found, the authors could not conclusively rule out either null alleles or a possible Wahlund effect.

*Haploid diversity*

Gene diversity (*h*) also can be calculated for haploid data. Estimates of genetic diversity based on mitochondrial data, for example, often use *h* as a measure of haplotype diversity. In this context, *h* describes the numbers and frequencies of different mitochondrial haplotypes and is essentially the heterozygosity equivalent for haploid loci. However, the haplotype diversity of relatively rapidly evolving genomes such as animal mtDNA will often approach 1.0 within a population if a high proportion of individuals have unique haplotypes. It can be more informative, therefore, to consider the number of nucleotide differences between any two sequences as opposed to simply determining whether or not they are different. This can be done by calculating **nucleotide diversity** (π; Nei, 1987), which quantifies the mean divergence between sequences. Nucleotide diversity is calculated as:

$$\pi = \Sigma f_i f_j p_{ij} \qquad (3.7)$$

where $f_i$ and $f_j$ represent the frequencies of the *i*th and *j*th haplotypes in the population, and $p_{ij}$ represents the sequence divergence between these haplotypes. By factoring in both the frequencies and the pairwise divergences of the different sequences, *p* calculates the probability that two randomly chosen homologous nucleotides will be identical.


## Choice of marker

When comparing populations, it is important to realize that estimates of genetic diversity will vary depending on which molecular markers are used. This is because, as noted in earlier chapters, mutation rates vary both within and between genomes, and rapidly evolving markers such as microsatellites will generally reflect higher levels of diversity than more slowly evolving markers such as allozymes. Furthermore, comparisons between nuclear and organelle genomes may be influenced by past demographic histories; recall from Chapter 2 that the relatively small effective population sizes of mtDNA and cpDNA mean that mitochondrial and chloroplast diversity will be lost more rapidly than nuclear diversity following either permanent or temporary reductions in population size.

   Discrepant estimates of genetic diversity were found in a study that used several different markers to compare European populations of the common carp (*Cyprinus carpio*) (Kohlmann *et al.*, 2003). According to data from 22 allozyme loci, $H_o$ = 0.066, $H_e$ = 0.062 and A = 1.232. Substantially higher values of $H_o$ = 0.788, $H_e$ = 0.764 and A = 5.75 were obtained from four microsatellite loci. An even greater difference was found in the mitochondrial genome. Mitochondrial haplotypes identified using PCR-RFLP revealed haplotype and nucleotide diversity estimates of zero. Genetic diversity in European common carp therefore ranges from non-existent when estimated from mitochondrial markers to highly variable when estimated from microsatellite markers. This does not, however, mean that organelle markers always will be less diverse than nuclear markers. Red pine (*Pinus resinosa*) populations in Canada showed no allozyme variation and very little RAPD variation, but a survey of nine chloroplast microsatellite loci revealed 25 alleles and 23 different haplotypes in 159 individuals (Echt *et al.*, 1998). Table 3.4 gives some other examples of genetic diversity estimates that vary depending on which markers were used.

**Table 3.4**  Comparisons of within-population variation, measured as $H_e$, based on several different types of markers. Microsatellite loci often are more variable than either allozyme or dominant markers

| Species | $H_e$ | Reference |
| --- | --- | --- |
| Gray mangrove (*Avicennia marina*) | AFLP: 0.19 <br> Microsatellites: 0.78 | Maguire, Peakall and Saenger (2002) |
| Russian couch grass (*Elymus fibrosus*) | RAPD: 0.10 <br> Allozymes: 0.008 <br> Microsatellites: 0.25 | Sun *et al.* (1998) |
| Wild and cultivated soybean (*Glycine soja* and *G. max*) | AFLP: 0.32 <br> RAPD: 0.31 <br> Microsatellites: 0.60 | Powell *et al.* (1996) |
| Wild barley (*Hordeum spontaneum*) | AFLP: 0.16 <br> Microsatellites: 0.47 | Turpeinen *et al.* (2003) |
| Lodgepole pine (*Pinus contorta*) | RAPD: 0.43 <br> Microsatellites: 0.73 | Thomas *et al.* (1999) |
| Chinese native chickens (*Gallus gallus domesticus*) | Allozymes: 0.221 <br> RAPD: 0.263 <br> Microsatellites: 0.759 | Zhang *et al.* (2002) |
| Pink ling, a marine fish (*Genypterus blacodes*) | Allozymes: 0.324 <br> Microsatellites: 0.823 | Ward *et al.* (2001) |
| Roe deer (*Capreolus capreolus*) | Allozymes: 0.213 <br> Microsatellites: 0.545 | Wang and Schreiber (2001) |

Regardless of how variable they are, the effective number of loci being screened will be the same as the actual number only if they are in **linkage equilibrium**, which will be true only if they segregate independently of each other during reproduction. Non-random association of alleles among loci is known as **linkage disequilibrium**; this can occur for a number of reasons, the most common being the proximity of two loci on a chromosome. When analysing data from multiple loci it is always necessary to test for linkage disequilibrium before ruling out the possibility that there are fewer independent loci for genetic analysis than anticipated. Linkage disequilibrium may also cause loci to behave in an unexpected manner, for example neutral alleles that are linked to selected alleles will appear non-neutral and are unlikely to be in HWE even if the population is large and mating is random.

# What Influences Genetic Diversity?

Genetic diversity is influenced by a multitude of factors and therefore varies considerably between populations. In this section we shall look at some of the most important determinants of genetic diversity, including **genetic drift**, population bottlenecks, natural selection and methods of reproduction. While reading about these, it is important to keep in mind that no process acts in isolation, for example the rate at which a population recovers from a bottleneck will depend in part on its reproductive ecology. Furthermore, it is difficult to predict the extent to which a particular factor will influence genetic diversity because no two populations are the same. Nevertheless, several factors have a universal relevance to genetic diversity, and these will comprise the remainder of this chapter.

## Genetic drift

Genetic drift is a process that causes a population's allele frequencies to change from one generation to the next simply as a result of chance. This happens because reproductive success within a population is variable, with some individuals producing more offspring than others. As a result, not all alleles will be reproduced to the same extent, and therefore allele frequencies will fluctuate from one generation to the next. Because genetic drift alters allele frequencies in a purely random manner, it results in non-adaptive evolutionary change. The effects of drift are most profound in small populations where, in the absence of selection, drift will drive each allele to either fixation or extinction within a relatively short period of time, and therefore its overall effect is to decrease genetic diversity. Genetic drift will also have an impact on relatively large populations but, as we shall see later in this chapter, a correspondingly longer time period is required before the effects become pronounced. Genetic drift is an extremely influential force in population genetics and forms the basis of one of the most important theoretical measures of a population's genetic structure: **effective population size** ($N_e$). Because genetic drift and $N_e$ are inextricably linked, we will now spend some time looking at how $N_e$ differs from **census population size** ($N_c$), how it is linked to genetic drift, and what this ultimately means for the genetic diversity of populations.

### What is effective population size?

A fundamental measure of a population is its size. The importance of population size cannot be overstated because, as we shall see throughout this text, it can influence virtually all other aspects of population genetics. From a practical point of view, relatively large populations are, all else being equal, more likely to survive than small populations. This is why the World Conservation Union (IUCN) uses population size as a key variable, considering a species to be critically endangered if it consists of a population that numbers fewer than 50 mature individuals. Taken in its simplest form, population size refers simply to the number of individuals that are in a particular population – this is the census population size ($N_c$). From the point of view of population genetics, however, a more relevant measure is the effective population size ($N_e$).

The $N_e$ of a population reflects the rate at which genetic diversity will be lost following genetic drift, and this rate is inversely proportional to a population's $N_e$. In an ideal population $N_e = N_c$, but in reality this is seldom the case. If an actual population of 500 individuals is losing genetic variation through drift at a rate that would be found in an ideal population of 100 individuals, then this population would have $N_c = 500$ but $N_e = 100$, in other words it will be losing diversity much more rapidly than would be expected in an ideal population of 500. An $N_e/N_c$ ratio of $100/500 = 0.2$ would not be considered unusually low; one review calculated the average ratio of $N_e/N_c$ in wild populations, based on the results of nearly 200 published results, as approximately 0.1 (Figure 3.3; Frankham, 1995). We will now look at three of the most common reasons why $N_e$ is often much smaller than $N_c$: uneven sex ratios, variation in reproductive success, and fluctuating population size. At the end of this section we will return to an explicit discussion of the relationship between $N_e$, genetic drift, and genetic diversity.

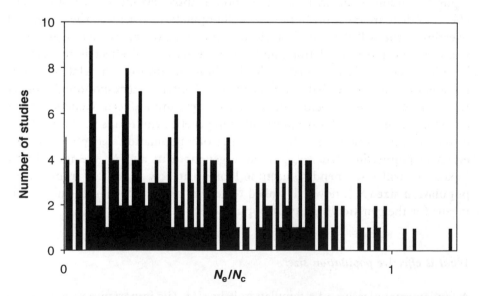

**Figure 3.3**  A review of published studies revealed a range of $N_e/N_c$ values in insects, molluscs, fish, amphibians, reptiles, birds, mammals and plants. Note that although $N_e$ is often much less than $N_c$, it is a theoretical measurement and under some conditions can be greater than $N_c$ (data from Frankham, 1995, and references therein)

## What influences $N_e$?

*Sex ratios*   Unequal sex ratios generally will reduce the $N_e$ of a population. An excess of one or the other sex may result from adaptive parental behaviour. Although the mechanisms behind this are not well understood, there is increasing evidence for parental manipulation of offspring sex ratios in a number of taxonomic groups, including some bird species, which may be responding to environmental constraints such as a limited food supply (Hasselquist and Kempenaers, 2002). Even if the overall sex ratio in a population is close to 1.0, the sex ratio of breeding adults may be unequal, and it is the relative proportion of reproductively successful males and females that ultimately will influence $N_e$. In elephant seal populations, for example, fighting between males for access to harems is fierce. This intense competition means that within a typical breeding season only a handful of dominant males in each population will contribute their genes to the next generation, whereas the majority of females reproduce. This disproportionate genetic contribution results in an effectively female-biased sex ratio.

The effect of an unequal sex ratio on $N_e$ is approximately equal to:

$$N_e = 4(N_{ef})(N_{em})/(N_{ef} + N_{em}) \tag{3.8}$$

where $N_{ef}$ is the effective number of breeding females and $N_{em}$ is the effective number of breeding males. The importance of the sex ratio can be illustrated by a comparison of two hypothetical populations of house wrens (*Troglodytes aedon*), which tend to produce an excess of females when conditions are harsh (Albrecht, 2000). Each of these populations has 1000 breeding adults. In the first population, conditions have been favourable for several years and so the $N_{ef}$ of 480 was comparable to the $N_{em}$ of 520. The $N_e$ therefore would be:

$$N_e = 4(480)(520)/(480 + 520) = 998$$

The second population, however, has been experiencing relatively harsh conditions for some time. As a result, the $N_{ef}$ is 650 but the $N_{em}$ is only 350. The $N_e$ in this population is:

$$N_e = 4(650)(350)/(650 + 350) = 910$$

In this example, the $N_e/N_c$ in the first population, which had almost the same number of males and females, was $998/1000 = 0.998$. The $N_e/N_c$ in the second population, with its disproportionately large number of females, was $910/1000 = 0.910$. Although the $N_e/N_c$ ratio was smaller in the second population, the reduction in $N_e$ that is attributable to uneven sex ratios was actually relatively low in both of these hypothetical populations compared to what we would find in many wild populations. According to one survey of multiple taxa, uneven sex ratios cause effective population sizes to be an average of 36 per cent lower than census population sizes (Frankham, 1995), although not surprisingly there is considerable variation both within and among species.

*Variation in reproductive success*    Even if a population had an effective sex ratio of 1:1, not all individuals will produce the same number of viable offspring, and this **variation in reproductive success** (**VRS**) will also decrease $N_e$ relative to $N_c$. In some species the effects of this can be pronounced. Genetic and demographic data were obtained from a 17-year period for a steelhead trout (*Oncorhynchus mykiss*) population in Washington State, and variation in reproductive success was found to be the single most important factor in reducing $N_e$ (Ardren and Kapuscinski, 2003). When this trout population is at high density, i.e. when $N_c$ is large, females experience increased competition for males, spawning sites and other resources. The successful competitors will produce large numbers of off-spring whereas the less successful individuals may fail to reproduce. In other species, variation in reproductive success may have relatively little influence on $N_e$. The relatively high $N_e/N_c$ ratio in balsam fir (*Abies balsamea*; Figure 3.4) has been attributed partly to overall high levels of reproductive success in this wind-pollinated species (Dodd and Silvertown, 2000).

The effects of reproductive variation on $N_e$ can be quantified if we know the VRS of a population. Reproductive success reflects the number of offspring that each individual produces throughout its lifetime and therefore can be determined from a single breeding season in short-lived species, although individuals with multiple breeding seasons must be monitored for the requisite number of years. Long-term monitoring of a population of Darwin's medium ground finch (*Geospiza fortis*) on the Galápagos archipelago provided an estimated VRS of 7.12 (Grant and Grant, 1992a). The effects of VRS on $N_e$ can be calculated as follows:

$$N_e = (4N_c - 2)/(VRS + 2) \tag{3.9}$$

If the census population size of *G. fortis* is 500 on a particular island, then the influence of variation in reproductive success on $N_e$ will be:

$$N_e = [4(500) - 2]/(7.12 + 2) = 219$$

Therefore, even if the sex ratio is equal, the variation in the number of chicks that each individual produces will cause $N_e$ to be substantially smaller than $N_c$.

VRS may be highest in clonal species. In the freshwater bryozoan (moss animal) *Cristatella mucedo* (Figure 3.5), clonal selection throughout the growing season means that some clones are eliminated whereas others reproduce so prolifically

**Figure 3.4** Balsam fir (*Abies balsamea*). Wind pollination in this species helps to maintain overall high levels of reproductive success, and this helps to keep the $N_e/N_c$ ratios high within populations. Photograph provided by Mike Dodd and reproduced with permission

that the $N_c$ of a population may be in the tens of thousands by the end of the growing season (Freeland, Rimmer and Okamura, 2001). Because clonal selection is decreasing the proportion of unique genotypes throughout the summer (Figure 3.6), the VRS must be substantial, with some clones producing no offspring and others producing large numbers of young. In the most extreme scenario, some populations of bryozoans and other clonal taxa may become dominated by a single clone that experiences all of the reproductive success within that population, and when this happens the effective population size is virtually one (Freeland, Noble and Okamura, 2000). If this occurs in a population with a large $N_c$, the $N_e/N_c$ ratio will approach zero.

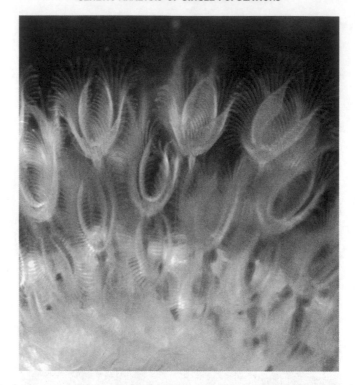

**Figure 3.5**  A close-up photograph showing a portion of a colony of the freshwater bryozoan *Cristatella mucedo*. These extended tentacular crowns are approximately 0.8 mm wide and capture tiny suspended food particles. Photograph provided by Beth Okamura and reproduced with permission

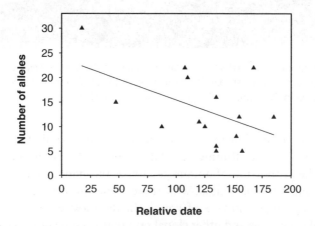

**Figure 3.6**  Linear regression of ln-relative date (sampling date represented as number of days after 1 January) versus total number of alleles in a UK population of the freshwater bryozoan *Cristatella mucedo* (redrawn from Freeland, Rimmer and Okamura, 2001). Clonal selection has reduced the genetic diversity of this population throughout the growing season, even though the number of colonies increased during this time. This leads to a reduction in the $N_e/N_c$ ratio

*Fluctuating population size* Regardless of a species' breeding biology, fluctuations in the census population size from one year to the next will have a lasting effect on $N_e$. A survey of multiple taxa suggested that fluctuating population sizes have reduced the $N_e$ of wild populations by an average of 65 per cent, making this the most important driver of low $N_e/N_c$ ratios (Frankham, 1995). This is because the long-term effective population size is determined not by the $N_e$ averaged across multiple years, but by the **harmonic mean** of the $N_e$ (Wright, 1969). The harmonic mean is the reciprocal of the average of the reciprocals, which means that low values have a lasting and disproportionate effect on the long-term $N_e$. A population crash in one year, therefore, may leave a lasting genetic legacy even if a population subsequently recovers its former abundance. A population crash of this sort is known as a bottleneck and it may result from a number of different factors, including environmental disasters, over-hunting or disease.

Because fluctuations in population size have such lasting effects on genetic diversity, we will take a more detailed look at bottlenecks later in this chapter. For now, we will limit ourselves to looking at how fluctuating population sizes influence $N_e$, which can be calculated as follows:

$$N_e = t/[(1/N_{e1}) + (1/N_{e2}) + (1/N_{e3}) \cdots + (1/N_{et})] \qquad (3.10)$$

where $t$ is the total number of generations for which data are available, $N_{e1}$ is the effective population size in generation 1, $N_{e2}$ is the effective population size in generation 2, and so on.

The fringed-orchid (*Platanthera praeclara*) is a globally rare plant that occurs in patches of tallgrass prairie in Canada. The $N_e$ of most populations is substantially reduced by fluctuations in population size from one year to the next. If a population had a census size of 220, 70, 40 and 200 during each of the past four years, and we assume that $N_e/N_c = 1.0$, then the effects that these fluctuations would have had on the $N_e$ can be calculated as:

$$N_e = 4/[(1/220) + (1/70) + (1/40) + (1/200)]$$
$$N_e = 82$$

Even though this population rebounded from the bottleneck that it experienced in years 2 and 3, this temporary reduction in $N_c$ means that the current $N_e/N_c$ ratio is only $82/200 = 0.41$. Note that we have limited our example to a 4-year period for the sake of simplicity, although a longer period is needed for an accurate estimation of $N_e$.

So far we have looked at how individual factors – sex ratios, VRS, and fluctuating population sizes – can influence $N_e$. In each of the preceding sections we calculated the effects of a single variable on $N_e$, but in reality all of these variables can simultaneously influence a population's $N_e$. We are highly unlikely to have enough information to calculate individually the reduction in $N_e$ that is attributable to each relevant variable. In the next section, therefore, we will move away from examining the effects of single variables and instead look at how we can calculate a population's overall $N_e$ regardless of which factors have caused the biggest reduction in $N_e$.

### Calculating $N_e$

There are three general approaches for estimating $N_e$. The first of these, based on long-term ecological data, requires accurate census sizes and a thorough understanding of a population's breeding biology, neither or which are available for most species. A second approach is based on some aspect of a population's genetic structure at a single point in time, e.g. heterozygosity excess (Pudovkin, Zaykin and Hedgecock, 1996) or linkage disequilibrium (Hill, 1981). The application of mutation models to parameters such as these can provide estimates of $N_e$, although this approach is not used widely because it makes many assumptions about the source of genetic variation and can be influenced strongly by demographic processes such as immigration (Beaumont, 2003).

The third approach, which is considered by many to be the most reliable, requires samples from two or more time periods that are separated by at least one generation. Several different methods can then be used to calculate $N_e$ from the variation in allele frequencies over time. At this time, the most widely used method is based on Nei and Tajima's (1981) method for calculating the variance of allele frequency change ($F_c$) as follows:

$$F_c = 1/K\Sigma(x_i - y_i)/[(x_i + y_i)^2/(2 - x_iy_i)] \tag{3.11}$$

where $K =$ the total number of alleles and $i =$ the frequency of a particular allele at times $x$ and $y$, respectively. This value then can be used to calculate $N_e$ while correcting for sample size and $N_c$ by using the following equation (after Waples, 1989):

$$N_e = t/2[F_c - 1/(2S_0) - 1/(2S_t)] \tag{3.12}$$

where $t =$ generation time, $S_0 =$ sample size at time zero and $S_t =$ sample size at time $t$.

The temporal variance in allele frequencies was used to calculate the $N_e$ of crested newt (*Triturus cristatus*) populations that were sampled from ponds in western France. Researchers first were able to obtain an accurate census size of these populations using a standard mark–recapture method. As they were counted, individuals were marked by removing toes, which then were used as sources of DNA for deriving genetically based estimates of $N_e$. The census

**Table 3.5** Some estimates of $N_e/N_c$. In all these examples, $N_e$ was calculated using a method based on the temporal variance in allele frequencies

| Species | $N_e/N_c$ | Reference |
|---|---|---|
| Steelhead trout (*Oncorhynchus mykiss*) | 0.73 | Ardren and Kapuscinski (2003) |
| Domestic cat (*Felis catus*) | 0.40–0.43 | Kaeuffer, Pontier and Perrin (2004) |
| Red drum, a marine fish (*Sciaenops ocellatus*) | 0.001 | Turner, Wares and Gold (2002) |
| Crested newt (*Triturus cristatus*) Marbled newt (*T. marmoratus*) | 0.16 0.09 | Jehle *et al.* (2001) |
| Shining Fungus beetle (*Phalacrus substriatus*) | 0.021 | Ingvarsson and Olsson (1997) |
| Carrot (*Daucus carota*) | 0.71 | Le Clerc *et al.* (2003) |
| Grizzly bear (*Ursus arctos*) | 0.27 | Miller and Waits (2003) |
| Apache silverspot butterfly (*Speyeria nokomis apacheana*) | 0.001–0.030 | Britten *et al.* (2003) |
| Pacific oyster (*Crassostrea gigas*) | $<10^{-6}$ | Hedgecock, Chow and Waples (1992) |
| Giant toad (*Bufo marinus*) | 0.016–0.008 | Easteal and Floyd (1986) |

population size in one pond was approximately 77 newts in 1989 and 73 newts in 1998. The variance in allele frequencies between 1989 and 1998, based on eight microsatellite loci, provided an $N_e$ estimate of approximately 12 and an $N_e/N_c$ ratio of 0.16 (Jehle *et al.*, 2001). Other examples of $N_e/N_c$ ratios that have been calculated from temporal changes in allele frequencies are given in Table 3.5.

Estimating $N_e$ from the variance in allele frequencies can be logistically challenging because of the time and expense involved in sampling the same population in multiple years. Obtaining samples from museums is one answer to this, although museum specimens are a finite resource and not all species will have sufficient representation. Furthermore, some taxa such as soft-bodied invertebrates are not amenable to preservation in museums, and in many cases plants will be underrepresented. Practical limitations may also arise from the availability of markers; because it is based on allele frequencies, the temporal method ideally should be done with data from co-dominant loci. Dominant data such as AFLPs can also be used, although, as noted earlier, accompanying estimates of allele frequencies will assume Hardy–Weinberg equilibrium, which may be unrealistic.

Perhaps the biggest drawback to estimating $N_e$ from the temporal variance in allele frequencies is the assumption that all changes in allele frequencies are a result of genetic drift. This does not allow for the possibility that immigrants from other populations are introducing new alleles and therefore altering allele frequencies through a process that is completely separate from genetic drift. As we will see in the next chapter, most populations receive immigrants with some regularity, and therefore this assumption is unlikely to be met. This problem has been partially addressed by a recently developed **maximum likelihood** (**ML**) approach that estimates $N_e$ from temporal changes in allele frequencies in a way that partitions the effects of both immigration and genetic drift (Wang and Whitlock, 2003).

Maximum likelihood is a general term for a statistical method that first specifies a set of conditions underlying a particular data set, and then determines the likelihood that these particular conditions would have given rise to the data in question. In the case of $N_e$, conditions may include a particular evolutionary history of the alleles in question, and maximum likelihood would be used to calculate the probability that different scenarios would have resulted in the observed variance in allele frequencies (Berthier et al., 2002). Maximum likelihood is a powerful approach, although it is computationally demanding and analytically complex. For these reasons it has avoided the mainstream so far, although its popularity is increasing as computers become more powerful and software becomes more user-friendly, and it may soon become the analytical method of choice for several aspects of molecular ecology including estimates of $N_e$.

Wang and Whitlock's (2003) method is an extremely promising development in the quest for accurate estimates of $N_e$. However, it does require data from a sufficient number of variable markers to allow the detection of even relatively small changes in allele frequencies; this may be particularly demanding when $N_e$ is relatively large and migration rates are relatively small. In addition, it requires allele frequency data from both the population under investigation (focal population) and the populations from which immigrants may be originating (potential source populations). Assuming that the latter can be identified, one option is to pool data from all possible source populations and estimate the extent to which their collective contribution of migrants to the focal population has influenced the variance in allele frequencies that might otherwise be attributed entirely to drift. This method was applied to a metapopulation of newts (*Triturus cristatus* and *T. marmoratus*) in France. The $N_e/N_c$ ratios ranged from 0.07 to 0.51 when researchers assumed that changes in allele frequencies were solely a result of drift, and were 0.05 – 0.65 when they allowed for the effects of immigrants (Jehle et al., 2005). Because it aims to separate the effects that genetic drift and migration have on changing allele frequencies, this approach marks a significant step forward in the quantification of $N_e$. Although none is perfect, methods for estimating $N_e$ have become increasingly refined in recent years, and this trend will undoubtedly continue because accurate estimates of $N_e$ are crucial for understanding many different aspects of population genetics and evolution.

### Effective population size, genetic drift and genetic diversity

We started this section by identifying genetic drift as one of the key processes that influences the genetic diversity of populations. We will now return to that concept by looking at the specific relationship between $N_e$, genetic drift and genetic diversity. The genetic diversity of a population will be reduced whenever an allele reaches fixation (attains a frequency of 1.0) because, when this occurs, the population has only one allele at that particular locus. The probability that a novel mutation will become fixed in a population as a result of genetic drift is $1/(2N_e)$ for diploid loci, in ohter words it is inversely proportional to the population's $N_e$ (Figure 3.7). Since the rate at which alleles drift to fixation also represents the rate at which all other alleles at that locus will be lost, $1/(2N_e)$ can be considered as the rate at which genetic variation will be lost within a population as a result of genetic drift.

The predictable relationship between $N_e$ and genetic drift means that if we know the effective size of a population and its current genetic diversity (measured as expected heterozygosity), and if we assume that the population size remains essentially constant, we can calculate what the heterozygosity will become after a given time period as:

$$H_t = [1 - 1/(2N_e)]^t H_0 \qquad (3.13)$$

where $H_t$ and $H_0$ represent heterozygosity at time $t$ and time zero, respectively. Time intervals refer to generations, not years (although they will of course be the same if the generation time is 1 year). The predicted heterozygosity at time $t$ is represented more commonly as a proportion of the heterozygosity at time zero:

$$H_t/H_0 = [1 - 1/(2N_e)]^t \qquad (3.14)$$

This tells us what proportion of the initial heterozygosity will be remaining after $t$ generations. We can use this equation to compare the expected changes in heterozygosity in two hypothetical populations of crested newts that have a generation time of 1 year. The first population lives in a lake and retains an effective population size of approximately 200 for a period of 10 years. The second population inhabits a small pond and has an $N_e$ of approximately 40 for the same time period. From Equation 3.14 we can estimate the proportional change in heterozygosity as:

$$H_t/H_0 = [1 - 1/(2 \times 200)]^{10} = [0.9975]^{10} = 0.975$$

for the lake population, and as:

$$H_t/H_0 = [1 - 1/(2 \times 40)]^{10} = [0.9875]^{10} = 0.882$$

for the pond population. This means that the lake population will lose approximately 2.5 per cent of its initial heterozygosity in ten generations, whereas the smaller pond population will lose around 12 per cent of its heterozygosity.

The rate of drift does not depend solely on a population's $N_e$; it is also influenced by the population sizes of the genome in question (Table 3.6 and Figure 3.7). Since the population sizes of plastids and mitochondria are effectively

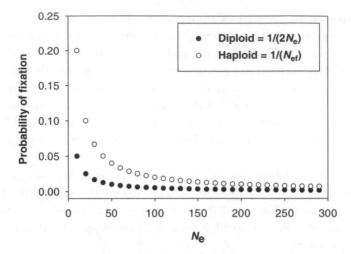

**Figure 3.7** The probability that a neutral mutation will reach fixation in any given generation reflects the rate at which genetic variation will be lost following genetic drift. In this figure, the probability of fixation is given for diploid loci, calculated as $1/(2N_e)$, and for mitochondrial haplotypes, which is calculated as $1/N_{ef}$ (here we have assumed that half of the breeding population is female; see Table 3.6). Note that the probability of fixation (and the accompanying loss of alleles) following genetic drift is inversely proportional to the effective population size

a quarter those of nuclear genomes in diploid species, they will lose genetic variation at a faster rate than most nuclear genes (Figure 3.7). Returning to our example of crested newts, we know that the rate at which genetic variation is lost from diploid nuclear genes is $1/2N_e$ per generation, which is around $1/400 = 0.0025$ in the lake population of 200 individuals. Table 3.6 tells us that in the same population, assuming that half of the $N_e$ is female, mitochondrial variation will be lost at the much faster rate of approximately $1/100 = 0.01$ per generation.

**Table 3.6** The rate at which genetic variation is lost each generation following genetic drift will depend on the population size of the locus in question. Adapted from Wright (1969)

| Relative population size | Rate at which variation is lost each generation |
| --- | --- |
| Haploid | $1/N_e$ |
| Diploid | $1/(2N_e)$ |
| Tetraploid | $1/(4N_e)$ |
| Plastid DNA | $1/N_{ef}$[a] |
| mtDNA | $1/N_{ef}$[a] |

[a]This is true for taxa in which plastid DNA (including cpDNA) and mtDNA are maternally inherited, because $N_{ef}$ is the effective number of females in a population.

## Population bottlenecks

We have noted already that a population bottleneck will reduce the effective size of a population and hence its overall level of genetic diversity. We are now returning to bottlenecks for a more in-depth discussion because, as mentioned previously, they are one of the most important determinants of a population's genetic diversity and therefore merit further discussion. The methods outlined previously for quantifying the effect of bottlenecks on $N_e$, and hence genetic diversity, required long-term demographic data. Because such data are seldom available, potential relationships between bottlenecks and genetic variation are more likely to be inferred from situations in which biologists discover a genetically depauperate population for which no records exist, and then have to try to decide whether or not a past bottleneck can explain why current genetic diversity is so low.

Detecting past reductions in population size is seldom straightforward. For one thing, the severity of any population bottleneck depends on both the size that a population is reduced to and the speed at which it recovers. In general, the initial loss of alleles is proportional to the reduction in population size. Genetic drift means that diversity will continue to deplete while the population size remains low, and as a result populations that take longer to rebound will generally lose more genetic diversity than populations that recover rapidly.

The detection of bottlenecks is confounded further by the fact that not all measures of genetic diversity will show a uniform decrease. On the one hand, allelic diversity usually decreases after a bottleneck because rare alleles will be lost. This often is associated with a drop in $H_e$ because fewer alleles typically lead to reduced expectations of heterozygosity under HWE. At the same time, $H_o$ values may not deplete and in fact there may be a temporary increase in observed heterozygosity compared with expectations under HWE. In some cases, observed heterozygosity excess can be a useful indicator of past bottlenecks, although this test requires a large number of polymorphic loci and can detect only relatively recent bottlenecks (Luikart and Cornuet, 1998).

If samples are available from two or more generations then we may be able to infer past bottlenecks from the variation in allele frequencies between generations. We know that if a population has undergone a bottleneck its $N_e$ will be reduced. Because the rate of genetic drift is inversely proportional to a population's $N_e$, a bottleneck will accelerate drift. This will lead to an inflated variance in allele frequencies, which may be taken as evidence that a bottleneck occurred between two sampling periods. Simulations suggest that the temporal variance method provides an 85 per cent probability of detecting a bottleneck after a single generation, provided that data can be obtained from at least five highly variable loci and at least 30 individuals from each sampling period (Luikart et al., 1998). However, the same study also suggested that the temporal variance test may not be useful for populations that have undergone prolonged bottlenecks (>3–5 generations) because these populations will continue to lose alleles. Once an allele is lost, its frequency must remain at zero and therefore it cannot exhibit any variance between generations. Prolonged bottlenecks mean that a relatively large proportion of alleles will go extinct and therefore will show no change in frequency over time, and as a result $N_e$ will be overestimated.

Several methods for inferring past bottlenecks were evaluated in a series of experimental populations of the western mosquitofish *Gambusia affinis* (Spencer, Neigel and Leberg, 2000). The authors of this study set up multiple populations that were founded by 2–16 individuals, and after two or three generations they quantified genetic diversity in a number of different ways. The most useful measure was the temporal variance in allele frequencies, because the degree to which allele frequencies changed was inversely proportional to the size of the founding population. Allelic diversity and $H_e$ also proved to be suitable for detecting bottlenecks (Figure 3.8), but neither $H_o$ nor the proportion of polymorphic loci showed any relationship with the severity of the bottleneck.

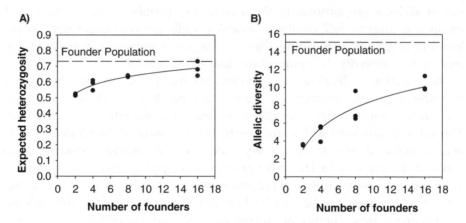

**Figure 3.8** Relationship between the size of a founding population and (A) expected heterozygosity ($H_e$) and (B) allelic diversity ($A$), both of which were measured after two or three generations in experimental populations of the western mosquitofish *Gambusia affinis*. Dashed lines show the relevant measures of genetic diversity in the founding population (data from Spencer, Neigel and Leberg, 2000)

It is important to keep in mind that the results from the experimental western mosquitofish populations will not be transferable to all populations. The most appropriate indicator of past bottlenecks will depend on a number of factors, including the variability of the marker that is used and both the recency and severity of the bottleneck. A related study of the eastern mosquitofish (*G. holbrooki*), based on allozyme data, found that temporal variance in allele frequencies may be misleading if only a few loci are characterized or if the population lost many alleles during a bottleneck (Richards and Leberg, 1996). In this study, changes in allelic diversity were tied more closely to bottlenecks than were changes in allele frequencies. It is therefore advisable, whenever possible, to obtain multiple estimates of genetic diversity before inferring past bottlenecks (see also Box 3.4).

Conflicting results have also been found when nuclear and organelle markers are used to evaluate the effects of bottlenecks. The relatively low population size of organelle genomes means that they are more sensitive to bottlenecks and may provide exaggerated estimates of post-bottleneck reductions in genetic diversity. This is illustrated by Scandinavian brown bears (*Ursus arctos*), which underwent a well-documented population bottleneck between the mid-1800s and 1930, during which time the population of around 5000 plummeted to approximately 130 bears distributed among four subpopulations. Today, the total population is thriving (approximately 1000 bears), but there has been some concern about current levels of genetic diversity.

A recent investigation revealed only two mtDNA haplotypes in 380 bears, which likely reflects the loss of most haplotypes during the bottleneck of the late 19th century. At the same time, data from 19 microsatellite loci revealed estimates of $H_e$ comparable with those found in North American brown bear populations that have not undergone severe bottlenecks (Waits *et al.*, 2000) (Table 3.7 and Figure 3.9). Although the bottleneck in Scandinavia was severe enough for most mtDNA haplotypes to be lost, multiple microsatellite alleles appeared to be maintained within each of the four subpopulations. In this case, mitochondrial data suggest that bottlenecks have had a lasting impact on genetic diversity in Scandinavian brown bears, whereas nuclear data show that the impact on highly polymorphic markers has been minimal.

**Table 3.7** Comparison of $H_e$ values (based on microsatellite data) from four post-bottleneck subpopulations of Scandinavian brown bears (Waits *et al.*, 2000) and five North American brown bear populations that have not experienced severe bottlenecks (Paetkau *et al.*, 1998); $N_c$ is the estimated census size of each population at the time that samples were taken for genetic analysis

| Population | $N_c$ | $H_e$ |
| --- | --- | --- |
| Scandinavian subpopulation 1 | 108 | 0.70 |
| Scandinavian subpopulation 2 | 29 | 0.69 |
| Scandinavian subpopulation 3 | 156 | 0.68 |
| Scandinavian subpopulation 4 | 00 | 0.67 |
| Kluane, Yukon | 50 | 0.76 |
| Flathead River, British Columbia/Montana | 40 | 0.69 |
| Richardson Mountains, Northwest Territories | 119 | 0.75 |
| East Slope, Alberta | 45 | 0.67 |
| Yellowstone, Montana/Wyoming | 57 | 0.55 |

**Figure 3.9**   A North American brown bear (*Ursus arctos*). The genetic diversity of North American and Scandinavian brown bears is comparable at microsatellite loci, even though the latter underwent a severe bottleneck in the late 19th century. This bottleneck did, however, drastically deplete mitochondrial haplotype diversity, which is now much lower in Scandinavian compared with North American populations. Author's photograph

### Box 3.4   The long-term decline of the Hawaiian néné

Knowledge of a past bottleneck, coupled with currently low levels of genetic variation, does not necessarily imply cause and effect because genetic variation may have been reduced by either multiple bottlenecks or a gradual and prolonged decline in population size. Long-term genetic data produced some surprising results in a study of the population genetic history of the Hawaiian goose, or néné (*Branta sandvicensis*). The néné once occurred on most of the main Hawaiian islands, but by the time Captain Cook arrived in 1778 it was found only on the island of Hawaii. By the middle of the 20th century there were fewer than 30 surviving néné.

In order to assess the impact of this recent decline on genetic diversity, Paxinos *et al.* (2002) compared sequence variation in the mitochondrial control region from samples representing four different time periods: current, museum specimens (dating from 1833 to 1928), archaeological samples (160–500 radiocarbon years before present) and palaeontological samples (500–2540 radiocarbon years before present). All but one of the current, museum and archaeological samples had the same haplotype, but 7 haplotypes were found in the 14 palaeontological samples. This translates into haplotype diversity values of zero in the current population, 0.067 in the archaeological and museum specimens combined and 0.802 in the palaeontological samples. The researchers concluded that the néné lost most of its haplotype diversity during a period of prehistoric human population growth, and as a result had very little mitochondrial diversity even before its recent population decline. A direct approach such as this can provide valuable insight into how historical processes have affected genetic diversity, although unfortunately obtaining data from multiple time periods is seldom a practical option.

## Founder effects

The effects that both the size of a bottleneck and the speed of the population's recovery can have on the long-term genetic diversity of a population have been well illustrated by a number of studies on species that have been introduced by humans into new geographical areas such as islands. These introductions involve a particular type of bottleneck known as a **founder effect**, so-called because the founders of a new population will carry only a portion of the genetic diversity that was present in the source population.

The advantage of using human-mediated introductions to investigate the effects of bottlenecks is that they often occurred on known dates and involved a founding population of a known size. On the Hawaiian archipelago, the Mauna Kea silversword (*Argyroxiphium sandwicense* ssp *sandwicense*) has undergone at least two bottlenecks in recent times. The first known bottleneck occurred approximately 100–150 years ago when the spread of introduced ungulates caused the population to plummet to fewer than 50 plants. Microsatellite data showed that this bottleneck did not significantly deplete genetic diversity according to either allelic diversity ($A$) or heterozygosity ($H_e$) values (Friar *et al.* 2002). This retention of genetic diversity in a population of <50 individuals is partly attributable to the species' reproductive ecology; silverswords often grow for 30–50 years before flowering and reproducing, and therefore recovery was rapid because the bottleneck lasted for only one or two generations

The second known Mauna Kea silversword bottleneck occurred in a reintroduced population that originated from two or three transplanted individuals in 1973. Today this population numbers more than 1500 individuals, but the 1973 bottleneck was so extreme that its effects on genetic diversity could not be ameliorated by rapid population growth and long generation times, and levels of $A$ and $H_e$ remain very low in the current population (Friar *et al.*, 2002). This example illustrates how a population's rapid recovery can ameliorate a moderately severe, but not an extremely severe, bottleneck.

The effects that founding population size and post-bottleneck growth rates had on genetic variation were also investigated in a review of several studies of avian introductions onto islands that used allozyme data to compare the genetic diversity of mainland (source) and island populations (Baker, 1992). These data showed that introduced island populations do not necessarily have lower levels of genetic diversity than the source mainland populations. The genetic diversity of island populations was correlated with the size of the founding population and also with the rate at which the population grew (Figure 3.10), a finding that supports the idea that bottlenecks may reduce overall genetic diversity only if severe and prolonged.

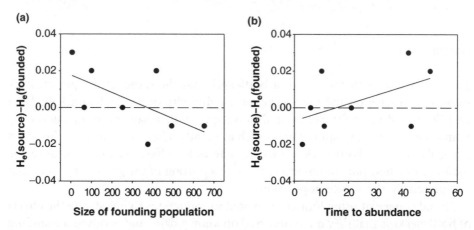

**Figure 3.10** The effects of (a) size of founding population and (b) time to abundance on the $H_e$ of founded (island) populations compared with their source (mainland) populations. The values of $H_e$ were calculated from allozyme data representing populations of starlings (*Sturnus vulgaris*), greenfinches (*Carduelis chloris*) and chaffinches (*Fringilla coelebs*) that were introduced to New Zealand, and common mynas (*Acridotheres tristis*) and house sparrows (*Passer domesticus*) that were introduced to both New Zealand and Australia. Time to abundance refers to the number of years between the founding event and the time when populations were first described as abundant. Dashed lines represent parity between $H_e$ of the source and founded populations. Data points above dashed lines represent founded populations with lower $H_e$ than source populations; data points below dashed lines represent founded populations with higher $H_e$ than source populations. Generally, the $H_e$ of the founded population is positively correlated with the number of founders and the population growth rate. Data from Merila, Björklund and Baker (1996) and references therein

## Natural selection

In the previous sections we have been looking at some of the ways in which non-adaptive processes can affect the genetic variation of populations. Genetic diversity is affected also by natural selection, a process that leads to 'the differential reproduction of genetically distinct individuals or genotypes within a population' (Li, 1997). Natural selection can alter allele frequencies in a number of different ways that can either increase or decrease overall genetic variation. Both **stabilizing selection** and **directional selection** will generally decrease genetic diversity. The former reduces diversity by favouring the average phenotype (and associated genotype) over either extreme. Directional selection generally reduces diversity either through **negative** or **purifying selection**, which is selection *against* any mutation that reduces the fitness of their carriers, or through **positive selection**, which selects *for* a particular mutation that increases the fitness of carriers. During the time in which positive selection is increasing the frequency of a particular allele, diversity will increase temporarily; this is known as **transient polymorphism**. Once the selected allele reaches fixation, that locus becomes monomorphic and genetic variation is lost. The type of selection most likely to increase variation is **disruptive** (or **diversifying**) selection, which occurs when two or more phenotypes are fitter than the intermediate phenotypes that lie between them. Figure 3.11 summarizes the three main modes of selection.

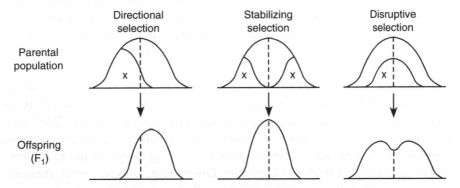

**Figure 3.11** Three common modes of selection that can influence the level of phenotypic variation – and the variability of underlying genotypes – within a population. In each diagram, phenotype frequency is illustrated by the distribution curve, the dashed line represents the mean value of the phenotypic character in the parental generation and the cross identifies the portion of individuals that have a relatively low fitness in the parental generation. Directional selection favours an extreme phenotype, stabilizing selection favours an intermediate phenotype, and disruptive selection favours two or more phenotypes at the expense of any intermediate phenotypes. Redrawn from Endler (1986)

**Balancing selection** is a general term given to selective processes that maintain genetic diversity because particular alleles are selected for in some situations and selected against in others. In large populations, where drift plays a relatively small role in shaping allele frequencies, balancing selection can influence the genetic diversity of some non-neutral loci. One form of balancing selection is **heterozygote advantage**, also known as **overdominance**. This occurs when heterozygotes have a higher fitness than homozygotes, and ensures that both alleles will be maintained within a population. One of the best-known examples of single-locus haeterozygote advantage occurs at the $\beta$-haemoglobin locus in some African and Mediterranean human populations. These populations have two alleles at this locus: the normal haemoglobin allele (A) and the sickle-cell haemoglobin allele (S). SS individuals suffer from sickle-cell anaemia and usually die young. SA individuals suffer from slight anaemia but they also have a much higher resistance to the type of malaria caused by the parasite *Plasmodium falciparum* than AA individuals, who do not suffer from anaemia. Therefore, in regions where malaria is widespread, SA individuals have an advantage over both AA and SS individuals, and for that reason both alleles are maintained within populations.

The second common form of balancing selection that can influence genetic diversity is **frequency-dependent selection**. This occurs when the fitness of a genotype depends on the frequency of that genotype within a population. Populations of the tropical butterfly *Heliconius erato* have different-coloured forms that are subject to frequency-dependent selection. The main line of defence that this butterfly has against bird predation is its colouration, which serves as a warning of its unpalatability. If a butterfly from one population is introduced into a new population in which an alternative phenotype predominates, it will appear conspicuous and probably will be eaten despite its unpalatability. This is positive frequency-dependent selection because the fittest genotype is the one that is most common within a given population.

The converse of this is negative frequency-dependent selection, which occurs if a trait is more advantageous when it is relatively rare. An example of this is the maintenance of both yellow and purple flowers in populations of the Elderflower orchid *Dactylorhiza sambucina*. This species relies on insect pollinators but does not reward them with any nectar. If yellow is the most common colour within a population, insects are likely to visit yellow flowers first. However, when their first visit is not rewarded, their next visit will generally be to a purple flower. As a result, the relatively rare purple flowers will receive a large proportion of insect visits and therefore should have higher fitness values than the yellow flowers. This relatively high fitness means that purple flowers eventually will become more common than yellow flowers, at which point negative frequency-dependent selection will favour yellow flowers until such time as the situation is reversed once more (Gigord, Macnair and Smithson, 2001).

### The major histocompatibility complex

The potential effects of balancing selection on genetic polymorphism can be illustrated further using the example of the **major histocompatibility complex** (**MHC**). This is a large multigene family in vertebrates that is involved in the immune response and is therefore extremely important in fighting disease.

Molecules encoded by the MHC will bind to antigens (peptides) and transport them to the cell surface membrane. Here, T-cell receptors will initiate an immune response if the antigens are recognized as foreign but will fail to initiate an immune reaction if the antigens are recognized as 'self'. The human MHC, which is usually referred to as the human leucocyte antigen (HLA), is found on chromosome 6 and spans a region more than 4 million bp long that contains over 100 genes. The MHC genes and their products are grouped into three classes on the basis of their chemical structure and biological properties. In general, MHC genes are extremely polymorphic, for example in humans the class I HLA-B locus has 149 alleles and the class II DRB locus has 179 alleles (Hedrick and Kim, 2000).

There is little doubt that MHC polymorphism is a result of selection. For one thing, particular MHC alleles can persist within populations for tens of millions of years, whereas neutral alleles are not expect to persist within a population for more than $2N_e$ generations (Takahata and Mei, 1990). One outcome of this long-term persistence is that some alleles are maintained in multiple species. This was found in a comparison between Arabian oryx (*Oryx leucoryx*) MHC sequences and homologous class II DRB sequences from bison, cattle, sheep and goats. Three of the oryx MHC sequences were more closely related to MHC sequences from other species than they were to each other, a finding that is difficult to explain unless selection is maintaining the same alleles for a period of time that transcends the evolution of multiple species (Hedrick *et al.*, 2000).

Evidence that selection maintains a diversity of MHC genes also comes from the pattern of nucleotide substitutions, since non-synonymous substitutions often outnumber synonymous substitutions. In one study a 254 bp region of an MHC class II B gene was sequenced from 666 wild Atlantic salmon (*Salmo salar*), and 40 nucleotide polymorphisms within 18 different alleles showed that non-synonymous substitutions were nearly three times more common that synonymous substitutions (Landry and Bernatchez, 2001); recall from Chapter 1 that under the neutral theory of evolution, synonymous substitutions are expected to predominate.

Although there is general agreement that balancing selection maintains MHC polymorphism, there is some debate about whether the precise mechanism is frequency-dependent selection or overdominance. An individual that has two alleles at one locus should be able to bind twice the number of foreign peptides as an individual with only one allele, in which case heterozygote individuals will be at an advantage and overdominance will occur. On the other hand, negative frequency-dependent selection would mean that individuals would be at an advantage if they were carrying rare alleles to which pathogens are not adapted.

Support for both mechanisms has come from studies of laboratory and wild populations. One recent study on great reed warblers (*Acrocephalus arundinaceus*) compared the temporal variance in frequencies of 23 MHC class I alleles and 23 putatively neutral microsatellite loci over nine successive years (Westerdahl *et al.*,

2004). The variation in allele frequencies was significantly higher for the MHC alleles than for the microsatellite alleles, which provided evidence for selection because MHC and microsatellite allele frequency variations should be comparable if they were both caused by drift. Frequency-dependent selection could be maintaining MHC polymorphism in this species, because the frequencies of different blood parasite strains are known to vary between years.

## Reproduction

No discussion on the factors that influence genetic diversity is complete without a reference to mode of reproduction. Sexual reproduction has long been viewed as something of a paradox because it predominates within multicellular eukaryotic taxa and yet entails a cost. In most sexual species half of the offspring produced are male, whereas an asexual female will produce only female offspring. Males are costly because their only contribution to the next generation is the fertilization of females. Sexual females therefore have a reduction in fitness compared with asexual females because they pass on only 50 per cent of their genes to each offspring and, because they produce both females and males, their offspring multiply at half the rate of clonal offspring. This result is often expressed as the 'twofold cost of sex' (Maynard Smith, 1978). There must be some advantage, therefore, that can explain the predominance of sex, and the most commonly cited explanation is recombination, which continuously generates novel genotypes that have the potential to adapt to changing environments. This rapid generation of genetic diversity through recombination means that sexually reproducing populations usually have higher levels of genetic diversity than asexual populations (see also Box 3.5).

Comparisons between the genetic diversity of sexual and asexual populations are most reliable when they involve a single species because this reduces the potentially confounding effects that variables other than reproduction may have on genetic variation. One example of this is the aphid *Sitobion avenae*, which tends to reproduce sexually in areas of Europe that have harsh winters, because the sexually produced eggs can survive low temperatures. In milder areas, parthenogenetic lineages predominate, because adults can survive the winters and have the advantage of high reproductive output. A microsatellite-based comparison between largely asexual populations in western France and sexual populations in Romania showed that the Romanian population had much higher genetic diversity than the French population – 94 per cent of aphids sampled in Romania had unique genotypes compared with only 28 per cent in France (Papura *et al.*, 2003). The low diversity in France was largely attributable to two dominant genotypes that were found in more than 60 per cent of samples.

Although relatively low levels of genetic diversity have been found in many asexually reproducing populations, it is important to note that not all asexual

**Table 3.8** Average heterozygosity values in plants that are grouped according to their mode of reproduction. Calculations are based on RAPD and microsatellite data and *n* refers to the number of studies included in each category. Numbers within a column are significantly different from one another if they are denoted by different letters. Adapted from Nybom (2004) and references therein

| Mode of reproduction | RAPD data | | Microsatellite data | | | |
|---|---|---|---|---|---|---|
| | $n$ | $H_e$ | $n$ | $H_e$ | $n$ | $H_o$ |
| Selfing[a] | 10 | $0.12^b$ | 15 | $0.41^b$ | 4 | $0.05^b$ |
| Mixed (selfing and outcrossing) | 8 | $0.18^b$ | 15 | $0.60^c$ | 13 | $0.51^c$ |
| Outcrossing | 38 | $0.27^c$ | 71 | $0.65^c$ | 60 | $0.63^d$ |

[a]Selfing refers to the practice of **self-fertilization**.

populations are genetically depauperate. Seven microsatellite loci revealed a mean of 6.6 alleles per locus and an average $H_e$ of 0.60 in asexual populations of the aphid *Rhopalosiphum padi*, an appreciable level of diversity that was nevertheless lower than that found in sexual populations, which had a mean of 11.3 alleles per locus and an average $H_e$ of 0.642 (Delmotte *et al.*, 2002). In flowering plants, a survey of 653 studies of allozyme variation showed no significant difference between the mean level of gene diversity within populations of sexually reproducing plants (0.114) compared with populations having a mixed (sexual and asexual) reproductive mode (0.103) (Hamrick and Godt, 1990). However, this finding may be an artefact of relatively invariant markers, because a later review of studies based on RAPD and microsatellite data revealed overall higher levels of genetic diversity in outcrossing plants (Nybom, 2004; Table 3.8).

## Box 3.5 Hybridization and genetic diversity

Another aspect of reproduction that can influence genetic diversity is hybridization. Interspecific hybridization can result in the transfer of genes, or even entire genomes, from one species to another. Provided that there are no fitness costs associated with these episodes of genetic introgression (and there often are), hybridization can increase the genetic diversity of populations and species. In many cases, phenotypic variation in a hybrid population not only will be greater than that found in each parent population, but also will exceed the combined variation of both parent populations; this is known as **transgressive segregation**. The novel genetic diversity often will be maladaptive, resulting in hybrids that are inviable or have relatively low fitness. At times, however, the genetic reshuffling may result in stable populations with unique phenotypes, and when this happens a new species may emerge.

The extremely high levels of genetic diversity in populations of the outcrossing monkey flower *Mimulus guttatus* have been attributed to long-term introgression of genes from the selfing congeneric *M. nasutus* into *M. guttatus* (Sweigart and Willis, 2003). Another example of this phenomenon has been found in populations of Darwin's ground finches (genus *Geospiza*), which experience regular bottlenecks often caused by environmental extremes such as drought or excessive rainfall. Despite these bottlenecks, populations show little evidence of depleted genetic diversity, and this seems to be at least attributable partly to ongoing hyridization among all six species (Freeland and Boag, 1999). The generation of genetic diversity following hybridization has been cited as one reason for the rapid evolutionary change that has led to **adaptive radiation** after species invade new environments, two examples of this being Darwin's finches on the Galápagos archipelago and African cichlid fish in Lake Malawi (Seehausen, 2004). The acquisition of new genes through introgression can also be of practical concern because bacteria frequently exchange antibiotic resistance genes. In an unusual case, there is evidence that some of these bacterial genes have even crossed a kingdom boundary and been incorporated into a fungal genome (Penalva *et al.*, 1990).

## Inbreeding

Inbreeding occurs when individuals mate with their relatives. Inbreeding does not alter the allele frequencies within a population but it does increase the proportion of homozygotes at all loci, thereby decreasing levels of *individual* genetic diversity. Inbreeding therefore will reduce a population's diversity when measured as $H_o$, although it will not immediately affect $H_e$; however, as we will see in Chapters 6 and 7, inbreeding is a potential concern because it can reduce fitness levels, and if this happens it will deplete all measures of genetic diversity through increased mortality rates.

Historically, the only way to estimate inbreeding was from detailed pedigree records, a method that is completely impractical for most wild populations. Fortunately, the advent of molecular markers has provided alternative methods. Key to these methods is the identification of alleles that are **identical by descent**, i.e. alleles that are copies of a single allele that existed in a relatively recent ancestor. Individuals that have two alleles that are identical by descent are **autozygous** and must, by definition, be homozygous at the locus in question. This contrasts with **allozygous** individuals, which have two alleles that are not identical by descent (Figure 3.12). Note that allozygous individuals can be either homozygous or heterozygous at the locus in question because an allozygous individual may have

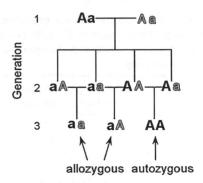

**Figure 3.12** Diagram showing how both allozygous and autozygous individuals can be generated within the same family. Two unrelated heterozygotes in generation 1 produced four offspring, all with different parental alleles (generation 2). Inbreeding occurs among the siblings of generation 2, resulting in two allozygous and one autozygous individual in generation 3. Note that the allozygous individuals include both a heterozygote and a homozygote

two copies of the same allele that did not originate in a recent common ancestor. For this reason, estimates of inbreeding are not exactly the same as measures of observed homozygosity.

As noted earlier, the first clue that a population may be inbred is a deviation from Hardy–Weinberg equilibrium caused by $H_o$ deficits. Observed and expected heterozygosity values form the basis of the inbreeding coefficient ($F$), which can be calculated as:

$$F = (H_s - H_I)/H_s \qquad\qquad (3.15)$$

where $H_I$ is the frequency of heterozygous genotypes in a population at the time of investigation (*individual* heterozygosity) and $H_S$ is the frequency of heterozygotes that would be expected if the population was in HWE (populations are commonly referred to as subpopulations in the literature surrounding the theory of $F$, hence *subpopulation* heterozygosity). In other words, $F$ measures the reduction in heterozygosity compared with the heterozygosity that we would expect to find in a randomly mating population with the same allele frequencies. When $F = 0$ there is no inbreeding at all. At the other extreme, when $F = 1$, all individuals within a population are homozygous and there may be complete inbreeding. It is, however, worth reiterating that a deficit of heterozygotes is not always due to inbreeding because population substructure (Wahlund effect), null alleles or natural selection may also contribute. Table 3.9 provides some estimates of $F$ in wild populations.

In vertebrates, which (with a few exceptions) are relatively invariant in their reproductive modes, inbreeding is more likely to occur in small, isolated

**Table 3.9** Some examples showing the range of average population inbreeding coefficients (*F*) across a variety of taxonomic groups. The examples in this table also show how *F* varies *within* a species, depending on whether the population reproduces predominantly by sexual or asexual means, or whether *F* is calculated for adults or offspring

| Species | Marker | Inbreeding coefficient ($F$) within populations | Reference |
|---------|--------|-------------------------------------------------|-----------|
| Aphids (*Rhopalosiphum padi*) | Allozymes | Asexual: 0.339<br>Sexual: 0.020 | Delmotte *et al.* (2002) |
| | Microsatellites | Asexual: −0.499<br>Sexual: 0.182 | |
| Aphids (*Sitobion avenae*) | Microsatellites | Asexual: 0.033<br>Sexual: 0.244 | Simon *et al.* (1998) |
| Lemurs (*Propithecus verreauxi*) | Microsatellites | Offspring: −0.068<br>Adults: 0.003 | Lawler, Richard and Riley (2003) |
| Greenfinch (*Carduelis chloris*) | Allozymes | Native (continental): 0.133<br>Introduced (island): 0.124 | Merilä, Björklund and Baker (1996) |
| Alpine pennycress (*Thlaspi caerulescens*) | Allozymes | Metal-tolerant: 0.217 and 0.501<br>Metal-intolerant: 0.578 and 0.704 | Dubois *et al.* (2003) |
| Common ash (*Fraxinus excelsior*) | Microsatellites | Seeds: 0.163<br>Adults trees: 0.292 | Morand *et al.* (2002) |

populations because of the increased likelihood of mating with a relative even when mating is random. In other taxa that have two or more methods of reproduction, such as many plants and invertebrates, inbreeding may be influenced further by a species' ecology. The succulent shrub *Agave lechuguilla* that grows in the Chihuahan Desert can undergo either clonal or sexual reproduction. Sex can involve either outcrossing, which occurs when pollinators transfer pollen between plants, or self-pollination (selfing), which is the most extreme form of inbreeding. The main pollinators of *A. lechuguilla* are nocturnal hawk moths (*Hyles lineata*) and diurnal large bees (*Bombus pennsylvanicus* and *Xylocopa californica*). A comparison of *A. lechuguilla* populations from different latitudes found that populations further north have fewer visits by pollinators compared with those in the southern part of its range (Silva-Montellano and Eguiarte, 2003a). They also had relatively low $H_o$ values, although $H_e$ values were comparable across latitudes. The authors of this study concluded that, in populations with fewer pollinators, self-pollination is more common and this results in inbreeding and therefore relatively high levels of $H_o$ (Silva-Montellano and Eguiarte, 2003b; Figure 3.13).

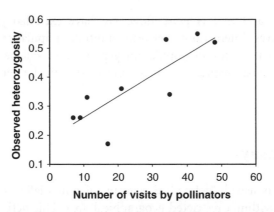

**Figure 3.13** Relationship between the number of visits by pollinators and the levels of inbreeding in *Agave lechuguilla* populations. Populations at higher latitudes are visited less frequently by pollinators than those further south. With fewer pollinators, selfing becomes more common and the observed heterozygosity ($H_o$) decreases. Data from Silva-Montellano and Eguiarte (2003a,b)

## Overview

Genetic diversity is arguably the most important aspect of the genetic analysis of single populations. In this chapter we have looked at methods for quantifying genetic diversity and also at some of the factors that influence genetic diversity. An overview of the main factors that influence the genetic diversity of populations is shown in Figure 3.14. Note that although factors that reduce genetic diversity outnumber those that increase it, levels of genetic diversity within most populations remain high. It should be evident from Figure 3.14 that we have not yet discussed one key process that increases within-population

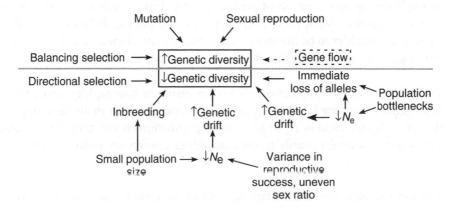

**Figure 3.14** An overview of some of the main factors that influence levels of genetic diversity within populations

genetic diversity, and that is **gene flow**. We have excluded gene flow so far because it involves interactions between multiple populations, but it is an extremely important factor in population genetics and will be a main topic of discussion in the next chapter, which is concerned with the genetics of multiple populations.

## Chapter Summary

- A population is usually defined as a group of potentially interbreeding individuals living within a restricted geographical area. This definition has limited application to asexual species, and even in sexually reproducing species the population boundaries are often difficult to identify.

- Genetic diversity is key to the long-term survival of populations. Estimates of genetic diversity are usually calculated as one or more of the following: allelic diversity ($A$), proportion of polymorphic loci ($P$), observed heterozygosity ($H_o$), gene diversity ($h$) or nucleotide diversity ($\pi$). Gene diversity is often equivalent to the heterozygosity that is expected under Hardy–Weinberg equilibrium ($H_e$).

- Genetic diversity estimates are usually higher when based on microsatellite data compared with dominant data, which in turn tend to be more variable than allozyme data. Organelle genomes may yield lower estimates of diversity than nuclear genomes.

- Genetic drift can rapidly diminish genetic diversity in small populations through the random fixation of alleles. Genetic drift is such a powerful determinant of genetic diversity and non-adaptive evolution that it forms the basis of what many consider to be the most important theoretical measurement in population genetics, and that is the effective size of a population ($N_e$).

- The effective size of a population ($N_e$) is a measure of how rapidly a population loses genetic diversity following genetic drift, and although in an ideal population $N_e = N_c$, the ratio of $N_e$ to $N_c$ in most populations is substantially less than 1.0 owing to several possible factors, including uneven sex ratios, variation in reproductive success, and fluctuating population size.

- The most reliable way to calculate $N_e$ is from the variance in allele frequencies, although this can be done only if we have allele frequency data from the same population representing at least two different time periods that are separated by at least one generation.

- Population bottlenecks, which include founder effects, often reduce genetic diversity; however, not all measures of diversity will be affected in the same ways, and the long-term impact will depend on both the severity and duration of the bottleneck.

- Natural selection, which includes directional and balancing selection, can either decrease or increase genetic diversity.

- Sexually reproducing populations usually have higher levels of genetic diversity than asexual populations because of the generation of new genotypes through recombination. Inbreeding in sexual populations does not directly alter allele frequencies but it does reduce observed heterozygosity ($H_o$).

## Useful Websites and Software

- Hardy–Weinberg equilibrium calculator: http://www.changbioscience.com/genetics/hardy.html

- Bottleneck. Detects recent effective population size reductions from allele data frequencies: http://www.montpellier.inra.fr/URLB/bottleneck/bottleneck.html

- Analysis of molecular variance (AMOVA) is a method for studying molecular variation within a species: http://www.bioss.ac.uk/smart/unix/mamova/slides/frames.htm

- FSTAT. Calculates genetic diversity and tests for differences in genetic diversity between groups of samples: http://www2.unil.ch/izea/softwares/fstat.html

- NeEstimator. Estimates the effective population sizes ($N_e$) from allele frequency data: http://www.dpi.qld.gov.au/fishweb/13887.html

- MLNE. A program for calculating maximum likelihood estimates of effective population size ($N_e$): http://www.zoo.cam.ac.uk/ioz/software.htm#MLNE

## Further Reading

### Books

Hartl, D.L. 2000. *A Primer of Population Genetics*. Sinauer Associates, Sunderland, Massachusetts.
Maynard Smith, J. 1998. *Evolutionary Genetics* (2nd edn). Oxford University Press, Oxford.

## Review articles

Frankham, R. 1995. Effective population size/adult population size ratios in wildlife: a review. *Genetical Research* **66**: 95–107.

Hughes, A.L. and Yeager, M. 1998. Natural selection at major histocompatibility complex loci of vertebrates. *Annual Review of Genetics* **32**: 415–435.

Nybom, H. 2004. Comparison of different nuclear DNA markers for estimating intraspecific genetic diversity in plants. *Molecular Ecology* **13**: 1143–1155.

Simon, J.-C., Rispe, C. and Sunnucks, P. 2002. Ecology and evolution of sex in aphids. *Trends in Ecology and Evolution* **17**: 34–39.

van Tienderen, P.H., de Haan, A., van der Linden, C. and Vosman, B. 2002. Biodiversity assessment using markers for ecologically important traits. *Trends in Ecology and Evolution* **17**: 577–582.

# Review Questions

**3.1.** The following data show the number of individuals in a population of *Drosophila polymorpha* that had the three alternative abdomen colour genotypes (after Cunha, 1949):

*EE* (dark): 3969

*Ee* (intermediate): 3174

*ee* (light): 927

    (i)    Calculate the frequency of each allele.

    (ii)    Calculate the genotype frequencies that would be expected if this population was in Hardy–Weinberg equilibrium.

    (iii)    Calculate the number of individuals of each genotype that would be expected if this population was in Hardy–Weinberg equilibrium.

    (iv)    Calculate the $\chi^2$ value for a comparison between observed and expected genotype frequencies.

**3.2.** Tawny owls (*Strix aluco*) produce more females in their broods when vole densities are high. Uneven sex ratios meant that Population 1 consisted of 41 breeding males and 68 breeding females, whereas Population 2, which was of the same size, had a more even sex ratio of 52 breeding males and 57 breeding females. Assuming that both populations consist entirely of breeding adults, calculate the $N_e / N_c$ ratio in these two populations.

**3.3.** A population goes through a bottleneck, and as a result its effective population size over six generations is $10^4$, $10^4$, $10^4$, $10^3$, $10^4$ and $10^4$.

    (i)    Calculate this population's long-term effective population size.

    (ii)    Calculate the current $N_e/N_c$ ratio.

**3.4.** In a small population of medium ground finches consisting of ten males and ten females, what is the rate at which genetic diversity would be lost each generation following genetic drift (assuming that population size remains constant) from:

(i)   the nuclear genome?
(ii)  the mitochondrial genome?

**3.5.** Provide five possible explanations why the $H_o$ of a population can be significantly lower than its $H_e$.

# 4

# Genetic Analysis of Multiple Populations

## Why Study Multiple Populations?

In Chapter 3 we learned that by quantifying the genetic diversity of single populations we can gain considerable insight into processes as varied as bottlenecks, reproduction and natural selection. We must understand these before we can interpret genetic data in an ecological context, but the genetics of populations are influenced by both intrapopulation and interpopulation processes, and therefore the next step is for us to investigate how gene flow (the transfer of genes from one population to another) influences the evolution of populations and species. We will start this chapter by looking at how we can quantify population subdivision and gene flow. However, we cannot predict the effects of gene flow without adding genetic drift and natural selection into the equation, because these are the three processes that jointly determine the extent to which populations will genetically diverge or converge. In the second part of this chapter, therefore, we shall revisit genetic drift and natural selection, but this time the emphasis will be on the ways in which they interact with gene flow to influence the genetic structure of populations across a species' range.

## Quantifying Population Subdivision

With the exception of rare or endangered species, which have limited distributions, virtually every species investigated to date has revealed some level of genetic differentiation between populations. A lack of population differentiation would mean that all populations had the same allele frequencies. This would be possible only if the entire species constituted a single group of randomly mating

*Molecular Ecology* Joanna Freeland
© 2005 John Wiley & Sons, Ltd.

individuals, which will not occur in species that comprise multiple populations. For many years, a frequently cited exception to this rule was European eels (*Anguilla anguilla*). These eels overwinter in numerous, geographically distinct sites across Europe, but regardless of their overwintering site, all individuals migrate to the Sargasso Sea in the summer to reproduce. This common mating ground provides a mechanism for species-wide panmixia, a hypothesis that was supported by a complete lack of population genetic structure across Europe according to either allozyme or mitochondrial data (Avise *et al.*, 1986).

More recently, however, data from seven variable microsatellite loci revealed weak but significant genetic differences between groups of eels from several over-wintering sites. These genetic differences suggest that eels from across Europe do not, after all, form a single panmictic population in the Sargasso Sea, possibly because eels migrating from different latitudes reproduce at different times, in which case reproduction should occur more often between eels that overwinter at similar latitudes (Wirth and Bernatchez, 2001). Population divisions such as these are often very difficult to detect without the aid of appropriate molecular markers; bearing this in mind, we shall look now at some methods of analysing genetic data that allow us to determine whether or not populations are genetically distinct from one another.

## Genetic distance

One way in which we can measure the genetic similarly of two populations is by estimating the genetic distance between them. There are many different ways in which this can be done, one of the most common being Nei's (1972) standard genetic distance, $D$. To calculate this, we first need to know Nei's measure of genetic identity ($I$), which reflects the genetic similarity of populations. For a given locus, this is calculated as:

$$I = \sum_{i=1}^{m} \left( p_{ix} p_{iy} \right) \Big/ \left[ \left( \sum_{i=1}^{m} p_{ix}^{2} \right) \left( \sum_{i=1}^{m} p_{iy}^{2} \right) \right]^{0.5} \tag{4.1}$$

where $p_{ix}$ is the frequency of allele $i$ in population $x$, $p_{iy}$ is the frequency of allele $i$ in population $y$ and $m$ is the number of alleles at the locus (see Box 4.1).

Values of $I$ range from zero to one. Once calculated, they can be used to obtain Nei's $D$ as follows:

$$D = -\ln I \tag{4.2}$$

Values of $D$ range from zero to infinity. If two populations have similar allele frequencies, i.e. if $p_{ix} \approx p_{iy}$, genetic similarity ($I$) between the two will approach one and genetic distance ($D$) will approach zero. At the other extreme, if two populations have no alleles in common, $I$ will be zero and $D$ will be infinity.

**Table 4.1** Allele frequencies at the *Pgm* locus in two *D. pulex* populations (data from Crease, Lynch and Spitze, 1990)

| *Pgm* allele | Illinois population | Indiana population |
|:---:|:---:|:---:|
| F | 0.146 | 0.491 |
| M | 0.818 | 0.106 |
| S | 0.036 | 0.403 |

## Box 4.1  Calculating Nei's genetic distance

The data in Table 4.1 represent the allele frequencies of a single allozyme locus (phosphoglucomutase, or *Pgm*) in two populations of the freshwater zooplankton *Daphnia pulex* in Illinois and Indiana, USA. We will denote Illinois as population $x$ and Indiana as population $y$. Since $I = \sum_{i=1}^{m}(p_{ix}p_{iy})/[(\sum_{i=1}^{m}p_{ix}^2)(\sum_{i=1}^{m}p_{iy}^2)]^{0.5}$, we first need to calculate the sum of the products of the allele frequencies as follows:

$$\sum_{i=1}^{m}(p_{ix}p_{iy}) = (0.146)(0.491) + (0.818)(0.106) + (0.036)(0.403)$$

$$= 0.173$$

We then need to calculate the sums of the squared allele frequencies, which for the Illinois population is:

$$\left(\sum_{i=1}^{m}p_{ix}^2\right) = 0.146^2 + 0.818^2 + 0.036^2 = 0.692$$

and for the Indiana population is:

$$\left(\sum_{i=1}^{m}p_{iy}^2\right) = 0.491^2 + 0.106^2 + 0.403^2 = 0.415$$

Therefore:

$$I = 0.173/[(0.692)(0.415)]^{0.5} = 0.173/\sqrt{[(0.692)(0.415)]}$$

$$= 0.173/0.536 = 0.323$$

and because $D = -\ln I$, in this case the genetic distance between the two populations is:

$$D = -\ln(0.323) = 1.13$$

## F-Statistics

Perhaps the most common method for quantifying the genetic differentiation between populations is based on **F-statistics**, which were developed by Wright (1951). F-Statistics use inbreeding coefficients to describe the partitioning of genetic variation within and among populations and can be calculated at three different levels. Note that in describing these levels we will follow the convention set by the literature underlying the theory of F-statistics, and refer to spatially discrete samples as subpopulations instead of populations; elsewhere we will revert to the more common practice of referring to discrete breeding units as populations. The first F-statistic, $F_{IS}$, measures the degree of inbreeding within individuals relative to the rest of their subpopulation. This reflects the probability that two alleles within the same individual are identical by descent, and is the same as the inbreeding coefficient F that was introduced in Chapter 3. It is calculated as:

$$F_{IS} = (H_S - H_I)/H_S \qquad (4.3)$$

where $H_I$ is the observed heterozygosity in a subpopulation at the time of investigation (*individual* heterozygosity) and $H_S$ is the heterozygosity that would be expected if the subpopulation was in HWE (*subpopulation* heterozygosity)

The second F-statistic is $F_{ST}$ (also known as the fixation index), and this provides an estimate of the genetic differentiation between subpopulations. It is a measure of the degree of inbreeding within a subpopulation relative to the total population (total population here meaning all of the subpopulations combined), and reflects the probability that two alleles drawn at random from within a subpopulation are identical by descent. It is calculated as:

$$F_{ST} = (H_T - H_S)/H_T \qquad (4.4)$$

where $H_S$ is the same as in Equation 4.4 and $H_T$ is the expected heterozygosity of the *total* population. The third F-statistic, which is used much less frequently than the other two, is $F_{IT}$. This provides an overall inbreeding coefficient for an individual by measuring the heterozygosity of an individual relative to the total population. $F_{IT}$ is therefore influenced by both non-random mating within a subpopulation ($F_{IS}$) and population subdivision ($F_{ST}$), and is calculated as:

$$F_{IT} = (H_T - H_I)/H_T \qquad (4.5)$$

where $H_T$ and $H_I$ are the same as in Equations 4.4 and 4.5. The relationship between the three statistics is given as:

$$F_{IT} = F_{IS} + F_{ST} - (F_{IS})(F_{ST}) \qquad (4.6)$$

Since $F_{ST}$ measures the extent to which populations have differentiated from one another, this is the $F$-statistic with which we are most concerned in this chapter (see Box 4.2).

**Table 4.2** Frequencies of alternative wing genotypes in two hypothetical subpopulations of the scarlet tiger moth

|                  | No. of individuals | | | Genotype frequency | | |
|------------------|------|------|------|-------|-------|-------|
|                  | *AA* | *Aa* | *aa* | *AA*  | *Aa*  | *aa*  |
| Subpopulation 1  | 352  | 63   | 12   | 0.824 | 0.148 | 0.028 |
| Subpopulation 2  | 312  | 77   | 27   | 0.750 | 0.185 | 0.065 |

---

**Box 4.2    Calculating *F*-statistics**

We will use some hypothetical data from two scarlet tiger moth subpopulations to calculate *F*-statistics. Recall from Chapter 3 that moths with white spotting were assigned the genotype *AA*, moths with little spotting were *aa* and moths with intermediate spotting were the heterozygotes, *Aa*. The frequencies of the alternative genotypes from two subpopulations are given in Table 4.2, and from these data we need to calculate $H_I$, $H_S$ and $H_T$.

First we calculate $H_I$, which is the average observed heterozygosity across populations:

$$H_I = (0.148 + 0.185)/2 = 0.167$$

Next we calculate the subpopulation heterozygosity, $H_S$, which is the heterozygosity that would be expected if the subpopulations were in HWE. We know that under HWE heterozygosity is equal to $2pq$, therefore we need to first calculate $p$ and $q$. In subpopulation 1 there are 427 moths and therefore 854 alleles, so:

$$p = [(2)(352) + 63]/854 = 0.898$$
$$q = [(2)(12) + 63]/854 = 0.102$$

In subpopulation 2 there are 416 moths and therefore 832 alleles, so:

$$p = [(2)(312) + 77]/832 = 0.843$$
$$q = [(2)(27) + 77]/832 = 0.157$$

The expected heterozygosity is therefore:

$2pq = 2(0.898)(0.102) = 0.183$ in subpopulation 1 and

$2pq = 2(0.843)(0.157) = 0.265$ in subpopulation 2. This means that:

$H_S = (0.183 + 0.265)/2 = 0.224$.

Next, we calculate $H_T$ by using the average allele frequencies from the two subpopulations to calculate the expected heterozygosity if the total population was in HWE. In this case:

$$p = (0.898 + 0.843)/2 = 0.871$$
$$q = (0.102 + 0.157)/2 = 0.130$$

Therefore:

$$H_T = 2(0.871)(0.130) = 0.226.$$

We can now calculate inbreeding within populations as:

$$F_{IS} = (H_S - H_I)/H_S = (0.224 - 0.167)/0.224 = 0.254$$

and inbreeding due to population differentiation as:

$$F_{ST} = (H_T - H_S)/H_T = (0.226 - 0.224)/0.226 = 0.009$$

with overall inbreeding (both within and differentiation among populations) as:

$$F_{IT} = (H_T - H_I)/H_T = (0.226 - 0.167)/0.226 = 0.261$$

### Interpreting $F_{ST}$

If two populations have identical allele frequencies they will not be genetically differentiated and therefore $F_{ST}$ will be zero. At the other extreme, if two populations are fixed for different alleles then $F_{ST}$ will be equal to one (see Table 4.3 for some examples of $F_{ST}$ values and Box 4.3 for analogues of $F_{ST}$). Within that range, $F_{ST}$ values of 0–0.05 are generally considered to indicate little genetic differentiation; values of 0.05–0.25 indicate moderate genetic differentiation; and values of $>0.25$ represent pronounced levels of genetic differentiation. However, these are only approximate guidelines, and in reality even very low levels of $F_{ST}$ may represent important levels of genetic differentiation. In the European eel example referred to at the beginning of this section, the global $F_{ST}$ value was extremely low at 0.0017 but was nevertheless significant ($P = 0.0014$). The significance of $F_{ST}$ estimates is based on a permutation procedure that shuffles genotypes among populations thousands of times, and an $F_{ST}$ value is calculated for each permutation. The $P$ value of the test is based on the number of times that these $F_{ST}$ values are equal to or larger than that calculated from the actual data set.

**Table 4.3**  Some examples of $F_{ST}$ values that have been calculated for various taxa

| Species | Distance (km)[a] | Molecular marker | $F_{ST}$ (or analogue)[b] | Reference |
|---|---|---|---|---|
| Stonefly (*Peltoperla tarteri*) | 3.5 | Mitochondrial sequence | 0.004–0.21 | Schultheis, Weigt and Hendricks (2002) |
| Nematode (*Heterodera schachtii*) | 175 | Microsatellites | 0–0.107 | Plantard and Porte (2004) |
| Collembola (*Orchesella cincta*) | 10 | AFLPs | 0–0.09 | van der Wurff *et al.* (2003) |
| Red seaweed (*Gracilaria gracilis*) | 5 | Microsatellites | 0–0.031 | Engel, Restombe and Valero (2004) |
| Canada thistle (*Cirsium arvense*) | 5 | AFLPs | 0.63 | Solé *et al.* (2004) |
| Common frog (*Rana temporaria*) | 1600 | Microsatellites | 0.24 | Palo *et al.* (2003) |
| Plunkett Mallee tree (*Eucalyptus curtisii*) | 500 | Microsatellites | 0.22 | Smith, Hughes and Wardell-Johnson (2003) |
| Island fox (*Urocyon littoralis*) | 13 | Microsatellites | 0.11 | Roemer *et al.* (2001) |

[a]Maximum distance between pairs of populations.
[b]Single values represent the $F_{ST}$ values averaged across all populations. See Box 4.3 for analogues of $F_{ST}$.

---

### Box 4.3    Analogues of $F_{ST}$

$F_{ST}$, which was developed by Wright (1951), was the original method for estimating population differentiation from allele frequencies. Since then several variations have been developed, with each measuring population differentiation in a slightly different way. One of these is $G_{ST}$, developed by Nei (1973), which is equivalent to $F_{ST}$ when there are only two alleles at a locus. In the case of multiple alleles, $G_{ST}$ is equivalent to the weighted average of $F_{ST}$ for all alleles. A similar measurement is $\theta$ (Weir and Cockerham, 1984), which is often preferred because it takes into account the effects of uneven sample sizes and the number of sampled populations. More recently, $R_{ST}$ was developed by Slatkin (1995) specifically for the analysis of microsatellite data. $R_{ST}$ differs from the other measurements because it assumes a stepwise mutation model (SMM; Chapter 2) that may be more realistic for microsatellite data than the infinite alleles model (IAM).

It is beyond the scope of this text to provide a comprehensive discussion on the many different measures of population differentiation. The theoretical literature surrounding these measurements is copious and often opaque; interested readers are referred to recommended reading at the end of this chapter. Nevertheless, even if few of us can understand all of the mathematics that underly estimates of genetic differentiation, it is important to understand that these estimates often will vary depending on which method is used. This may be particularly important if we are comparing the results of multiple studies, for example we may be interested in whether populations of perennial plants show greater differentiation than populations of annual plants, but such a comparison will be valid only if it is based on estimates of genetic differentiation that were calculated in the same way. This was illustrated by a study that compared 227 measurements of $G_{ST}$ that had been calculated using both the method of Nei (1973) and the method of Hamrick and Godt (1990). The main difference between the two methods is that in the former $G_{ST}$ is calculated from the mean values of $H_T$ and $H_S$ after they have been averaged across all loci, whereas in the latter the $G_{ST}$ values are calculated separately for all loci and then averaged. A comparison of the two methods revealed that, although results often were comparable, $G_{ST}$ values in 15 per cent of all studies differed by $>0.10$ (Culley *et al.*, 2002; Figure 4.1)

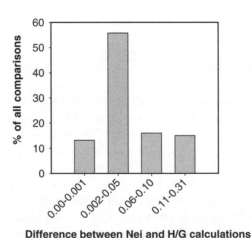

**Difference between Nei and H/G calculations**

**Figure 4.1**   Range of differences between $G_{ST}$ values calculated using both Nei's (1973) and Hamrick and Godt's (1990) methods (H/G), based on 227 comparisons. Data from Culley *et al.* (2002)

It can be extremely difficult to make predictions about the level of differentiation that may be expected between any pair of populations because the range of population divergence is remarkable both within and among species. $F_{ST}$ values are affected by many aspects of a species' ecology and demographic history. Later in this chapter we shall look at how genetic drift and natural selection influence population differentiation, but we shall start with a discussion of gene flow, a process that is of fundamental importance because without gene flow a group of populations could not be maintained as a cohesive species.

# Quantifying Gene Flow

Before looking at ways to quantify gene flow, we have to understand exactly what constitutes gene flow and how it differs from dispersal and migration. Dispersal generally refers to the movement of individuals or propagules between discrete locations or populations, whereas migration refers to periodic movement to and from a specific geographical area, often occurring seasonally and along a consistent route. Although either dispersal or migration must precede gene flow, neither results in gene flow unless individuals successfully reproduce once they have arrived at their new location. Nevertheless, dispersal is sometimes used as a surrogate for gene flow and we need to be aware of this potential limitation when evaluating different methods of quantification. Understanding the movements of individuals, gametes (e.g. pollen) and genes between populations is fundamental to studies of ecology and population genetics because, as we shall see later in the chapter, gene flow can profoundly influence a host of relevant variables, including population size, genetic diversity, local adaptation and ultimately speciation. Unfortunately, quantifying the movements of organisms and their genes is seldom straightforward. In this section we shall look at several methods that are commonly used to estimate ongoing dispersal and gene flow. These methods can be divided into three main categories: direct, indirect and **assignment tests**.

## Direct methods

The colonization of new habitats, such as the invasion of new ponds by freshwater zooplankton (Jenkins, 1995), provides direct, incontrovertible evidence for dispersal, although such observations impart very little information on the extent or frequency of movements between established populations. A direct method that can provide more detailed estimates of dispersal is mark–recapture. Individuals are first marked in one of a number of different ways that include rings on birds' legs or paint markings on insect wings or carapaces (Figure 4.2). Direct evidence of dispersal is obtained when marked individuals are later recaptured some distance from their source population. A major drawback to this approach is that individual marking is a very time-consuming exercise and often yields limited returns because many of the marked individuals will never be recaptured.

**Figure 4.2** A ruddy darter (*Sympetrum sanguineum*) that can be identified from the label on its wing as individual K15. This dragonfly was marked using indelible ink and subsequently recaptured as part of a study on the dispersal behaviour of odonates (Conrad *et al.*, 1999). Photograph provided by Kelvin Conrad and reproduced with permission

Even more intensive, although potentially a greater source of information, is the use of radio tracking or, in some cases, radar tracking. Radio tags have been used to monitor migratory movements of a number of taxa, including birds and butterflies (Figure 4.3). Although the expense and laboriousness associated with radio tracking means that relatively few individuals can be monitored in this way, it does have the strong advantage of providing data on all movements, not just on the sites of recapture. This can be particularly useful if we are interested in individual foraging behaviour or wish to reconstruct a migratory route. Other studies have used stable isotopes to mark individuals, one example of this being the use of $^{15}$N to mark more than 1.5 million aquatic stonefly larvae (*Leuctra inermis*). Subsequent screening of individuals at adjacent streams revealed only a few isotopically enriched adults (Briers *et al.*, 2004). In this case, traditional mark–recapture methods, which for logistical reasons would have targeted far fewer individuals, would probably not have provided any evidence for dispersal.

Radio tracking and mark–recapture studies provide estimates of dispersal but not gene flow, because they provide no information on whether or not immigrants will reproduce. Other direct measures of dispersal that accurately reflect gene flow are based on parentage studies and have been used most commonly in plants. One study used this approach to determine how far animal frugivores were dispersing seeds in a Spanish population of the cherry tree *Prunus mahaleb* (Godoy and Jordano, 2001). Endocarp tissue in this species is maternally derived and therefore

**Figure 4.3**  A female willow ptarmigan (*Lagopus lagopus*) in northwest British Columbia, Canada, with a radio collar around her neck. Ptarmigan chicks are precocial, which means that they are mobile almost as soon as they hatch, and without radio tracking it is almost impossible to monitor adults and chicks as they move around the tundra. Author's photograph

is genetically identical to the source tree. By comparing the genotypes of 180 adult trees with the genotypes of endocarp tissue from 95 seeds that had been dispersed naturally, the authors were able to match seeds with maternal trees in 78 cases, and in each case the seed had originated from within the population. The remaining 17 seeds did not match the genotypes of any trees within the population and therefore must have been transported by frugivores from elsewhere. A similar approach has been used in other studies that have estimated pollen-mediated gene flow by comparing the genotypes of seedlings to the genotypes of potential pollen-donors (e.g. Dunphy, Hamrick and Schwagerl, 2004).

Estimating gene flow by comparing the genotypes of offspring and putative parents can provide a comprehensive picture of dispersal, but this approach is impractical or impossible in large populations or in species that often travel long distances. In fact, a common denominator to all direct studies of dispersal is that they are costly, time-consuming, and generally can be conducted only on a relatively small number of individuals. Furthermore, with the exception of parentage analysis, direct evidence for dispersal does not necessarily translate into evidence of gene flow. For these reasons researchers in recent years have turned to alternatives, and we shall look now at the most common of these: indirect estimates of gene flow based on genetic differentiation among populations.

## Indirect methods

After Wright (1951) developed $F$-statistics he went on to demonstrate that there is a simple relationship between the genetic divergence of two populations, measured as $F_{ST}$, and the amount of gene flow between them, which is given as:

$$F_{ST} = 1/(4N_em + 1) \qquad (4.8)$$

where $N_e$ is the effective size of each population and $m$ is the migration rate between populations, and therefore $N_em$ is the number of breeding adults that are migrants. From Equation 4.8 we can calculate $N_em$ as:

$$N_em = (1/F_{ST} - 1)/4 \qquad (4.9)$$

In the scarlet tiger moth example that was given earlier in this chapter, the $F_{ST}$ value for the two populations was 0.009. This would translate into a very high indirect estimate of gene flow since $N_em = (1/0.009 - 1)/4 = 27.5$ individuals migrating between the two populations each generation.

$N_em$ is a popular method for estimating gene flow from genetic data, in part because of the ease with which it can be calculated. There are, however, several problems associated with estimates of $N_em$. For one thing, it is based on what is known as the island model of population structure. This model assumes no selection or mutation and an infinite number of populations, all of which are the same size and have an equal probability of exchanging migrants. $N_em$ also assumes that populations are in migration–drift equilibrium, which occurs when the increase in genetic differentiation caused by drift is equal to the decrease in genetic differentiation caused by gene flow. Populations reach equilibrium only when population sizes and migration rates remain more or less constant; equilibrium is therefore negated by a number of demographic processes, including recent range expansions, habitat fragmentation, and population bottlenecks.

The extent to which a lack of migration–drift equilibrium can influence $N_e m$ estimates was illustrated by a study of the pond-dwelling whirligig beetle *Dineutus assimilis*. Gene flow estimates based on $F_{ST}$ values were compared with dispersal estimates that were based on mark–recapture. Four sites within a few kilometres of each other yielded an average $F_{ST}$ estimate of 0.0949, which translated into an average $N_e m$ value of 2.12. However, the mark–recapture study showed that up to eight immigrants arrived at a pond in a single season. An explanation for these contradictory results emerged after a closer investigation showed that regular bottlenecks were occurring within the ponds. These bottlenecks would accelerate genetic drift and thus prevent the populations from reaching migration–drift equilibrium. As a result, $F_{ST}$ values would be inflated, which would lead to a corresponding reduction in $N_e m$. Since the ecology of this species meant that a high proportion of immigrants were likely to reproduce in their new populations (i.e. dispersal and gene flow should be comparable), a lack of equilibrium provided the most plausible explanation for the discrepancy between $N_e m$ and direct dispersal estimates in this study (Nürnberger and Harrison, 1995).

There has been an ongoing debate in the literature over just how useful $N_e m$ values are as estimates of gene flow. On the one hand, $N_e m$ values in numerous studies are comparable with direct estimates of dispersal. A meta-analysis of 230 studies of phytophagous (plant-feeding) insects revealed estimates of $N_e m$ that generally agreed with direct estimates of dispersal that were based on mark–recapture studies or observations of new colonizations, since highly vagile species showed higher levels of gene flow ($N_e m$) than sedentary species (Peterson and Denno, 1998b). Similar findings were obtained in another meta-analysis, this time based on 333 species of vertebrates and invertebrates from terrestrial, freshwater and marine habitats (Bohonak, 1999). Nevertheless, the $F_{ST}$ values between populations are influenced by factors other than gene flow, and the unrealistic assumptions of the island model of population structure upon which $N_e m$ estimates are based mean that they should be interpreted with caution (Whitlock and McCauley, 1999).

## Assignment tests

A more recently developed suite of methods for quantifying dispersal from genetic data is based on what are known as assignment tests. These assign individuals to their most likely population of origin by comparing their genotypes to the genetic profiles of various populations. This approach differs from the indirect methods of assessing gene flow ($N_e m$) because it identifies individuals that have dispersed from their natal population, as opposed to comparing overall genetic similarities between populations. The original assignment tests use a maximum likelihood method to calculate the probabilities that a given genotype arose from alternative populations based on the allele frequencies in those populations (Paetkau *et al.*,

1995). Individuals are then assigned to the population from which they have the highest probability of originating, unless all of the probabilities are low, in which case the true natal population may not have been sampled. This approach assumes that all populations are in Hardy–Weinberg equilibrium and that the loci being characterized are in linkage equilibrium.

In recent years, various modifications have been made to improve the accuracy of assignment tests. One refinement has been the application of a **Bayesian** method (Pritchard, Stephens and Donnelly, 2000; Wilson and Rannala, 2003). The Bayesian approach to statistical inference is based on subjective statements of probability. Data are assumed to be fixed, and prior information is used to test the likelihood that various parameters can explain the data. Unlike classical statistics, which typically provide a single $P$ value, Bayesian statistics often provide multiple probabilities, and this means that numerous scenarios (such as several candidate populations of origin) can be compared simultaneously. Simulation studies and applications to actual data sets have found that the Bayesian approach often outperforms the original frequency-based method (Cornuet *et al.*, 1999; Manel, Berthier and Luikart, 2002), although this approach does assume that in each case the true population of origin has been sampled, which may not be the case.

Assignment tests generally perform better when detailed genetic profiles of populations are obtained from a relatively large number of individuals using multiple (up to 20) highly polymorphic loci. For this reason, assignment tests have been most commonly based on microsatellite data, although they can also be applied to allozyme, microsatellite, RAPD and RFLP data. A number of simulation studies have suggested that these tests work best when candidate populations are genetically distinct from each other (Cornuet *et al.*, 1999), in which case dispersal must be relatively infrequent, although a recent study on the grand skink (*Oligosoma grande*) demonstrated that these conditions need not always be met. The grand skink is a large territorial lizard that lives in New Zealand in groups of around 20 that inhabit free-standing rock outcrops separated by 50–150 m of inhospitable vegetation. Mark–recapture studies have revealed frequent dispersal among sites, but assignment tests still managed to correctly assign between 65 and 100 per cent of migrant individuals to their natal population (depending on which assignment method was used), even when $F_{ST}$ values were as low as 0.04 (Berry, Tocher and Sarre, 2004).

Although assignment tests often provide valuable information, their performance does depend on a number of variables, including the extent of population differentiation, the sample size of individuals and loci, and the variability of markers. Under some conditions they may be rather imprecise, as illustrated by a study of Montana wolverines that used four different methods of Bayesian and frequency-based assignment tests to investigate dispersal. A total of 89 individuals were genotyped and, although 25 of these were classified as migrants according to at least one method, only nine individuals were classified as migrants by all four methods (Cegelski, Waits and Anderson, 2003). Nevertheless, assignment tests have tremendous potential as a method for tracking dispersal and it is likely that future refinements will increase their use. We will return to assignment tests in later chapters when we look at some specific applications of these methods to questions in behavioural ecology and wildlife forensics.

## Gene flow discrepancies

Since no method for calculating gene flow is infallible, conclusions should, whenever possible, be based on estimates that have been obtained using two or more different methods. If these are more or less in agreement then we can be reasonably confident that our findings are correct; however, discrepant results are not uncommon and there are several possible reasons for this. We have noted already that indirect estimates may not reflect the actual extent of gene flow because the calculation of $N_e m$ is based on an unrealistic model. Although assignment tests are becoming an increasingly popular alternative, there is still a need for rigorous testing of different models against various combinations of sample sizes, marker polymorphisms, and demographic parameters. At the same time, direct estimates often suffer from their inability to detect more than a very small proportion of dispersing individuals, for example dispersal estimates based on mark–recapture will be flawed if attempts at recapture are made at the wrong time or in the wrong place. An example of how mark–recapture can lead to underestimates of dispersal was provided by a study of yellow-bellied marmots (*Marmota flaviventris*), in which many of the dispersal events (65 per cent of males and 42 per cent of females) that were picked up by radio tracking were missed completely by mark–recapture (Koenig, van Vuren and Hooge, 1996). Parentage studies will also remain inconclusive unless the number of potential parents is very small. Table 4.4 summarizes some of the attributes of the different methods for estimating dispersal and gene flow.

On a final note, it is worth pointing out that the estimation of dispersal, as opposed to gene flow, from direct methods and assignment tests is not necessarily a drawback because dispersing individuals can have an impact on an invaded ecosystem even if subsequent reproduction does not occur. Parasites and microbes, for example, can decimate a population through disease even if they do not subsequently reproduce. We could argue, therefore, that estimates of both dispersal and gene flow are needed before we can have a truly comprehensive picture of population dynamics.

# What Influences Gene Flow?

There are both advantages and disadvantages to dispersal. Dispersing individuals may benefit by avoiding inbreeding, locating a new site with relatively few

**Table 4.4** A summary of the attributes of different estimates of gene flow

|  | Indirect ($N_e m$) | Direct | Assignment tests |
| --- | --- | --- | --- |
| **Assumptions** | Population equilibrium, marker neutrality, island model | Capture–mark–recapture, radio tracking: negligible. Parentage: individual-specific genetic profiles | Linkage equilibrium, Hardy–Weinberg equilibrium |
| **Number of populations feasibly studied** | Many | Few | Many |
| **Dispersal or gene flow** | Gene flow | Mark–recapture, radio tracking = dispersal. Parentage = gene flow | Dispersal |
| **Likelihood of detecting long-distance dispersal** | Low if long-distance dipersal is rare | Radio tracking: high. Mark-recapture, parentage: low | Theoretically high, as long as migrants' source populations are genotyped |
| **Dependence on sample size** | Moderate | Radio tracking: low. Mark–recapture, parentage: high | High |
| **Direction of dispersal identified?** | No | Yes | Yes |

competitors, or escaping from pathogens, parasites or predators. On the other hand, they may be unable to locate a suitable new site or mate, or they may be preyed upon *en route*. From an evolutionary perspective, dispersal should occur only if the benefits outweigh the risks, and this cost–benefit analysis is part of the reason why the frequency, mode and distance of dispersal vary tremendously between species.

Dispersal patterns are particularly complex in plants, fungi and invertebrates because often they have multiple dispersal mechanisms. The complexity of this is illustrated by the freshwater bryozoan *Cristatella mucedo*, which is a sessile, colonial, **benthic** species that can reproduce vegetatively and also through the production of larvae and seed-like propagules (statoblasts). The adult colonial stage can disperse only very slowly by creeping along substrate, and an individual colony is unlikely to move more than a few centimetres during the growing season. Sperm are broadcast into the water column, providing a means for local dispersal of the paternal genome. Similarly, larvae are free-swimming, but even if they are swept along in a current both sperm and larvae are short-lived and are unlikely to

**Figure 4.4** Scanning electron micrograph of statoblasts (approximately 0.8 mm in diameter) from the bryozoan *Cristatella mucedo* that have become attached to a feather with the aid of small hooks. This is an important mechanism for the dispersal of this species between lakes and ponds. Photograph provided by Beth Okamura and reproduced with permission.

disperse further than their natal lake or pond. Statoblasts, on the other hand, are resistant to dessication and therefore can survive overland dispersal. They have small hooks that facilitate their attachment to feathers or fur (Figure 4.4) and there is evidence that they can be transported 600 km or further on migratory waterfowl (Freeland *et al.*, 2000). Other examples of species whose dispersal mechanisms vary throughout their life cycles are shown in Table 4.5.

Another important reason for relatively complex dispersal mechanisms in plants, fungi and invertebrates is the ability that some of them have to disperse through time. Many plants have seed banks, which are formed when seeds become covered up by substrate and enter a period of dormancy that, depending on the species, may last anywhere from a couple of years to 100 years or longer. Delayed germination causes genotypes that may have been absent for some years to be reintroduced into a population, resulting in temporal gene flow. In one extreme example,

**Table 4.5** Examples of species in which dispersal mechanisms vary throughout their life cycles

| Species | Pre-adult dispersal | Adult dispersal |
|---|---|---|
| Stonefly (*Peltoperla tarteri*) | Local: larvae that crawl or are swept by current | Local and long-distance: flying or wind-borne |
| Trematode (*Microphallus* sp.) | Local: larvae ingested by freshwater snail host (*Potamopyrgus* spp.) | Local and long-distance: dispersed by waterfowl hosts |
| Water flea (*Daphnia pulex*) | Local and long-distance: eggs swept along by currents. Long-distance: eggs transported by animal vectors, boats | Local: swimming |
| Honey mushroom (*Armillaria bulbosa*) | Local and long-distance: spores blown by wind | Local: underground hyphae |
| Red seaweed (*Gracilaria gracilis*) | Local and long-distance: male gametes and spores swept by current | Local and long-distance: floating torn thallus fragments |
| Nudibranch (*Adalaria proxima*) | Local and long-distance: larvae swept along by ocean currents | Local and long-distance: rafting on drifting algae |

seeds from sacred lotus plants (*Nelumbo nucifera*) were recovered from a lake bed in the Liaoning Province, China. Although these were dated at between 1200 and 330 years old, the majority germinated and grew (Shen-Miller *et al.*, 1995).

Some fungi produce propagules by asexual means that can remain dormant for several years. Sclerotia are multicellular propagules produced by the ascomycete fungi in the genus *Sclerotinia*, and these can remain viable in the soil for at least 4–5 years (Adams and Ayers, 1979). A number of invertebrate species, including some rotifers, bryozoans and copepods, have banks of diapausing eggs or stato-blasts that are analogous to seed banks and also result in temporal gene flow. In some of these species the eggs can remain viable for more than 300 years (Hairston, Van Brunt and Kearns, 1995). In other habitats, juvenile sugar-beet cyst nematodes (*Heterodera schachtii*) can survive encysted in the soil for many years, and the nematode *Caenorhabditis elegans* responds to unfavourable environmental conditions by diapausing as a dauer larva that can last for several times the normal life span. Many insect species are also capable of prolonged diapause, such as the Yucca moth *Prodoxus y-inversus*, whose larvae developed into adults following a 30-year diapause in laboratory winter-like conditions (Powell, 2001).

The complexity of dispersal mechanisms, and the small size of many dispersing units (e.g. pollen, spores) in plants, fungi and invertebrates have made direct observation of movement very difficult in these taxonomic groups. Even when esti-mates of dispersal have been quantified using direct methods such as seed traps, they may bear little relation to gene flow. The contribution of molecular data to our understanding of dispersal and gene flow in these species has therefore been particularly noteworthy, as we shall see in the following sections that will focus on some of the most important factors that influence gene flow: dispersal ability, barriers to dispersal, reproductive mode, habitat patchiness, and interspecific interactions. In this chapter the emphasis will be on invertebrates, plants and microbes, but in Chapter 7 we shall add to our discussion by looking at patterns of dispersal in a number of vertebrate species.

## Dispersal ability

Although dispersal does not always lead to gene flow, it is a necessary precursor and there is, as we would expect, a tendency for highly mobile species to show accordingly high levels of dispersal and gene flow. For example, one review estimated the mean maximum dispersal distance as 148.1 km in 77 bird species, compared with 74.6 km in 40 mammal species (Sutherland et al., 2000), a result that is not particularly surprising given the flight capacity of birds. Similarly, in this chapter's earlier section on indirect estimates of gene flow, we saw that a correlation between dispersal ability and $N_e m$ has been used to defend the accuracy of gene flow estimates that have been obtained in this way. Although this may be considered a somewhat circular argument, it gains support from an additional correlation between dispersal ability and gene flow that is based on more complex patterns of dispersal: **isolation by distance**.

### Isolation by distance

So far we have been comparing dispersal ability with gene flow estimates that were based on a single (usually an average) $F_{ST}$ value. Reducing complex patterns of gene flow to a single variable is logistically attractive but it does mean that a lot of detailed information about dispersal patterns will be lost. For example, in many species the amount of gene flow between populations is inversely proportional to the geographical distances between them because individuals are most likely to disperse to nearby sites. This is known as isolation by distance (IBD; Wright, 1943), and is most commonly assessed by regressing log-transformed estimates of gene flow between pairs of populations against the appropriate log-transformed geographic distances. The significance of this relationship can be assessed using a **Mantel test**, which tests for a correlation between genetic and geographical distances. The Mantel test is appropriate for this comparison because it does not assume that the population pairwise comparisons are independent. The slope and the intercept of the IBD regression can be used to test the strength of the relationship.

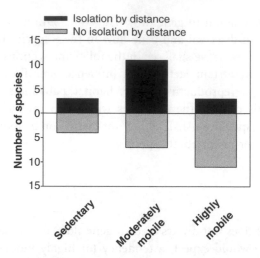

**Figure 4.5** The presence and absence of isolation by distance in 39 phytophagous insect species with varying dispersal abilities. Isolation by distance is defined as a significant relationship ($P \leq 0.05$) between gene flow and geographical distances, and is most prevalent in species with moderate mobility. Data from Peterson and Denno (1998a)

In a review of phytophagous insects, Peterson and Denno (1998a) compared dispersal ability with the patterns of IBD revealed by allozyme data from 39 species. Based on the significance of associations between gene flow and geographical distance (see Figure 4.5), and also on the slopes and the intercepts of the IBD regressions, the authors concluded that from distances of tens of kilometres to more than 1000 km, the IBD was more pronounced in moderately mobile insects compared with both sedentary and highly mobile species. This pattern, which has also been found in studies on other taxonomic groups, can be explained by the high levels of genetic differentiation that may exist between even neighbouring populations of sedentary species, and the low levels of genetic differentiation that may be found between even distant populations of highly mobile species.

It therefore seems that, whether we are looking at single estimates of gene flow or at patterns of isolation by distance, dispersal ability can be a useful predictor of the extent to which genes are exchanged between populations. We are, however, talking about generalities here, and there are many exceptions to this rule. To give just one example, the soil collembolan *Orchesella cincta* is a small, wingless insect that inhabits soil and leaf litter and is generally considered to be stationary. The $F_{ST}$ values calculated from AFLP data revealed no significant genetic differentiation between several pairs of *O. cincta* populations that were separated by distances of approximately 10 km, which suggests surprisingly high levels of gene flow (van der Wurff *et al.*, 2003). Conversely, in a comparison of gene flow within different species of carabid beetles, a fully winged species showed similar levels of population differentiation (and hence inferred gene flow) to those of a species with

vestigial wings (Liebherr, 1988). Clearly, dispersal ability is only a partial explanation for gene flow, and we shall now take a look at the importance of barriers.

## Barriers to dispersal

Barriers to dispersal may be relatively easy to recognize on land, for example water
bodies and mountain ranges limit the distribution of many ground-dwelling
species, and dry land is a barrier for many riverine species. In contrast, robust
physical barriers are relatively uncommon in the ocean, and therefore oysters,
clams, starfish, sea urchins and other species whose **planktonic** larvae can remain
free-living for days, weeks or even months are theoretically capable of being
transported across vast expanses of open ocean. Following on from this, it is
intuitively appealing that long-lived larvae should have a higher potential for
dispersal than short-lived larvae, and a number of comparative studies have
supported this prediction. One example of this is sea urchins in the genus
*Heliocidaris. H. tuberculata* has a planktonic larval stage of several weeks, and
showed very little genetic differentiation between populations separated by 1000
km of ocean, whereas *H. erythrogramma* has a planktonic larval stage of only 3–4
days and showed much higher levels of genetic differentiation across a similar
spatial scale (McMillan, Raff and Palumbi, 1992). This trend was summarized
further by a study in which data from 19 species of benthic marine invertebrates
revealed a significant correlation between the duration of the larval stage and the
distances across which individuals dispersed (Figure 4.6).

It is tempting to conclude from data such as those summarised in Figure 4.6 that
most, if not all, species with long larval periods will disperse widely on ocean
currents, but although this is sometimes true, an increasing number of molecular
studies have shown that trans-oceanic dispersal patterns are actually much more

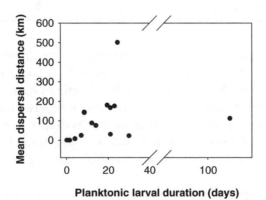

**Figure 4.6** Comparison between the length of the planktonic larval stage and the mean dispersal
distance (inferred from genetic differentiation) estimated for 19 species of marine benthic
invertebrates. These data show that species with relatively long larval durations tend to disperse
over relatively broad distances. Data from Siegel *et al.* (2003) and references therein

complicated than this. For one thing, several taxonomic groups show little or no correlation between larval duration and gene flow, such as coral species in the Great Barrier Reef (GBR), which can be divided into species that brood their larvae (brooders) and species that broadcast their larvae into the water (broadcast spawners). Since the larvae of brooding corals typically begin to settle within 1–2 days of release, whereas the larvae of broadcast spawning corals usually drift for 5–7 days before settling, the latter may be expected to show relatively high levels of gene flow and accompanying low levels of genetic differentiation among populations. This, however, was disproved by one study that found similar levels of genetic differentiation and gene flow among populations of both brooders and broadcast spawners (Ayre and Hughes, 2000; Table 4.6).

**Table 4.6** $F_{ST}$ values calculated between different reefs within the Great Barrier Reef for nine species of coral. Four of the species are broadcast spawners and five are brooders. Note that there is considerable overlap in the $F_{ST}$ values of brooders and spawners. Adapted from Ayre and Hughes (2000)

| Species | Mean $F_{ST}$ | Mean $N_em$ |
|---|---|---|
| Brooders | | |
| *Pocillopora damicornis* | 0.01 | 25 |
| *Acropora palifera* | 0.02 | 12 |
| *A. cuneata* | 0.05 | 5 |
| *Stylophora pistillata* | 0.09 | 3 |
| *Seriatopora hystrix* | 0.15 | 1 |
| Broadcast spawners | | |
| *Acropora millepora* | 0.01 | 25 |
| *A. valida* | 0.02 | 12 |
| *A. cytherea* | 0.03 | 8 |
| *A. hyacinthus* | 0.05 | 5 |

Other molecular studies of marine invertebrates have shown surprisingly low levels of gene flow between populations of species that have a high potential for dispersal. These seemingly paradoxical results sometimes have been attributed to previously unidentified physical barriers. Barriers to dispersal may be as simple as an expanse of open ocean. Five common and widespread coral species were sampled from several reefs within the GBR and also from Lord Howe Island. Within the GBR, gene flow was moderate to high across a distance of 1700 km. However, although Lord Howe Island is only 700 km further south, a lack of gene flow showed that dispersal there from the GBR was essentially non-existent (Ayre and Hughes, 2004). The authors of this study concluded that movements between the connected reefs of the GBR occurred in a stepping-stone manner, whereas the expanse of ocean between the GBR and Lord Howe Island was a barrier to dispersal. Similarly, the mantis shrimp *Haptosquilla pulchella* disperses across regions of semicontiguous coastlines that span thousands of kilometres, but populations separated by only 300 km across open ocean may experience very low levels of gene flow (Barber *et al.*, 2002).

Barriers can also be formed by currents such as upwellings, which occur when winds move surface water offshore, causing deeper water to flow in the opposite direction. The cold-water upwelling close to the tip of South Africa has provided a barrier to long-term gene flow in a number of marine species, including shallow water sea urchins in the genus *Tripneustes* (Lessios, Kane and Robertson, 2003). There are also many large-scale ocean currents that influence dispersal, such as the Almería–Oran oceanographic front, which is a well-defined hydrographic boundary caused by a strong current running between Almería in Spain and Oran in Algeria. This has created a barrier to gene flow at the junction between the Atlantic Ocean and the Mediterranean Sea for a number of species that disperse as planktonic larvae, including the mussel *Mytilus galloprovincialis* (Ladoukakis *et al.*, 2002) and the crustacean *Meganyctiphanes norvegica* (Zane *et al.*, 2000). Barriers to dispersal may also be created by changing environments, for example the Rio de la Plata estuary in Uruguay hinders the dispersal of pelagic larvae of the crab *Armases rubripes* because the salinity and temperature are suboptimal for larval development and survival (Luppi, Spivak and Bas, 2003).

Tracking the dispersal of small species in such an expansive area as the open ocean was extremely difficult before the advent of molecular markers. It is becoming clear that, even if they are not readily apparent to the human eye, barriers will influence gene flow across virtually every type of habitat. In the next section we will return to terrestrial species to look at how reproductive ecology can influence patterns of dispersal and gene flow.

## Reproduction

The way in which a species reproduces can have a strong influence on gene flow. Morjan and Rieseberg (2004) obtained estimates of gene flow from the literature by reviewing all relevant studies published in the journal *Molecular Ecology* between 1992 and 2002. They found a significant relationship between gene flow and the method of reproduction in both plants and animals. In both groups, populations that reproduce by either outcrossing or a mixture of outcrossing and selfing/cloning generally showed lower levels of differentiation and higher levels of gene flow than populations that reproduced solely by selfing/clonal reproduction (Table 4.7).

The relationship between reproduction and dispersal has been particularly well studied in plants, where it can be explained largely by different mechanisms of pollen dispersal. Outcrossing plants are either wind- or animal-pollinated, and therefore their pollen has the potential to travel relatively long distances. Plants that self-fertilize, on the other hand, generally lack any mechanism for pollen dispersal, and therefore pollen-mediated gene flow is low in these species. Both

**Table 4.7** Estimates of gene flow in plants and animals, grouped according to reproductive mode (Morjan and Rieseberg, 2004). In each category, *n* refers to the number of studies that were included in the comparisons. Note that $F_{ST}$ and $N_e m$ values in this table are not directly comparable as they are based on slightly different sets of comparisons

|  | $F_{ST}$ | | | $N_e m$ | | |
|---|---|---|---|---|---|---|
|  | *n* | Mean | Median | *n* | Mean | Median |
| **Plants** |  |  |  |  |  |  |
| Outcrossing | 74 | 0.29 | 0.14 | 73 | 1.38 | 1.47 |
| Mixed | 174 | 0.30 | 0.18 | 170 | 2.99 | 1.17 |
| Selfing or clonal | 22 | 0.43 | 0.36 | 22 | 0.43 | 0.45 |
| **Animals** |  |  |  |  |  |  |
| Outcrossing | 216 | 0.22 | 0.11 | 219 | 4.67 | 2.09 |
| Mixed | 14 | 0.40 | 0.16 | 13 | 7.55 | 1.50 |
| Selfing or clonal | 11 | 0.24 | 0.22 | 11 | 2.99 | 0.90 |

outcrossing and self-fertilizing species produce seeds, but seeds generally travel shorter distances than pollen. This has been demonstrated by numerous studies that estimated population differentiation and gene flow from at least two of the three plant genomes (nuclear, plastid and mitochondria) having different patterns of inheritance (Chapter 2). For any given species we can compare estimates of $F_{ST}$ (or analogue) that have been calculated from nuclear and organelle data to obtain a pollen/seed migration ratio (Ennos, 1994; Hamilton and Miller, 2002). A recent review of 93 studies calculated a mean pollen/seed gene flow ratio of 17, which shows that pollen is the main agent of gene flow in many plants (Petit *et al.*, 2005; see also Table 4.8). As a result, pollen dispersal mechanisms play an extremely

**Table 4.8** Some examples of the estimated ratios of pollen-mediated gene flow to seed-mediated gene flow. For each species the ratio was based on a comparison of $F_{ST}$ values (or analogue) that were calculated from two genomes that are inherited in different ways, typically nuclear versus plastid or mitochondrial data. A ratio of >0 means that pollen-mediated gene flow exceeds seed-mediated gene flow; the prevalence of this was reported recently in a review of 93 different studies in which the authors calculated a mean pollen/seed gene flow ratio of 17 (Petit *et al.*, 2005)

| Species | Pollen/seed gene flow | Reference |
|---|---|---|
| Oak (*Quercus petraea*) | 500 | El Mousadik and Petit (1996) |
| Oak (*Q. robur*) complex | 286 | El Mousadik and Petit (1996) |
| A tropical canopy tree, *Corythophora alta* | 200 | Hamilton and Miller (2002) |
| Lodgepole pine (*Pinus contorta*) | 28 | Ennos (1994) |
| White campion (*Silene alba*) | 3.4 (large scale)–124 (fine scale) | McCauley (1997) |
| Wild barley (*Hordeum spontaneum*) | 4 | Ennos (1994) |
| Rowan tree (*Sorbus torminalis*) | 2.21 | Oddou-Muratorio *et al.* (2001) |
| Helleborine orchid, *Epipactis helleborine* | 1.43 | Squirrell *et al.* (2001) |

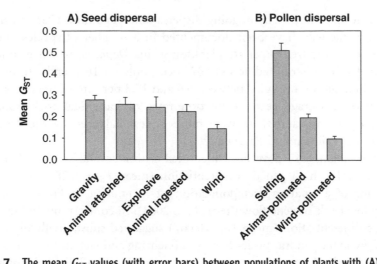

**Figure 4.7**  The mean $G_{ST}$ values (with error bars) between populations of plants with (A) different seed dispersal mechanisms, none of which are significantly different from each other, and (B) different pollen dispersal mechanisms, all of which are significantly different from each other. Adapted from Hamrick and Godt (1990)

important role in determining the genetic structure among plant populations (Figure 4.7).

## Fragmented habitats and metapopulations

Another factor that influences gene flow is the distribution of suitable habitats across the landscape. Species that live in temporary habitats may frequently disperse between sites, for example the coastal dune spider (*Geolycosa pikei*) that lives in ephemeral inlets showed very little genetic differentiation among ten seemingly isolated populations (Boulton, Ramirez and Blair, 1998). On the other hand, fairy shrimp (*Branchinecta sandiegonensis*) populations that inhabit ephemeral pools on coastal mesas in San Diego County, USA, showed substantial levels of genetic differentiation ($F_{ST}$ typically >0.25) that could be attributable to low gene flow (Davies, Simovich and Hathaway, 1997). It therefore seems that patchy or fragmented habitats lead to high gene flow in some species and low gene flow in others, although of course some gene flow is necessary for the survival of species that inhabit only ephemeral sites.

If a species lives in a fragmented habitat and regularly disperses between suitable sites it may form what is known as a **metapopulation**. The concept of a metapopulation dates back to the 1930s, although the actual term is attributable to Levins (1969) who presented an explicit model of a metapopulation. Levins' model, which we know now as a classical metapopulation, refers to a 'population of populations' that exist in a balance between extinction and recolonization, and are linked to one another by ongoing dispersal and gene flow. Classical metapopulations in the wild have been documented in a number of species, including three species in the freshwater zooplankton genus *Daphnia*, which in one study were found to be distributed across 507 rock pools on 16 islands. In any given year, the three species occupied between 5.6 and 17.8 per cent of the pools. Over a 17-year period, average yearly extinction rates of individual rock pools were around 20 per cent, and these were more or less balanced by the number of colonizations (Pajunen and Pajunen, 2003).

Metapopulations also have been identified within several species of the plant genus *Silene*. The high genetic differentiation (mean $F_{ST} = 0.287$) between subpopulations of a riparian metapopulation of *S. tatarica* in Finland suggested relatively low levels of gene flow (Tero *et al.*, 2003). In contrast, much lower levels of genetic differentiation (mean $F_{ST} = 0.007$) suggested substantially higher levels of gene flow in the tundra species *S. acaulis* (Gehring and Delph, 1999), once again reflecting a lack of consistency in the extent to which subdivided populations exchange immigrants. We may conclude from these previous examples that the subdivision of habitats is not a useful prediction of dispersal, possibly because it is a phenomenon that transcends a host of ecological and geographical variables. However, we must also bear in mind that populations that are regularly experiencing extinctions and recolonizations are unlikely to have reached equilibrium, in which case we must be particularly cautious in using $F_{ST}$ values to infer indirect estimates of gene flow.

As the concept of a metapopulation has become more commonplace in the literature, several variations have emerged. A slight modification of the original concept, which required all subpopulations to have an equal probability of going extinct at any given time, is the island–mainland model. This describes a metapopulation in which a small number of regional populations (mainlands) remain relatively stable over time and act as sources of (re)colonizers for new populations (islands). More recently, the term has often been adapted broadly to refer simply to a series of conspecific populations that are connected to one another by dispersal. In accordance with this modified view, in which localized extinctions are not necessarily a prerequisite, a number of alternative metapopulation models have been created. There is, however, some debate in the literature over whether these simply represent a series of subdivided populations.

The process of repeated extinctions and recolonizations can affect the genetic structure of a metapopulation in different ways. Extinctions and recolonizations will often be accompanied by population bottlenecks, which, as we know from Chapter 3, lead to reduced effective population size ($N_e$), accelerated genetic drift, and depleted genetic variation. The $N_e$ will be substantially reduced if all extant populations within a metapopulation descended relatively recently from a single ancestral population (Figure 4.8). Extinctions and recolonizations will also affect the extent to which populations are genetically differentiated from one another. Depending upon the balance between genetic drift and gene flow, differentiation

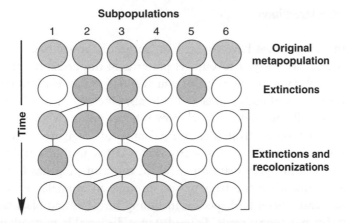

**Figure 4.8** An original metapopulation consisting of six subpopulations undergoes a series of extinctions and recolonizations. Shaded circles represent occupied sites and white circles represent extinct populations. Lines between circles indicate the source of colonizing individuals. In the most recent time period, the metapopulation consists of four subpopulations (2, 3, 4 and 5). All of these have descended recently from individuals that dispersed from a single subpopulation (subpopulation 3) and therefore the overall $N_e$ will be substantially lower than the $N_c$

among local populations may be enhanced if the founding of a new population is accompanied by a pronounced bottleneck, or diminished if gene flow remains high.

It is becoming increasingly important for us to understand the genetics of metapopulations, because habitat fragmentation is transforming a growing number of formerly continuous populations into subdivided units with varying degrees of connectivity. Some of the potential effects of habitat fragmentation were illustrated by a couple of studies that compared two sets of populations of the regal fritillary butterfly (*Speyeria idalia*). One set of populations was located in relatively continuous prairie and farmland habitat (Great Plains populations), whereas the other set was located in habitat that has been fragmented for approximately 150 years (Midwestern populations). The first study, which was based on mtDNA data, revealed little genetic structure among either the Great Plains or Midwestern populations, thereby suggesting that habitat fragmentation had little effect on population differentiation (Williams, 2002). However the second study, based on microsatellite data, revealed higher levels of among-population genetic differentiation and lower levels of gene flow and genetic diversity in the Midwestern populations compared with the continuous Great Plains populations (Williams, Brawn and Paige, 2003). Unfortunately the reduction of dispersal and genetic variation is a fairly common result of habitat fragmentation, making this a frequent matter of concern in conservation genetics. The regal fritillary butterfly studies also remind us how important it is to choose the correct molecular marker when evaluating gene flow and population differentiation.

## Interspecific interactions

So far in this chapter we have focused on some of the studies of *intra*specific populations that have helped to unravel patterns of dispersal and gene flow. We shall now look at some of the ways in which *inter*specific relationships can also influence movements of both plants and animals. One way in which this occurs is through plant–herbivore and predator–prey interactions, a classic example of this being plants that produce fruits that are palatable to frugivores. The rowan tree *Sorbus torminalis* is a temperate forest tree that is native to much of Europe. Seeds of this species will not germinate unless all of the fruit pulp has been removed, which often occurs when the fruit passes through an animal's intestine. Foxes, badgers, bears and several bird species, including thrushes and blackbirds, are all vectors that disperse rowan seeds. Long-distance dispersal is particularly likely to occur during the time (up to 12 h) that seeds remain in a bird's digestive tract (Oddou-Muratorio *et al.*, 2001). An example of a predator–prey interaction that can influence dispersal is the abandonment of hoarded sawfly cocoons (*Pristiphora erichsonii*) by small mammals such as shrews and voles (Buckner, 1959). However, the direct dispersal of prey by predators is not a widespread phenomenon, and in animals the role of predation may be more important in determining whether or not individuals disperse (i.e. to avoid predators) or where they disperse to (i.e. to a predator-free site).

### Parasites and hosts

Dispersal can also be influenced strongly by parasite–host relationships. From the point of view of the parasite, dispersal is often at the mercy of its host (Figure 4.9). In one study, high levels of gene flow based on PCR-RFLP analysis of three polymorphic loci were found in the nematode *Strongyloides ratti* over a range of approximately 300 miles in the UK. This is a parasite of wild rats (*Rattus norvegicus*), and it is most likely the dispersal of juvenile male rats from their natal sites that prevents substantial population subdivision of either *S. ratti* or their parasites (Fisher and Viney, 1998).

A more detailed comparison of host and parasite dispersal emerged from a study of five parasitic nematodes from three host species in the USA: *Ostertagia ostertagi* and *Haemonchus placei* from cattle, *H. contortus* and *Teladorsagia circumcincta* from sheep, and *Mazamastrongylus odocoilei* in white-tailed deer (Blouin *et al.*, 1995). Domestic sheep and cattle are moved regularly around the country, whereas white-tailed deer typically disperse over much shorter distances. Bearing in mind that these nematode parasites have no intermediate host and are unable to disperse in the free-living stage, it was not surprising that mitochondrial DNA revealed

**Figure 4.9** Scanning electron micrograph of a parasitic water mite (*Eylais* sp.) that is being transported by its host, a water boatman (*Sigara falleni*). The movements of their hosts greatly facilitate the dispersal of these mites between waterbodies. Photograph provided by David Bilton and reproduced with permission

evidence for high gene flow in the sheep and cattle parasites. In contrast, there was evidence of subdivision and isolation by distance among populations of the white-tailed deer parasites, patterns that are compatible with restricted gene flow in hosts and therefore their parasites.

Dispersal patterns of hosts and their parasites are less concordant when the parasites have multiple hosts. Liver flukes (*Fascioloides magna*) live as adults in several herbivorous species, including white-tailed deer. They produce eggs in the liver of their host that then pass through the bile duct into the host's intestinal tract. Eggs are expelled in faeces, and those that successfully reach an appropriate water body will hatch into free-swimming larval flukes known as miracidia. The next stage in their development requires the flukes to pass into the body of a suitable snail species. Inside the snail, the miracidia develop into cercaria that, after several stages of growth, are released from the snail and encyst on aquatic or wet terrestrial vegetation. These cysts are then ingested by deer or other herbivores that eat the vegetation, and so the cycle begins again (Figure 4.10). An investigation into the population genetics of white-tailed deer and liver flukes revealed significant genetic differentiation among populations of both species, but the complex life cycle of *F. magna* meant that there was little agreement in the extent to which fluke and host populations from the same geographical regions were genetically differentiated (Mulvey *et al.*, 1991).

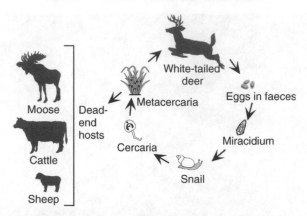

**Figure 4.10**   The life cycle of *Fascioloides magna*, a species of liver fluke that moves between several hosts throughout its development. This complex life cycle means that the dispersal patterns of this parasite show little concordance with those of its hosts.

While parasite dispersal often depends on the dispersal of the relevant hosts, it is also true that the dispersal of hosts may be influenced by the distribution of parasites. This is because a high parasite load at a particular site may cause hosts to move elsewhere. Male offspring of the common lizard *Lacerta vivipara* are more likely to disperse if their mothers harbour mites (Sorci and Clobert, 1995), and the kittiwake (*Rissa tridactyla*) is less likely to be faithful to a nesting site if there is a high level of tick infestation (Boulinier and Lemel, 1996). The great spotted cuckoo (*Clamator glandarius*) is a brood parasite, meaning that it lays its eggs in the nests of other bird species, which then rear the cuckoo's offspring, often at the expense of their own. In Europe, the main host of the great spotted cuckoo is the magpie (*Pica pica*). A study based on microsatellite markers revealed higher levels of gene flow among magpie populations living in sympatry with cuckoos compared with magpie populations living in allopatry (Martinez *et al.*, 1999). This may reflect a greater tendency for magpies to disperse from populations that are parasitised by cuckoos.

## Population Differentiation: Genetic Drift and Natural Selection

Gene flow is undoubtedly one of the most important processes in population genetics. In the absence of gene flow a combination of mutations and genetic drift will cause populations to diverge genetically from one another, whereas in the presence of gene flow populations can be held together as interconnected units that collectively contribute to a species' evolution. Nevertheless, gene flow does not necessarily make populations genetically homogeneous; populations may regularly exchange immigrants and still maintain a significant level of genetic divergence.

Population differentiation in the face of ongoing gene flow has been one of the most intensely debated subjects in evolutionary biology. How much gene flow is necessary to prevent differentiation by genetic drift? How is local adaptation possible if individuals are dispersing regularly among sites? What are the relative roles of drift and selection in promoting population differentiation? These questions are of fundamental importance to the evolution of populations and the cohesiveness of species. In this section we will look first at how gene flow and genetic drift interact, although this will be a relatively short discussion because genetic drift was covered in some detail in Chapter 3. We will then devote the rest of this section to the seemingly paradoxical co-existence of gene flow and local adaptation.

## Gene flow and genetic drift

In the absence of gene flow, conspecific populations will generally diverge from one another as a result of genetic drift. However, very little gene flow is necessary to reduce the rate of genetic drift and thereby prevent substantial population subdivision. Consider once more the relationship between gene flow and population differentiation ($F_{ST}$). Earlier in this chapter we learned that when populations are in equilibrium, $F_{ST} = 1/(4N_e m + 1)$. From this equation we can see that although populations are maximally divergent when $N_e m = 0$ and $F_{ST} = 1$, even small increases in gene flow ($N_e m$) will markedly reduce population differentiation ($F_{ST}$) (Figure 4.11). With only one migrant every fourth generation ($N_e m = 0.25$), $F_{ST}$ will be reduced to 0.5. If one migrant moves between a pair of populations every generation ($N_e m = 1$), then $F_{ST} = 0.20$. Since $F_{ST}$ is a measure of the degree

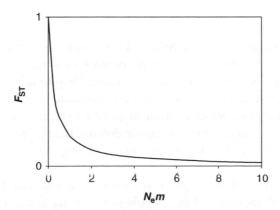

**Figure 4.11** Comparison of the genetic differentiation between populations ($F_{ST}$) and the accompanying indirect estimates of gene flow ($N_e m$), based on the relationship $F_{ST} = 1/(4N_e m + 1)$. The $F_{ST}$ values are reduced rapidly even with low levels of gene flow

of inbreeding within a local population relative to the collective population, and reflects the probability that two alleles drawn at random from within a population are identical by descent, an $F_{ST}$ value of 0.2 means that the local population is only 20 per cent more inbred than the collective population, even though gene flow is still very low. In fact, theory predicts that $N_e m$ values as low as one per generation may be sufficient to prevent the differentiation of populations by genetic drift (Wright, 1931).

Since gene flow reduces the rate of genetic drift, it stands to reason that, all else being equal, isolated populations will have a lower $N_e$, a higher rate of drift, and lower genetic variation compared with populations that receive immigrants (Table 4.9). Even low levels of immigration can introduce new genotypes and increase the effective size, and hence genetic diversity, of local populations (see Box 4.4). The importance of gene flow to the genetic diversity – and hence the survival – of threatened or endangered populations was illustrated by a study of Morelet's

**Table 4.9** Some of the ways in which gene flow (and lack thereof) can influence population genetics. Gene flow tends to be positively correlated with genetic diversity and $N_e$, and negatively with the rate of genetic drift within populations and the extent of genetic differentiation among populations

|  | Connected populations (gene flow) | Isolated populations (no gene flow) |
| --- | --- | --- |
| $N_e$ | ↑ | ↓ |
| Genetic drift | ↓ | ↑ |
| Genetic diversity | ↑ | ↓ |
| Population differentiation | ↓ | ↑ |

crocodile (*Crocodylus moreletii*) in Belize, Central America. In the 1960s, Morelet's crocodile declined to critically low levels because of intense hunting pressure, although populations now seem to be recovering. Despite this recent bottleneck, microsatellite analysis of crocodiles from seven populations in north-central Belize revealed surprisingly high levels of within-population genetic diversity ($H_e = 0.49$), plus low levels of among-population genetic differentiation. Both of these findings are at least partially attributable to high levels of migration (approximately five migrants per generation among localities). One of the populations included in this study, New River/New River Lagoon, is located in a relatively large and undisturbed area, and it appears that this population is acting as a genetic reservoir for the region. Regular dispersal from the New River Lagoon to other sites is increasing genetic diversity and reducing genetic drift within smaller regional populations (Dever *et al.*, 2002).

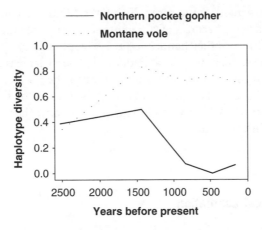

**Figure 4.12**  Changes in haplotype diversity over time for montane vole and northern pocket gopher populations. Adapted from Hadly *et al.* (2004)

## Box 4.4   Climate change, gene flow and genetic diversity

Despite its growing relevance, the effects that climate change may have on the genetic diversity of populations have seldom been empirically evaluated. Part of the difficulty lies with obtaining data from a sufficiently long time period. In one study, a group of researchers managed to obtain samples dating from between 2525 and 166 years before present (BP) for the montane vole (*Microtus montanus*) and the northern pocket gopher (*Thomomys talpoides*) (Hadly *et al.*, 2004). The authors of this study were interested in how the abundance and genetic diversity within populations of these two species had changed in Yellowstone National Park over the past 2500 years. They addressed this question by comparing abundance estimates from the fossil record with mtDNA sequences, an approach that effectively compared long-term population dynamics with genetic variation (measured as haplotype diversity). Both species prefer wet habitats, and this is reflected in their relative abundance in the fossil record during wet periods and their decline during dry periods. The decline of both species was particularly pronounced between 1438 and 470 years BP, a time that spanned the Medieval Warm Period. In the pocket gopher population this led to a decrease in haplotype diversity, whereas the opposite was true in the montane vole population (Figure 4.12).

These differences in genetic diversity have been attributed to contrasting levels of gene flow, because the two species differ in their dispersal patterns: montane voles move regularly between populations, whereas pocket gophers are highly territorial and seldom disperse. In the dispersive montane vole, gene flow increased during the Medieval Warm Period and as a result, a number of new haplotypes were introduced into the population. On the other hand, the pocket gopher population remained largely isolated, and a drop in abundance led to a decrease in haplotype diversity. Since genetic diversity is important to the long-term survival of populations, dispersal behaviour is likely to be one factor that determines which populations will survive climate change and other environmental disturbances in the future.

## Gene flow and local adaptation

Genetic drift is not the only force that causes populations to diverge from one another. Environments are not homogeneous, and therefore different genotypes will be selected for in different parts of a species' range; this is known as local adaptation. A commonly cited example of local adaptation is heavy metal tolerance in plants and fungi. Suilloid ectomycorrhizal fungi (*Suillus luteus* and *S. bovinus*), for example, were found to be highly zinc-tolerant when collected within 5 km of a zinc smelter, whereas those collected at least 15 km away from the pollution source were zinc-sensitive (Colpaert *et al.*, 2004). As with genetic drift, adaptation of populations to contrasting environmental conditions is a diverging force but, unlike drift, adaptation can occur even when gene flow is relatively high. The co-existence of these diverging and homogenizing forces has led to one of the most challenging questions in ecological genetics: how can adaptation occur in the face of gene flow?

The answer to this question involves a trade-off between the amount of gene flow and the strength of selection pressure acting on the trait in question. Selection pressures can be quantified on the basis of relative fitness levels. The phenotype (and its underlying genotype) that leaves the most descendants within a population is considered to have a **fitness value** (*w*) of one. Since all other genotypes have a relatively lower fitness, their fitness values must be less than one. Consider a hypothetical population in which genotype *A* has the highest fitness ($w = 1$) and the relative fitness of genotype *a* is 0.75. This means that for every 100 *A* genotypes that survive, only 75 *a* genotypes will survive. A related measure is the **selection coefficient**, which is the corollary of *w* and a measurement of the reduced probability of survival. It is calculated as:

$$s = 1 - w \tag{4.10}$$

The selection coefficient against genotype *a* is calculated as $s = 1 - 0.75 = 0.25$, i.e. survival of *a* genotypes is predicted to be 25 per cent lower than *A* genotypes in each generation. For any given genotype, the higher the selection coefficient, the stronger the selection pressure against that particular genotype will be, and the more likely that genotype is to become extinct. One of the best-known examples of selection coefficients is industrial melanism in the peppered moth *Biston betularia*. Although most populations were light coloured before the industrial revolution, dark forms of the moth became favoured in polluted areas because they are better camouflaged against blackened trees. The selection coefficient against light-coloured moths in these regions has been estimated to be 0.32, making their chances of survival approximately 32 per cent lower than dark-coloured moths (Haldane, 1924).

If selection pressure is high enough, local adaptations may persist despite high levels of gene flow (Endler, 1977). This has been demonstrated in a number of species, including the Dominican anole (*Anolis oculatus*), which is a morphologically variable member of the family Iguanidae that is endemic to Dominica. Morphological diversity in this species, which includes variable tail depth, scale size and body size, reflects adaptation to environmental variables such as rainfall and type of vegetation, both of which vary markedly across the island. Populations therefore show substantial levels of morphological differentiation, but at the same time microsatellite data revealed surprisingly high levels of gene flow, with only 7.8 per cent of population pairs revealing $F_{ST}$ values that were significantly greater than zero (Stenson, Halhotra and Thorpe, 2002). In another example, populations of Galápagos lava lizards (*Microlophus albemarlensis*) showed no evidence of genetic differentiation at six out of seven microsatellite loci, despite having differences in anti-predator behaviour that most likely reflect local adaptation (Jordan, Snell and Jordan, 2005). The anole and lava lizard studies provide just two of the numerous examples in which local adaptation has been maintained despite high levels of ongoing gene flow.

### Drift versus selection

Since gene flow can impede local adaptation only if the migration rate ($m$) exceeds the strength of selection ($s$), it follows that, if gene flow is low, even relatively weak selection pressures can accelerate the divergence of two populations from one another. This, however, is only part of the story; we must also remember the potential contribution of genetic drift to population differentiation. So how do drift and selection interact? Key here is the size of populations ($N_e$). If selection is strong relative to the population size, i.e. if $s$ is much greater than $1/(4N_e)$, then the effects of drift will be negligible. If, however, $s$ is much less than $1/(4N_e)$, then changes in allele frequencies will be largely attributable to genetic drift, although mutations also are likely to make a contribution. The interactions of gene flow, drift and selection are summarized in Figure 4.13.

There are a number of reasons why we may wish to know whether drift or selection is a more important driver of population differentiation. For one thing,

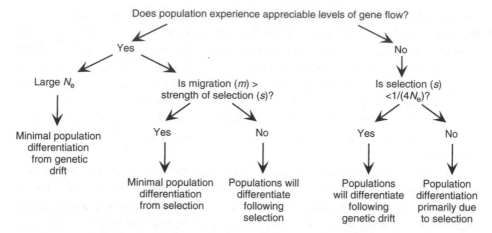

**Figure 4.13** Genetic differentiation of populations depends on a combination of gene flow, effective population size and natural selection

natural selection may mean that populations in one part of a species' range will not fare well if transplanted elsewhere, and this is an important consideration in reintroduction programmes in conservation biology. In practice, however, it is often extremely difficult to obtain estimates of $s$ and $N_e$ in wild populations, meaning that a comparison of these two variables is seldom a practical way of assessing the relative importance of selection. One alternative method for identifying local adaptation is through reciprocal transplant experiments, in which individuals are sampled from one or more populations and then transplanted to other sites of interest. Individuals then are monitored as they develop at both their home site and the sites to which they were relocated. At each population the relative fitness values of native and introduced individuals are compared, and a higher fitness in native plants provides evidence for local adaptation. Reciprocal transplant experiments have been used to great effect in many studies of plants and fungi, although they tend to be impractical when studying more mobile species. Molecular ecologists therefore often rely on other methods for inferring patterns of natural selection and local adaptation, some of which we will outline below.

### Patterns of molecular evolution

We know from Chapter 3 that one way to infer selection is to compare sequences from different populations and calculate the proportion of synonymous and non-synonymous substitutions. Synonymous substitutions do not alter the encoded amino acid and are therefore usually neutral, whereas non-synonymous substitutions are much more likely to cause phenotypic change. Patterns of substitution can therefore provide a way of identifying natural selection, because non-synonymous substitutions should be proportionately higher in selected genes.

**Table 4.10** Examples of genes in which selection has been inferred from a non-synonymous/synonymous substitution ratio that is greater than one. Adapted from Ford (2000) and references therein

| Taxa | Gene | Function |
|---|---|---|
| Salmonids | Transferrin | Resistance to bacterial infection |
| Humans, mice, fish | Major histocompatibility complex (MHC) | Immunity |
| Flowering plants (Solanaceae) | The S-locus system | Self-incompatibility |
| Filamentous fungi | The *het-c* locus (heterokaryon incompatibility) | Regulation of self/non-self-recognition during vegetative growth |
| Marine gastropods | Lysin genes | Proteins used by sperm to create a hole in the egg vitelline envelope |
| Sea urchins | Binding genes | Proteins that attach sperm to eggs during fertilization |
| *Drosophila* | *Acp26Aa* | Male ejaculate protein |
| House mouse | *Abpa* | Androgen-binding protein |
| Plasmodium | Surface protein genes | Detection of parasites by hosts |

Table 4.10 lists some examples in which the ratio of non-synonymous to synonymous substitutions has been used to infer natural selection.

One advantage to inferring selection from substitution ratios is that this method allows us to compare two categories of nucleotide substitutions along a single stretch of DNA, as opposed to comparing substitution rates from two or more gene regions that may have had different evolutionary histories. However, there are also some weaknesses associated with this method. For one thing it is not infallible; patterns of substitution will not reveal all instances of selection, in part because, as we saw in Chapter 1, sometimes only a single nucleotide change can have significant phenotypic consequences. Furthermore, we can use this method only if we know which gene has been selected for, and identifying candidate genes can be difficult. In the next section we will look at how a more general approach based on the comparison of data from multiple markers may give us clues about local adaptation.

### Discordant genetic differentiation

Migration and drift are expected to have approximately equal effects on all neutral loci, whereas the effects of selection will vary between neutral and non-neutral loci. We therefore expect all neutral loci to show similar levels of genetic divergence among populations, whereas non-neutral loci (or loci linked to non-neutral loci) are expected to show anomalous levels of divergence. These anomalous levels may be unusually high or unusually low, depending on the type of selection that the relevant genes have been subjected to; directional selection will increase population differentiation if different alleles are selected for in different populations, whereas balancing selection can decrease population differentiation by maintaining the same suite of alleles in multiple populations. By comparing multiple measures of population differentiation that are each based on a different locus, researchers may discover a marker that shows unusual levels of differentiation, and that may therefore indicate a genetic region that is under selection.

There are numerous examples of this in the literature. The dusky grouper (*Epinephelus marginatus*) inhabits coastal reefs around the Atlantic Ocean and the Mediterranean Sea. Researchers interested in the amount of population differentiation among different sites around the Mediterranean used data from nine allozyme loci to compare three populations. The average $F_{ST}$ was 0.214 but this result was influenced strongly by a single locus, *ADA* (adenosine deaminase), which yielded an $F_{ST}$ value of 0.713. This was seven times greater than the next highest $F_{ST}$ value, and once *ADA* was removed from the data set the average $F_{ST}$ value fell to 0.060. The high $F_{ST}$ value of *ADA* may be explained by directional selection at that locus (De Innocentiis *et al.*, 2001). In another example of discordant genetic markers in fish, four spawning populations of sockeye salmon (*Oncorhynchus nerka*) in Alaska were compared using data from allozymes, microsatellites and RAPDs (Allendorf and Seeb, 2000). The $F_{ST}$ values were comparable for all allozyme, microsatellite and RAPD loci with the exception of allozyme *sAH* (aconitate hydratase), which provided an extremely high measure of $F_{ST}$ (Figure 4.14). Once again, this discordant $F_{ST}$ value may be explained by natural selection.

Discordant genetic markers are often used to identify seemingly non-neutral genes, although they are seldom interpreted as conclusive evidence for natural

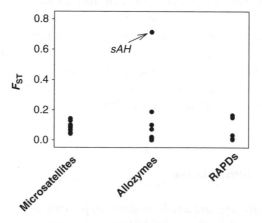

**Figure 4.14** The $F_{ST}$ values calculated from thirteen allozyme loci, eight microsatellite loci and five RAPD loci among sockeye salmon populations. The outlier is the $F_{ST}$ value for the *sAH* allozyme locus, marked by the arrow, which may be subject to selection (see text). Adapted from Allendorf and Seeb (2000)

selection because variation in $F_{ST}$ among different loci can result from stochastic variation in genetic drift. This randomly generated variance of $F_{ST}$ among loci may be particularly high if population sizes have fluctuated over time, since the loss of rare alleles during recent bottlenecks may have a greater impact on some loci than others. Care also must be taken when comparing $F_{ST}$ measures that are based on different types of genetic markers, because markers often mutate at different rates. The effects of mutation on $F_{ST}$ should be negligible if populations are in drift–migration equilibrium, but rapidly evolving markers may yield relatively high $F_{ST}$ values if new mutations are not dispersed between populations rapidly enough to attain equilibrium between gene flow and genetic drift. When all $F_{ST}$ values are higher in one set of markers (e.g. all microsatellites) compared with another set of markers (e.g. all allozymes), discrepancies are more likely to be attributable to different mutation rates than to natural selection.

## Clinal variations in allele frequencies

Another way to look for evidence of natural selection is to examine changes in allele frequencies along geographical **clines**. These are caused by a gradual change in one or more environmental variables, for example a change in photoperiod along a latitudinal gradient. If particular alleles are associated with environmental variables then this may be indicative of selection. Some of the best examples of this have been found in the fruitfly *Drosophila melanogaster*. One study compared alleles of the 70-kd heat shock protein (hsp70) along a latitudinal gradient in Australia (Bettencourt *et al.*, 2002). Heat shock proteins provide a mechanism that enables organisms to survive extreme temperatures and other environmental stresses. The authors of this study found a significant association between allele frequency and latitude. Since hsp70 is associated with heat resistance, and latitude is associated with both average temperatures and thermal extremes along the sampled transect, it seems possible that changes in temperature are causing natural selection to favour alternative hsp70 alleles. However, it is important to note that before selection can be invoked as an explanation for clinal variation in allele frequencies, the potential role of historical processes such as founder events and bottlenecks must be taken into account because changes in allele frequencies may simply reflect the genotypes of founding individuals.

Stochastic events such as bottlenecks become a less likely explanation for clinal variations in allele frequencies if similar genetic gradients are found in multiple geographically distinct regions. In *D. melanogaster*, concordant clines of alcohol dehydrogenase (*Adh*) alleles have been found in several different regions throughout the world. *D. melanogaster* feeds on rotting fruit, which ferments and produces alcohol. The *Adh* allele is important in the metabolism of this alcohol, and flies that lack *Adh* activity are extremely sensitive to the toxic effects of alcohol. A worldwide polymorphism in *D. melanogaster Adh* activity is attributable to a single amino acid replacement that determines whether or not individuals will have allele F (fast), which confers relatively high *Adh* activity, or allele S (slow), which confers relatively low *Adh* activity. The frequency of the F allele increases with latitude in several geographically distant parts of the world, including China (Jiang, Gibson and Chen, 1989), India (Parkash, Shamin and Vashist, 1992) and the Mediterranean (David *et al.*, 1989). This agreement between allele frequencies and latitude in several different areas cannot be easily explained by random neutral processes, therefore associations between latitude and the frequencies of *Adh* alleles present a compelling case for selection along environmental gradients.

### $F_{ST}$ *versus* $Q_{ST}$

Selection can also be inferred from comparisons between $F_{ST}$ and variations in **quantitative traits**, which are traits that are influenced by several different genes. Many important traits are quantitative, including height, weight and measures of reproductive fitness such as clutch size in birds and time to flowering in plants. These contrast with **qualitative traits**, which are discrete and controlled by one or a few loci, such as the round versus wrinkled seeds that were made famous by Mendel's experiments. Quantitative traits are controlled by a suite of genes known as **quantitative trait loci** (QTL). They are often influenced strongly by environmental effects, e.g. time to flowering in plants will depend on external factors such as temperature and photoperiod as well as the relevant genes. The variance in quantitative traits can therefore be partitioned into the amount of variation that is due to genetic factors, which is known as the genotypic variance ($\sigma^2_g$), and that due to environmental factors, which is known as the environmental variance ($\sigma^2_e$).

Differentiating between the contributions made by environmental and genetic factors to the phenotypic variation of quantitative characters can be problematic because many of these characteristics are phenotypically plastic. In laboratory settings, breeding experiments and pedigree analyses can be used to good effect, but these are seldom practical for wild populations. An alternative and often more practical way to estimate the genetic component of a quantitative trait is by calculating its heritability ($h^2$). Heritabilities range from zero to one, with a heritability of one meaning that a characteristic is determined purely by genetics, with absolutely no environmental influence. One way to estimate heritability is from parent–offspring regressions, which provide a comparison of parent and offspring phenotypes. If parent and offspring phenotypes consistently show a strong similarity to one another regardless of environmental influences, then genotypic variance (and hence heritability) is high and environmental variance is low. Parent–offspring regressions were used to estimate the heritability of bill depth in the medium ground finch *Geospiza fortis* on one of the Galápagos Islands. By measuring the phenotypic variance of bill depth in parents and offspring, Boag (1983) calculated an estimated heritability of 0.90 for this trait. Estimates of herit-

**Table 4.11**  Some examples of heritability that have been obtained through comparisons of the phenotypes of known relatives, e.g. parent–offspring regressions. The extent to which a trait will be influenced by environmental variables is inversely proportional to its heritability

| Species | Trait | Heritability | Reference |
|---|---|---|---|
| Medium ground finch (*Geospiza fortis*) | Body weight | 0.91 | Boag (1983) |
| European starling (*Sturnus vulgaris*) | Tarsus length | 0.49 | Smith (1993) |
| Squinting Bush Brown butterfly (*Bicyclus anynana*) | Egg size | 0.4 | Fischer, Zwaan and Brakefield (2004) |
| Common frog tadpoles (*Rana temporaria*) | Body size | 0.12 | Pakkasmaa, Merila and O'Hara (2003) |
| Egyptian cotton leafworm (*Spodoptera littoralis*) | Haemolymph henoloxidase activity (immune response) | 0.69 | Cotter and Wilson (2002) |
| Mouse (*Mus musculus*) | Life span | 0.44–0.62 | Klebanov *et al.* (2000) |
| Cotton-top tamarin (*Saguinus oedipus*) | Body size | 0.35 | Cheverud *et al.* (1994) |
| Tree snail (*Arianta arbustorum*) | Shell width | 0.70 | Cook (1965) |

ability can also be obtained by comparing phenotypic variance in full-siblings, i.e. siblings with the same mother and father, or in half-siblings, i.e. siblings with only one parent in common. Further examples of heritability are given in Table 4.11.

Once heritability estimates have been obtained, the genotypic variance of QTLs can be partitioned within and between populations. This is designated as $Q_{ST}$ and is comparable to $F_{ST}$ because it represents the degree to which populations are genetically differentiated. $Q_{ST}$ is calculated as:

$$Q_{ST} = \sigma^2_{g(between)} \Big/ \left( \sigma^2_{g(between)} + 2\sigma^2_{g(within)} \right) \qquad (4.11)$$

where $\sigma^2_{g(between)}$ is the amount of genotypic variance between populations and $\sigma^2_{g(within)}$ is the amount of genotypic variance within populations.

We know that $F_{ST}$, when based on neutral genetic markers, estimates the degree to which populations have diverged from one another as a result of gene flow and genetic drift. The $Q_{ST}$ values should show similar levels of population differentiation if they are based on neutral quantitative traits, but estimates of $Q_{ST}$ and $F_{ST}$ are often discordant. A survey of the literature reveals three possible outcomes in comparisons of $Q_{ST}$ and $F_{ST}$ between populations of the same species. Under the first scenario $Q_{ST} > F_{ST}$, and this means that quantitative traits have differentiated to a greater extent than would be expected by genetic drift alone. This is often taken as evidence that directional selection is favouring different phenotypes in different populations. In the second scenario $Q_{ST} = F_{ST}$, which, as noted above, could mean that the quantitative trait is selectively neutral. It is important to note, however, that parity between $Q_{ST}$ and $F_{ST}$ does not necessarily mean that the quantitative trait is neutral, because in these situations we cannot distinguish between the forces of selection and drift. The third possible outcome is $Q_{ST} < F_{ST}$, which means that population differentiation is less than would be attributable to drift and therefore the same phenotype is being selected for in multiple populations.

To date, few studies have found instances in which $Q_{ST} = F_{ST}$. $Q_{ST}$ is occasionally less than $F_{ST}$, one example being an endemic Mediterranean plant species *Brassica insularis* (Petit *et al.*, 2001). Populations had average $F_{ST}$ values of 0.213, whereas overall $Q_{ST}$ values were only 0.023 in seedlings and 0.087 in adults. These relatively low $Q_{ST}$ values have been attributed to selection pressures that have decreased the phenotypic variability of this species. Reasons for the discrepant seedling and adult $Q_{ST}$ values are less clear, although this may be attributable to age-dependent selection. The majority of comparisons between $Q_{ST}$ and $F_{ST}$ have found $Q_{ST} > F_{ST}$ (Figure 4.15). An extreme example of this was found in Finnish populations of Scots Pine (*Pinus sylvestris*) along a latitudinal gradient from 60°N to 70°N. Each year, the timing of bud burst occurs around 21 days earlier in the northernmost populations compared with those further south. This is a quantitative trait with a strong genetic component that provides a $Q_{ST}$ value of 0.80. In contrast, $F_{ST}$ values calculated from allozymes, microsatellites, RAPDs and RFLPs all revealed very low population differentiation ($F_{ST} < 0.02$;

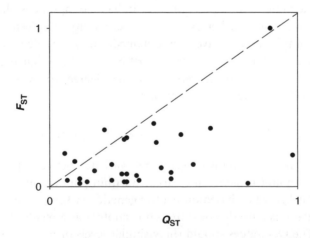

**Figure 4.15** The relationship between $F_{ST}$ and $Q_{ST}$ in 29 species (21 plants, 5 invertebrates, 3 vertebrates). The dashed line in the middle represents parity between the two measurements; points below this line mean that $F_{ST} < Q_{ST}$, and points above this line mean that $F_{ST} > Q_{ST}$. The $F_{ST}$ values were based on data from allozymes, microsatellites, RFLPs and RAPDs. The $Q_{ST}$ values were based on a mixture of morphological and life history traits. After Merilä and Crnokrak (2001), McKay and Latta (2002) and references therein

Karhu *et al.*, 1996). Evidently, directional selection is favouring a relatively early date of bud burst in northern populations.

## Overview

Broadly speaking, gene flow maintains populations as cohesive units, whereas genetic drift and natural selection can cause populations to diverge and eventually to split into distinct species. Throughout this chapter we have been concerned primarily with relatively recent adaptive and non-adaptive changes within populations. From time to time we have referred also to the more distant evolutionary histories of populations and species, which can have lasting impacts on genetic structure. This will be explored further in the next chapter, which is concerned primarily with the link between evolutionary histories and the current distributions of species.

## Chapter Summary

- Most species consist of multiple populations. The extent to which these populations are differentiated from one another can be quantified in a number of ways, including Nei's standard genetic distance ($D$).

- $F$-Statistics use inbreeding coefficients to partition genetic variation within and between populations: $F_{IT}$ reflects the total amount of inbreeding in a population; $F_{ST}$ reflects the amount of inbreeding in a population that is due to the differentiation of subpopulations; and $F_{IS}$ reflects the amount of inbreeding that occurs within subpopulations. The most widely used is $F_{ST}$, which is a measure of how genetically distinct populations are from one another.

- The amount of gene flow between populations is one important determinant of genetic differentiation. Dispersal and gene flow can be quantified using direct methods (such as radio tracking), indirect methods that provide estimates of $N_e m$, and assignment tests.

- Dispersal patterns vary widely across and within taxonomic groups. Dispersal is particularly complex in plants, fungi and invertebrates because many species within these groups have different dispersal mechanisms throughout their life cycle, and also may disperse through time following prolonged diapause.

- Although dispersal does not necessarily lead to gene flow, species with good dispersal abilities generally show correspondingly high levels of gene flow. Species with intermediate dispersal abilities may be most likely to show a pattern of isolation by distance.

- Barriers to dispersal will promote the genetic differentiation of populations. Molecular data can sometimes reveal previously unidentified barriers to dispersal and can be particularly useful in relatively inaccessible places such as the open ocean.

- Reproduction can also influence gene flow and population differentiation. In both plants and animals, outcrossing species have higher levels of gene flow than selfing/clonal species. In plants, this difference is largely attributable to mechanisms of pollen dispersal.

- Species living in ephemeral or fragmented habitats may be either regular or infrequent dispersers. If the former, they may exist within a metapopulation in which local populations are connected by gene flow and are subjected to repeated extinctions and recolonizations.

- The dispersal of species may also depend on the distribution of symbionts. Parasite dispersal often follows the movements of hosts, although hosts will sometimes disperse in response to parasites.

- Genetic drift and natural selection can cause populations to diverge. Gene flow tends to homogenize populations, although divergence between populations can occur despite ongoing gene flow if selection ($s$) is strong enough. Conversely, if the migration rate ($m$) exceeds the strength of selection ($s$), then local adaptation will be hindered by the continued introduction of alleles from other populations.

- Selection can be inferred from the proportion of non-synonymous substitutions, from discrepancies in the levels of differentiation that are revealed at different loci, or from clinal variations in allele frequencies.

- $Q_{ST}$ measures the genetic variance of quantitative trait loci (QTLs) within and among populations. If $Q_{ST}$ is either more or less than $F_{ST}$ then this too may be evidence of selection. So far, investigations have found that $Q_{ST}$ is usually greater than $F_{ST}$.

## Useful Websites and Software

- Genepop on the web, an online population genetics software package: http://wbiomed.curtin.edu.au/genepop/

- Arlequin software for population genetics data analysis: http://lgb.unige.ch/arlequin/

- Popgene. Analysis of genetic variation among and within populations using co-dominant and dominant markers: http://www.ualberta.ca/~fyeh/info.htm

- Structure. Uses multi-locus genotype data to investigate population structure: http://pritch.bsd.uchicago.edu/software.html

- Immanc. Assignment test to identify immigrants based on multilocus genotypes: http://www.rannala.org/labpages/software.html

## Further Reading

### Books

Futuyma, D.J. 2005. *Evolution*. Sinauer Associates, Sunderland, Massachusetts.
Lowe, A., Harris, S. and Ashton, P. 2004. *Ecological Genetics*. Blackwell Publishing, Oxford.

### Review articles

Bilton, D.T., Freeland, J.R. and Okamura, B. 2001. Dispersal in freshwater invertebrates. *Annual Review of Ecology and Systematics* **32**: 159–181.
Hanski, I. 1998. Metapopulation dynamics. *Nature* **396**: 41–49.
Latta, R.G. 2003. Gene flow, adaptive population divergence and comparative population structure across loci. *New Phytologist* **161**: 51–58.
Manel, S., O.E. Gaggiotti and R.S. Waples. 2005. Assignment methods: matching biological questions with appropriate techniques *Trends in Ecology and Evolution* **20**: 136–142.
Morjan, C.L. and Rieseberg, L.H. 2004. How species evolve collectively: implications of gene flow and selection for the spread of advantageous alleles. *Molecular Ecology* **13**: 1341–1356.
Neigel, J.E. 1997. A comparison of alternative strategies for estimating gene flow from genetic markers. *Annual Review of Ecology and Systematics* **28**: 105–128.
Petit, R.J., Duminil, J., Fineschi, S., Hampe, A., Salvini, D. and Vendramin, G.G. 2005. Comparative organisation of chloroplast, mitochondrial and nuclear diversity in plant populations. *Molecular Ecology* **14**: 689–702.
Weir, B.S. and Hill, W.G. 2002. Estimating F-statistics. *Annual Review of Genetics* **36**: 721–750.

## Review Questions

**4.1.** A study of six house sparrow (*Passer domesticus*) populations in Western Canada revealed an average $F_{ST}$ of 0.136 and an average $F_{IS}$ of 0.085. What is the overall inbreeding coefficient across all six populations?

**4.2.** From the data in question 1, calculate the average indirect measure of gene flow among the house sparrow populations.

**4.3.** An investigation into four populations of the damselfly *Coenagrion puella* revealed an average $N_em$ of 8.8 per generation, whereas mark–recapture studies identified only four immigrants during the 8-week breeding season. How could you explain this discrepancy?

**4.4.** What factors determine the rate at which two populations diverge genetically from one another?

**4.5.** The $F_{ST}$ values between two populations of the vernal sweetgrass *Anthoxanthum odoratum* were estimated from AFLP and microsatellite data as follows:

AFLP
| Locus 1 | 0.124 |
| Locus 2 | 0.098 |
| Locus 3 | 0.103 |
| Locus 4 | 0.699 |
| Locus 5 | 0.111 |
| Locus 6 | 0.107 |

Microsatellites
| Locus 1 | 0.327 |
| Locus 2 | 0.396 |
| Locus 3 | 0.421 |
| Locus 4 | 0.372 |
| Locus 5 | 0.345 |

Do these data suggest that natural selection is acting on any of the loci?

**4.6.** What conclusion might you reach about height QTL in ten populations of sunflowers (*Helianthus annus*) situated along a latitudinal gradient if $Q_{ST} < F_{ST}$?

# 5

# Phylogeography

## What is Phylogeography?

Current patterns of gene flow may bear little resemblance to the historical connections among populations, but both are relevant to the contemporary distributions of species and their genes. Understanding how historical events have helped to shape the current geographical dispersion of genes, populations and species is the major goal of **phylogeography**, a term that was introduced by Avise in 1987 (Avise *et al.*, 1987). Phylogeography can be defined as a '. . . field of study concerned with the principles and processes governing the geographic distributions of genealogical lineages, especially those within and among closely related species' (Avise, 2000). By comparing the evolutionary relationships of genetic lineages with their geographical locations, we may gain a better understanding of which factors have most influenced the distributions of genetic variation. Phylogeography therefore embraces aspects of both time (evolutionary relationships) and space (geographical distributions).

## Molecular Markers in Phylogeography

Phylogeography is concerned with the distribution of genealogical lineages, and we know from Chapter 2 that DNA sequences are the markers that are best suited for inferring genealogies. A looser interpretation of phylogeography does allow the use of markers such as microsatellites and AFLPs that provide information about the genetic similarity of populations based on allele frequencies or bandsharing, although strictly speaking such data do not comply with Avise's original definition of phylogeography. Nevertheless, as we saw in Chapter 4, allele frequencies can provide us with information on gene flow and the genetic subdivision of

*Molecular Ecology*   Joanna Freeland
© 2005 John Wiley & Sons, Ltd.

populations and therefore often make useful contributions to studies of phylogeography.

Over the years the markers of choice, at least when studying animals, have been mitochondrial sequences that were obtained through either direct sequencing or RFLP analysis; in fact, prior to 2000, approximately 70 per cent of all phylogeographic studies were based on analyses of animal mitochondrial DNA (Avise, 2000). As we noted in Chapter 2, the popularity of mtDNA is based on several factors, including the ease with which it can be manipulated, its relatively rapid mutation rate, and its presumed lack of recombination, which results in an effectively clonal inheritance. Futhermore, universal animal mitochondrial primers are readily available and this is an important reason why animal phylogeographic studies have historically outnumbered those of plants.

At the same time, mtDNA markers are limited by the fact that the mitochondrion effectively comprises a single locus. Reconstructing population histories from a single locus is less than ideal if that locus has been subjected to selection or some other process that may have given it an unusual history. In addition, mitochondrial data may be misleading if mtDNA has passed recently from one species to another following hybridization. Furthermore, the sensitivity of mtDNA to bottlenecks is not always an advantage, and there is also the possibility that its maternal mode of inheritance will lead to an incomplete reconstruction of population histories if males and females had different patterns of dispersal.

The only way to test whether a mtDNA genealogy accurately reflects population history is to look for concordance with genealogies that are inferred from DNA regions in other genomes. In plants we can compare data from mitochondria, plastids and nuclear regions, but in animals mtDNA data can be supplemented only with data from nuclear loci. However, analysing nuclear data is less straightforward than analysing organelle data because recombination is common in the nuclear genome of sexually reproducing taxa. If the rate of recombination at a particular locus is similar to the rate of nucleotide substitutions, any given allele will, in all likelihood, have more than one recent ancestor, which means that different parts of the same locus will have different evolutionary histories. Although we need to be aware of this complication, a review of several nuclear gene phylogeographies recently suggested that recombination need not be an insurmountable problem (Hare, 2001).

Recombination can be identified with appropriate software (e.g. Holmes, Worobey and Rambaut, 1999; Husmeier and Wright, 2001). Once identified, the easiest way to deal with recombination, provided that it is present at only a low level, is to remove the relevant sequence regions before doing the genealogical analyses. This was the approach used in a study of the plant parasitic ascomycete fungus *Sclerotinia sclerotiorum* and three closely related species, all of which are parasites of agricultural and wild plants. Researchers sequenced seven nuclear loci and, after aligning the sequences, detected a low level of recombination using a software program that generates compatibility matrices. By removing recombinant haplotypes they were able to control for the effects of recombination in their analyses, and subsequently found some informative patterns regarding the fragmentation of populations in response to ecological conditions and host availability. Their findings were strengthened by their use of data from multiple, independent loci (Carbone and Kohn, 2001).

So far, most phylogeographic studies that have used nuclear data have sequenced specific genes such as bindin, a sperm gamete recognition protein that has been used to compare sea urchin populations (genus *Lytechinus*; Zigler and Lessios, 2004). There is, however, a growing interest in using single nucleotide polymorphisms (SNPs) from multiple loci for reconstructing population histories because they represent the most prevalent form of genetic variation (Brumfield *et al.*, 2003). At this time SNPs have not been characterized adequately to provide useful markers in most non-model organisms, although a recent study that used 22 SNP loci to genetically characterize Scandinavian wolf populations suggests that the practical constraints associated with SNPs will soon be substantially reduced at which time we are likely to see a rapid increase in SNP-based studies (Seddon *et al.*, 2005).

Regardless of which molecular markers are employed, there are a number of analytical techniques relevant to phylogeography that we have not yet discussed, and we must understand these before we can start to unravel the evolutionary relationships of populations. We will start by looking at some of the more traditional methods, which include molecular clocks and phylogenetic reconstructions. We will then move on to look at some more recently developed methods that are specifically designed to accommodate the sorts of data that we are most likely to encounter in phylogeography.

## Molecular Clocks

One of the easiest ways to obtain information about the evolutionary relationships of different alleles is to calculate the extent to which two sequences differ from one another (generally referred to as sequence divergence). This is most easily presented as the percentage of variable sites, although more complex models take into account mutational processes, for example by differentially weighting transitions versus transversions, or synonymous versus non-synonymous substitutions (Kimura, 1980). The similarity of two sequences provides us with some information about how long ago they diverged from one another because, generally speaking, similar sequences will have diverged recently whereas dissimilar sequences have been evolutionarily independent for a relatively long period of time. We may be able to acquire even more precise information about the time since sequences diverged from one another if we apply what is known as a **molecular clock**.

The idea of molecular clocks was introduced in the 1960s (Zuckerkandl and Pauling, 1965), based on the hypothesis that DNA sequences evolve at roughly constant rates and therefore the dissimilarity of two sequences can be used to calculate the amount of time that has passed since they diverged from one another. Molecular clocks have been used to date both ancient events, such as the emergence of ancestral mammals several millions of years before dinosaurs became extinct (Kumar and Hedges, 1998), and also more recent events, such as the splitting of the circumarctic-alpine plant *Saxifraga oppositifolia* into two subspecies approximately 3–5 million years ago (Abbott and Comes, 2004).

The calibration of molecular clocks is based on the approximate date when two genetic lineages diverged from one another. This date should ideally be obtained from information that is independent of molecular data, for example the fossil record or a known geological event such as the emergence of an island. The next step is to calculate the amount of sequence divergence that has occurred since that time. By dividing the estimated time since the lineages diverged by the amount of sequence divergence that has since taken place, we obtain an estimate of the rate at which molecular evolution is occurring, in ohter words the rate at which the molecular clock is ticking. Molecular clocks are usually represented as the percentage of base pairs that are expected to change every million years. If we sequence a gene from two species that were separated 500 000 years ago and we find that 490 out of 500 bp are still the same, the molecular clock would be calibrated as 10/500 = 2 per cent per 500 000 years, or 4 per cent per million years.

The most widely cited molecular clock is a 'universal' mtDNA clock of approximately 2 per cent sequence divergence every million years (Brown *et al.*, 1982). This was originally calculated using data from primates and has since been extrapolated to a wide range of taxonomic groups. In recent years, however, it has become increasingly apparent that the idea of a 'universal' clock is something of a fallacy because evolutionary rates differ within DNA regions (e.g. synonymous versus non-synonymous substitutions), between DNA regions, and also between taxonomic groups. Different mutation rates have been calculated for numerous species that were separated by geological events of a known age, such as the emergence of the Isthmus of Panama that divided the Pacific Ocean from the Atlantic Ocean and the Caribbean Sea approximately 3 million years ago. Subsequent population divergence on either side of the Isthmus has led to a number of sister species known as **geminate species**. A comparison of sequences from geminate shark species that were separated by the Isthmus of Panama revealed nucleotide substitution rates in the mitochondrial cytochrome *b* and cytochrome oxidase I genes that are seven or eight times slower than in primates (Martin, Naylor and Palumbi, 1992). Although there are no set rules, mutation rates in mtDNA seem to vary according to a number of taxonomic variables, including thermal habit, generation time and metabolic rates (Martin and Palumbi, 1993; Rand, 1994). Researchers therefore now prefer to use a molecular clock that has been calibrated within the taxonomic group and gene region ·that

**Table 5.1** Some examples of molecular clocks that have been calculated for various genomic regions in a variety of species. Each of these clocks was calibrated from the amount of time that has passed since species diverged from one another, which in turn was inferred from independent data such as the timing of a known geological event

| Species | DNA sequence | Sequence divergence rate (% per million years) | Method of calibration | Reference |
|---|---|---|---|---|
| Sorex shrews (Soricidae) | Cytochrome *b* (mtDNA) | 1.36 | Fossil record | Fumagalli *et al.* (1999) |
| Diatoms (bacillariophyta) | Small subunit ribosomal RNA | 0.04–0.06 | Fossil record | Kooistra and Medlin (1996) |
| Taiwanese bamboo viper (*Trimeresurus stejnegeri*) | Cytochrome *b* (mtDNA) | 1.1 | Age of Taiwan | Creer *et al.* (2004) |
| Geminate marine fishes | ND2 (mtDNA) | 1.3 | Time since the Isthmus of Panama emerged | Bermingham, McCafferty and Martin (1997) |
| Hawaiian *Drosophila* | Alcohol dehydrogenase gene (*Adh*) | 1.2 | Age of Hawaiian islands | Bishop and Hunt (1988) |
| California newt (*Taricha torosa*) | Cytochrome *b* (mtDNA) | 0.8 | Fossil record | Tan and Wake (1995) |
| Marine gastropods *Tegula viridula* and *T. verrucosa* | Cytochrome oxidase subunit I (mtDNA) | 2.4 | Time since the Isthmus of Panama emerged | Hellberg and Vacquier (1999) |

they are studying, instead of a so-called universal clock. Some examples of the molecular clocks that appear in the literature are shown in Table 5.1.

Some of the best examples of molecular clocks come from species that are endemic to oceanic islands. The Hawaiian islands are volcanic in origin and their ages have been estimated using potassium–argon (K–Ar) dating. This method, which is accurate on rocks older than 100 000 years, relies on the principle that the radioactive isotope of potassium (K-40) in rocks decays to argon gas (Ar-40) at a known rate. The proportion of K-40 to Ar-40 in a sample of volcanic rock therefore provides an estimate of when this rock was formed. Such K–Ar dating has revealed that the islands in the Hawaiian archipelago are arranged from the oldest at the northwest of the array to the youngest at the far southeast. Within the main Hawaiian Islands, Hawaii is approximately 0.43 million years old, Oahu is around 3.7 million years old and Kauai emerged approximately 5.1 million years ago (Carson and Clague, 1995).

Fleischer, McIntosh and Tarr (1998) superimposed these geological ages onto phylogenetic trees to calibrate the rates of sequence divergence in several endemic taxa. This provided them with molecular clocks of 1.9 per cent per million years for the yolk protein gene in *Drosophila*, 1.6 per cent per million years for the cytochrome *b* gene in Hawaiian honeycreeper birds (Drepananidae), and a variable rate of 2.4–10.2 per cent per million years for parts of the mitochondrial 12S and 16S rRNA and tRNA valine in *Laupala* crickets. The authors stressed that these estimates were based on a number of assumptions, including the establishment of populations very near to the time at which individual islands were formed, and there having been very little subsequent movement between populations. The surprisingly high rates for a ribosomal-RNA encoding gene that were calculated for *Laupala* crickets suggested that in this species at least one or more of the assumptions were not met.

There are two final points worth noting about molecular clocks. First, the rate at which a sequence evolves is not necessarily constant through time; in some cases, mutation rates are relatively rapid in newly diverged taxa but then slow down over time (Mindell and Honeycutt, 1990). Second, although many of the estimates presented in this section may appear very similar, a difference in mutation rates of only 0.5 per cent per million years can have a significant impact on the estimated timing of evolutionary events. If the sequences of two species diverged by 5 per cent then this would translate into a 5-million-year separation according to a clock of 1 per cent per million years, but a 10-million-year separation according to a clock of 0.5 per cent per million years. Molecular clocks remain widespread in the literature but are also highly contentious. In fact, some researchers have argued that we may never achieve molecular clocks that are sufficiently reliable to allow us to date past events (Graur and Martin, 2004). Molecular clocks should therefore be interpreted with caution and ideally should be based on accurately dated geological events or fossils, and be calibrated specifically for the gene region and taxonomic group that is being studied.

## Bifurcating Trees

One appeal of molecular clocks is that they are relatively easy to use once the correct calibration has been done, but with a bit more work a great deal more information on the evolutionary relationships of genetic lineages can be obtained from DNA sequences through the reconstruction of phylogenies. Traditionally, most phylogenetic inferences have been depicted in the form of hierarchical **bifurcating** trees, in other words trees that reflect a series of branching processes in which one lineage splits into two descendant lineages. These trees can be based on morphological characters, although in this book we will limit our discussion to phylogenetic trees that are inferred from genetic characters. The positioning of organisms on a tree is generally based on their genetic similarity to one another.

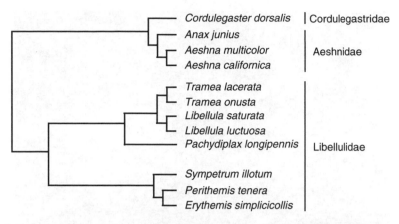

**Figure 5.1** A phylogeny of 13 dragonfly species based on the mitochondrial 12S ribosomal DNA gene. First species names, and then family names, are shown to the right of the tree. Note that congeneric species are closest together on the tree because they are genetically most similar to one another. Adapted from Saux, Simon and Spicer, (2003)

This is illustrated in Figure 5.1, which shows a tree that portrays the evolutionary relationships of some dragonfly species, genera and families. Congeneric species that diverged from a common ancestor relatively recently, such as *Libellula saturata* and *L. luctuosa*, will be close to each other on the tree. Confamilial genera, such as *Libellula* and *Erythemis* (Figure 5.2), are further apart on the tree because their common ancestor was more remote, and members of different families are even more widely spaced.

There are many different ways in which phylogenies can be reconstructed from genetic data, but most of them fall into one of four categories: **distance**, **parsimony, likelihood** and **Bayesian** methods. Note that the following discussion will focus on the phylogenies of closely related populations and species, and the limitations outlined below are not necessarily relevant to the phylogenies of more distantly related taxa.

Distance methods are based on measures of evolutionary distinctiveness between all pairs of taxa (Figure 5.3). These metrics may be calculated from the number of nucleotide differences if based on DNA sequence data or from estimates such as Nei's D (Chapter 4) if based on allele frequency data, such as that provided by allozymes or microsatellites. There are many different algorithms that can be used to reconstruct trees from genetic distances, the most common being the neighbour-joining method (Saitou and Nei, 1987). Details of these various methods are beyond the scope of this book; suffice it to say that the goal is to build a tree that accurately reflects how much genetic change has occurred – and therefore roughly how much time has passed – since lineages split from one other. Because branch lengths reflect the evolutionary distance between two points on a tree, this approach should ensure that neighbouring branches on a tree are

**Figure 5.2** An Eastern pondhawk (*Erythemis simplicicollis*). This is a common North American dragonfly that hunts for insects from low perches and often rests on the ground. Photograph provided by Kelvin Conrad and reproduced with permission

occupied by those lineages that have descended most recently from a common ancestor. When applied to closely related lineages, distance-based trees may be poorly resolved because a number of different lineages may be separated by the same distance, in which case decisions as to which lineages should be closest to each other on the tree are arbitrary.

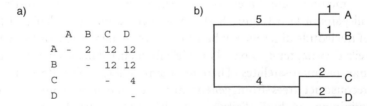

**Figure 5.3** A general distance method for reconstructing phylogenies. (a) The pairwise genetic distances between species A–D are provided in a matrix format, with the number referring to the percentage difference between any pair of species, e.g. the sequence from species A differs from that of species B sequence by 2%. (b) The genetic distances are then used to reconstruct a tree in which species that are separated by the smallest genetic distances are grouped together. Note that the branch lengths are proportional to the amount of genetic change that has occurred, and these add up to the total genetic distances that are given in (A)

(a)    (b)

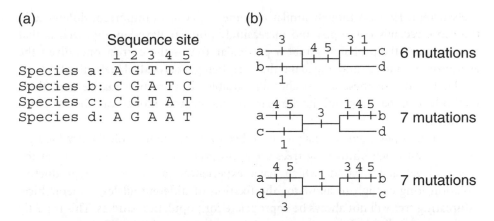

**Figure 5.4** A maximum parsimony (MP) phylogenetic analysis based on the DNA sequences shown in (a) of species a, b, c and d. Three possible trees are shown in (b). Vertical bars on branches represent the mutations that must have occurred at particular sequence sites. The tree that requires six mutations is more parsimonious than the trees that require seven mutations and therefore under MP analysis would be considered the correct tree

A maximum parsimony tree is the tree that contains the minimum number of steps possible, in other words the smallest number of mutations that can explain the distribution of lineages on the tree (Fitch, 1971; Figure 5.4). Parsimony is based on Ockham's Razor, the principle proposed by William of Ockham in the 14th century, which states that the best hypothesis for explaining a process is the one that requires the fewest assumptions. A maximum parsimony tree will maximize the agreement between characters on a tree. However, although intuitively appealing, parsimony trees may remain unresolved if data are insufficiently polymorphic, which is often the case in the recently diverged lineages that are typically found within and among populations. The small number of mutational changes that differentiate many conspecific haplotypes may mean that multiple, equally parsimonious trees exist, once again leading to a situation in which it may be impossible to determine which haplotypes should be adjacent to one another on the tree.

The third and fourth categories of phylogenetic analysis are maximum likelihood (ML; Chapter 3) and Bayesian approaches, both of which are based on specific models that describe the evolution of individual characters. Each model will make a particular set of assumptions, for example that all nucleotide substitutions are equally likely or, alternatively, that each nucleotide is replaced by each alternative nucleotide at a particular rate. Models are typically complex, for example they can accommodate different rates of transitions and transversions, and heterogeneous substitution rates, along a particular stretch of DNA. Once the assumptions have been established, ML determines the probability that a data set is best represented by a particular tree by calculating the likelihood of each possible phylogenetic tree occurring within a specified evolutionary model

(Felsenstein, 1981). Although similar in some respects, an important difference in the more recently developed and increasingly popular Bayesian approach is that it maximizes the probability that a particular tree is the correct one, given the evolutionary model and the data that are being analysed (Huelsenbeck *et al.*, 2001). In both of these approaches all variable sites are informative, and these methods can be powerful if the parameters of the model can be set with confidence.

Traditional phylogenetic analyses have been invaluable in evolutionary biology. However, although bifurcating trees are appropriate for taxonomic groups at the species level and beyond, which have experienced a period of reproductive isolation long enough to allow for the fixation of different alleles, a hierarchical bifurcating tree will not always be appropriate for population studies. This is partly because, as outlined above, there may be insufficient polymorphism in comparisons of conspecific sequences. In addition, bifurcating trees allow for neither the co-existence of ancestors and descendants nor the rejoining of lineages through hybridization or recombination (**reticulated evolution**), two processes that occur commonly at the population level. As a result, traditional phylogenetic trees are not always the most appropriate method for analysing the genealogies within and among conspecific populations, and in these cases can result in poorly resolved and sometimes misleading phylogenetic trees (Posada and Crandall, 2001). In recent years, this limitation has provided the impetus for researchers to develop a number of methods for phylogenetic anlaysis that are specifically tailored to accommodate the similar sequences that often emerge from comparisons of populations and closely related species.

## The Coalescent

With the exception of a small proportion of studies that use historical specimens from museums or other sources, phylogeographic studies typically use genetic information from current samples to reconstruct historical events. Inferences of past events are possible because most mutations arise at a single point in time and space. Assuming neutrality, the subsequent spread of each new mutation (allele) will be influenced by dispersal patterns, population sizes, natural selection and other processes that may be deduced from the contemporary distributions of these mutations. We may be able to make these deductions if we can determine when different alleles shared their **most recent common ancestor** (MRCA).

An MRCA can be identified using the **coalescent**, which is based on a mathematical theory that was laid out by Kingman (1982) to describe the genealogy of selectively neutral genes by looking backwards in time. If we apply the coalescent to the sequences of multiple alleles that have been identified at a particular locus, we can retrace the evolutionary histories of these alleles by looking back to the point at which they coalesce (come together). Although the

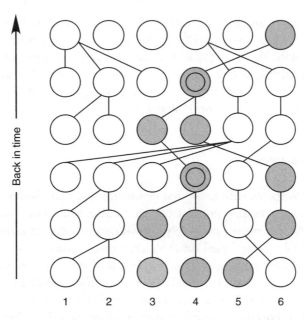

**Figure 5.5**   The evolutionary relationships of six haplotypes within a single population. Shaded circles are used to show how the lineages of haplotypes 3, 4 and 5 can be traced back to two coalescent events, which are indicated by double circles. Working backwards through time, the first of these coalescent events identifies the most recent common ancestor (MRCA) of haplotypes 3 and 4, whereas the second coalescent event identifies the MRCA of all three haplotypes

mathematical theory underlying the coalescent is too complicated for a detailed analysis in this book (see Hudson, 1990, for a review), the overall concept is relatively straightforward. This is illustrated by Figure 5.5, which shows us how we can work backwards through eight generations to reconstruct the history of six different genetic lineages within a particular population. Of the three lineages that have been highlighted in this example, haplotypes 3 and 4 coalesce relatively recently whereas the MRCA of all three lineages occurred in the more distant past.

If we go back far enough in time, all of the alleles within any population (discounting recent immigrants) should eventually coalesce to a single ancestral allele, but the time that this takes varies enormously and is influenced primarily by $N_e$. The importance of $N_e$ can be realized if we discount the possibility of natural selection (because this would preclude randomness) and think of haplotypes as randomly picking their parents as we go back in time (Rosenberg and Nordborg, 2002). Whenever two different haplotypes pick the same parents, they coalesce. Since there are fewer potential parents to choose from when $N_e$ is small, coalescence should occur relatively rapidly. If a population has a constant size of $N_e$ and individuals within this population mate randomly during each generation, then the likelihood that two different haplotypes pick the same parent in the preceding generation and coalesce is $1/2N_e$ for a nuclear diploid locus and, in most cases, $1/N_{ef}$ for mitochondrial DNA ($N_{ef}$ is the effective size of the female population). It must therefore follow that the probability of them picking different parents and remaining distinct is $1 - 1/2N_e$ or $1 - 1/N_{ef}$. The average time to coalescence of all gene copies in a population is $4N_e$ generations for diploid genes and $N_e$ generations for mitochondrial genes.

## Applying the coalescent

In reality, time to coalescence is affected by much more than simply $N_e$. A range of factors including fluctuating population sizes, natural selection and immigration tend to make coalescence an extremely convoluted process. As a result, statistical and mathematical models based on coalescent theory must be wide-ranging and able to accommodate numerous demographic, evolutionary and ecological parameters. Various mathematical models have used the coalescent successfully to analyse a number of different aspects of population genetics and molecular evolution, such as effective population sizes, past bottlenecks, selection processes, divergence times among populations, migration rates and mutation rates; note that coalescent theory has applications to traditional population genetics as well as to phylogeographic analysis e.g. (Coop and Griffiths, 2004; Wilkinson-Herbots and Ettridge, 2004; Degnan and Salter, 2005).

In one study, a coalescent-based approach was used to investigate why populations of the montane grasshopper *Melanoplus oregonensis* in the northern Rocky Mountains are genetically differentiated from one another. By using the coalescent to identify ancestral populations it became apparent that much of the genetic divergence dated back to the last Ice Age when populations were restricted to isolated geographical areas (Knowles, 2001). This finding has leant support to the idea that Pleistocene glaciations promoted speciation when ice sheets covered vast areas and populations became separated from one another for prolonged periods by inhospitable terrain. Another study used both traditional population genetics and coalescent theory to compare the distribution of mitochondrial haplotypes among yellow warbler (*Dendroica petechia*) populations across North America. In this species, eastern and western populations are genetically distinct from one another. A coalescent-based evolutionary model suggested that all western haplotypes are descended from an eastern lineage, and it therefore seems likely that western yellow warbler populations were established following infrequent colonizations from the east (Milot, Gibbs and Hobson, 2000).

The previous examples were based on the application of specific coalescent-based models to phylogeographic data, but the coalescent is also relevant to some recently developed general methods of phylogenetic reconstruction. Unlike the traditional bifurcating trees, these methods allow us to depict evolutionary relationships in the form of **multifurcating** trees in which a single haplotype can give rise to many haplotypes, thereby creating what is more commonly known as a **network**.

# Networks

Unlike many traditional phylogenetic trees, a graphical representation known as a network can be used to depict multifurcating, recently evolved lineages in a way that accommodates the co-existence of ancestors with descendants, and the reticulated evolution that accompanies hybridization and recombination (Table 5.2). There are several different ways to construct networks, most of which are distance methods that aim to minimize the distances (number of mutations) among haplotypes (reviewed in Posada and Crandall, 2001). Here we will limit our discussion to what has become one of the most commonly used methods in recent years, known as a **statistical parsimony network**.

A statistical parsimony network (Templeton, Crandall and Sing, 1992) links haplotypes to one another through a series of evolutionary steps. It is based on an algorithm that first estimates, with 95 per cent statistical confidence, the maximum number of base pair differences between haplotypes that can be attributed to a

**Table 5.2** Some characteristics of bifurcating trees versus network analysis, and the relevance of these characteristics to phylogeography

| Characteristic | Bifurcating trees | Network analysis | Relevance to phylogeography |
| --- | --- | --- | --- |
| **Branching pattern** | Assumes all lineages are bifurcating | Allows for multifurcating lineages | Population genealogies are often multifurcated |
| **Divergence** | Often requires numerous, variable characters | Can reconstruct genealogies from relatively little variation | Within species, sequences often show high overall similarity |
| **Ancestral haplotype** | Assumes ancestral haplotypes no longer exist | Allows for the co-existence of ancestral and descendant haplotypes | Ancestral and descendant haplotypes often coexist within populations |
| **Reticulated evolution** | Many algorithms assume no recombination or hybridization | Networks can reveal hybridization and some methods can allow for recombination | At the conspecific level, recombination and hybridization are often widespread |

series of single mutations at each site. This number is referred to as the parsimony limit. Haplotypes differing by a number of base pairs that exceeds the parsimony limit will not be connected to the network because **homoplasy** is likely to obscure their evolutionary relationships. Once the parsimony limit is calculated, the algorithm then connects haplotypes that differ by a single mutation, followed by haplotypes that differ by two mutations, three mutations and so on. As long as the parsimony connection limit is not reached, the final product is a single network showing the interrelationships of all haplotypes in a way that requires the smallest number of mutations.

The interpretation of parsimony networks draws on coalescent theory because the connections between haplotypes throughout the network represent coalescent events. By following some of the principles of coalescent theory, there are a number of predictions that we can make about parsimony networks, including:

1. High frequency haplotypes are most likely to be old alleles.

2. Within the network, old alleles are interior, whereas new alleles are more likely to be peripheral.

3. Haplotypes with multiple connections are most likely to be old alleles.

4. Old alleles are expected to show a broad geographical distribution because their carriers have had a relatively long time in which to disperse.

5. Haplotypes with only one connection (singletons) are likely to be connected to haplotypes from the same population because they have evolved relatively recently and their carriers may not have had time to disperse.

Figure 5.6A shows a statistical parsimony network of mitochondrial haplotypes from the migratory dragonfly *Anax junius* that was sampled from locations across North America spanning a maximum distance of approximately 8600 km between Hawaii and Nova Scotia (after Freeland *et al.*, 2003). Figure 5.6B shows the geographical locations of the different haplotypes. By comparing the network and the map, we can get some idea of whether the previously outlined predictions have been realized in this case. Haplotypes 1 and 25 are of the highest frequency, are central to the network, have more than one connection and show a broad geographical distribution. We cannot state unequivocally that these are the oldest alleles, but they meet the expectations of old alleles according to predictions 1–4. Although it is also true that, contrary to prediction 3, some of the haplotypes with more than one connection appear to be new alleles based on their low frequency and peripheral location in the network, haplotype 1 has considerably more connections (12) than any of the low-frequency haplotypes (maximum of 5).

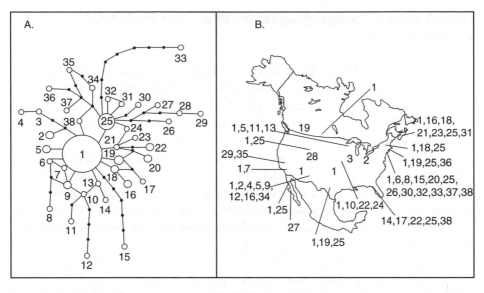

**Figure 5.6** (A) Statistical parsimony network of mitochondrial haplotypes that were identified from partial cytochrome oxidase I sequences for the common green darner dragonfly *Anax junius* in North America. Small dark circles represent missing or unsampled haplotypes, and each step along a lineage (marked by either a dark or an open circle) represents a single mutation. The sizes of the circles are proportional to the haplotype frequencies. (B) Map of North America showing the approximate sampling locations of the different haplotypes. Redrawn from Freeland *et al.* (2003)

Prediction 5, however, has not been met because there are many examples of singletons being connected to haplotypes that were found in distant locations, e.g. H3 and H4. Disjunctions such as these reflect the extremely high levels of gene flow in *A. junius*, which mean that mutations often spread before giving rise to new haplotypes. In fact, gene flow is so high in this migratory species that it shows essentially no phylogeographic structuring across a broad geographical range, despite high levels of genetic diversity (Freeland *et al.*, 2003).

While intuitively appealing and not without merit, it is important to note that network methods are not infallible. In one study, researchers investigating the phylogeography of dusky dolphins (*Lagenorhynchus obscurus*) compared the results that were obtained using four different methods of network construction (Cassens *et al.*, 2003). Although all four methods yielded networks that showed clear genetic differentiation between Pacific and Atlantic haplotypes, the evolutionary relationships within these two groups varied somewhat, depending on which network method was used. The authors of this study concluded that not all methods for constructing networks have been assessed rigorously under all evolutionary scenarios, and in some cases it may be appropriate to use multiple analytical methods so that any conflicting results can be identified and subsequently interpreted with caution.

# Nested Clade Phylogeographic Analysis and Statistical Phylogeography

Once we have established the genealogical relationships among haplotypes, the next step in phylogeography is to identify which historical and geographical factors may have influenced the current distributions of haplotypes. Traditionally, phylogeography has been based on the practice of gathering genetic data from samples collected across a geographical range and then looking for possible explanations for the genealogical patterns that are inferred; for example, a founder effect may explain pronounced genetic divergence between an island and a mainland population, and a mountain range in a north–south orientation may explain why eastern and western populations show independent evolutionary histories. This approach of seeking *post hoc* explanations for the current distribution of genetic variation has been an integral part of phylogeography since its inception, and may provide a useful initial assessment; at the same time, it is a largely descriptive approach that does not provide a rigorous framework within which specific hypotheses can be tested. For one thing, there is no way to determine whether or not the sample size of individuals and populations is large enough to rule out the possibility that the current distribution of genotypes resulted from chance alone.

In recent years, a number of increasingly rigorous methods based on statistical analyses and coalescent theory have been developed. One of these is nested clade phylogeographic analysis (NCPA; Templeton, Routman and Phillips, 1995), also known as nested clade analysis (NCA). The first step in NCPA is to construct a network such as the statistical parsimony network outlined in the previous section. NCPA then uses explicit rules to define a series of hierarchically nested clades within this network. The first level is made up of the clades that are formed by haplotypes that are separated by only one mutation. These one-step clades are then nested into two-step clades that contain haplotypes that are separated by two mutations, and so on. This is continued until the point when the next highest nesting level would result in a single clade encompassing the entire network. From our previous discussion on statistical parsimony networks we know that the oldest haplotypes should be central to the network and the newest haplotypes should be peripheral. As a result, the nested arrangement corresponds to evolutionary time, with higher nested levels corresponding to earlier coalescent events.

The next step is to superimpose geography over the clades, which then allows us to calculate two distance measures: $D_c$, which measures the mean distance of clade members from the geographical centre of the *clade*; and $D_n$, which measures the mean distance of nested clade members from the geographical centre of the *nested clade*. Permutation tests are then used to determine whether or not there is a non-random association between genetic lineages and geographical locations, in other words if there is an association between genotypes and geography. If the null hypothesis of no assocation between genotypes and geography can be rejected, an

**Figure 5.7** A North American bullfrog (*Rana catesbeiana*). This species is native to a wide area across eastern North America and is the largest true frog on that continent, weighing up to 0.5 kg. Photograph provided by Jim Austin and reproduced with permission

*a posteriori* inference key is used to determine which of several alternative scenarios, such as range expansion or allopatric fragmentation, is the most likely explanation for the patterns that have been revealed (Templeton, 2004).

An NCPA based on 41 haplotypes was used to test the hypothesis that the current distribution of genetic diversity in the North American bullfrog (*Rana catesbeiana*; Figure 5.7) was influenced by changing environmental conditions throughout the last Ice Age. Figure 5.8 shows the three nesting levels that were identified. Most haplotypes differed by a single mutation, although a notable exception was the connection between the eastern and western lineages (clades 3-1 and 3-2), which spanned at least five mutations. This greater than average divergence, together with the geographical distributions of these lineages either side of the Mississippi River, was interpreted as evidence for an early Pleistocene (last Ice Age) isolation of eastern and western populations. At the same time, widespread haplotypes within each of the two most divergent clades suggest that more recent levels of gene flow have been reasonably high on either side of the river (Austin, Lougheed and Boag, 2004).

NCPA is increasing in popularity because it allows researchers to test specific hypotheses about the geographical distribution of lineages based on both mitochondrial and nuclear sequence data. The power of nested analyses will, of course, be limited by the sampling regime, because the network upon which NCPA is based may be inaccurate if based on too few individuals or populations.

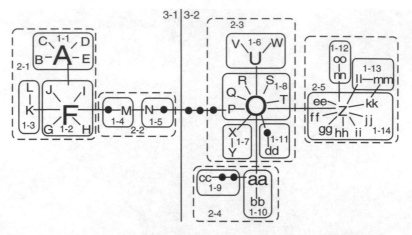

**Figure 5.8** A nested clade phylogeographic analysis based on DNA sequences from part of the mitochondrial cytochrome *b* gene of the North American bullfrog (*Rana catesbeiana*). The 41 haplotypes are labelled a – z and aa – oo. The size of the font is proportional to the frequency of the haplotype. One-step clades are prefixed with 1 (e.g. 1-1, 1-2) and are bounded by solid lines. Two-step clades are prefixed with 2 (e.g. 2-1, 2-2) and are bounded by dashed lines. The total network is divided into two three-step clades: clade 3-1, which occurs east of the Mississippi River, and clade 3-2, which occurs west of the river. Each line represents a single mutation change, and dark circles represent unsampled or extinct haplotypes. Redrawn by J. Austin from Austin, Lougheed and Boag (2004)

Nevertheless, a recent review of the performance of NCPA was conducted using 150 data sets that had strong *a priori* expectations based on known events such as post-glacial expansions or human-mediated introductions. The method generally performed well, although in a few cases it failed to detect an expected event (Templeton, 2004). Despite this track record, NCPA has been criticized for failing to provide any estimate of uncertainty along with its conclusions, because the *a posteriori* inference key provides only yes or no answers that have no confidence limits attached (Knowles and Maddison, 2002). This failing may be at least partially redressed by a suite of recently developed analytical methods that are known as **statistical phylogeography** (Rosenberg and Nordborg, 2002; Knowles, 2004).

The general approach of statistical phylogeography is to start with the development of specific hypotheses that may explain the current distribution of species. Models based on coalescent theory are then used for statistically testing these hypotheses by comparing the actual data set to the frequencies and distributions of alleles that we would expect to find under a variety of historical and ongoing scenarios. By using the coalescent to build models that reflect the complex demographic processes associated with alternative hypotheses, we should be able to accommodate all possible scenarios and hopefully identify specific historical events such as founder effects, geographical barriers to gene flow, and the relative roles of selection and drift.

At the moment, statistical phylogeography has great promise but is a newly emerging field that needs further development before applications become widespread. One difficulty lies with defining hypotheses that are simple enough to be tested but can nevertheless accommodate the complexities that are often associated with a species' evolutionary history. Parameters as varied as mutation rates, fluctuating population sizes, asymmetric migration, and geographical affiliations will often need to be accounted for. Models therefore may be highly complex, and detailed descriptions are beyond the scope of this textbook. This is nevertheless an area of investigation that should feature much more prominently in phylogeographic analysis in the coming years, and researchers in this field should be aware of the need to follow future developments in statistical phylogeography.

# Distribution of Genetic Lineages

So far in this chapter we have learned how to reconstruct evolutionary relationships, but we have done little more than allude to the processes that may have influenced the current distributions of genetic variation. We will now redress this imbalance by taking a more detailed look at what sorts of geographical and historical phenomena might have affected population sizes, population differentiation, gene flow and, ultimately, the distribution of species and their genes. We will begin this section by looking at some of the reasons why populations become isolated from one another, and we will then ask how long it takes for populations to become genetically distinct once reproductive isolation is complete. We will end this section with a discussion of the confounding influence that hybridization may have on our interpretation of past events.

## Subdivided populations

The distributions of species are extremely varied. No species that we know of has a truly worldwide distribution, although humans and some of their associates (dogs, rats, lice) come very close. Possibly the widest-distributed flowering plant is the common reed *Phragmites australis*, which is found on every continent except Antarctica. At the other end of the scale are many endemic species that have extremely restricted ranges, such as the giant Galápagos tortoises *Geochelone nigra*. Most of the 11 surviving subspecies are restricted to single islands within the archipelago, and in the case of *G. n. abingdonii* the entire subspecies is reduced to a single male known as Lonesome George who now lives at the Charles Darwin Research Station on the Island of Santa Cruz. All other species on Earth can be placed somewhere along the geographical continuum from humans to Lonesome George. Equally variable are species' patterns of distribution, with some forming essentially continuous populations throughout their range and others having extremely disjunct distributions. Examples of the former once again include humans, and examples of the latter include the strawberry tree *Arbutus unedo*, which is native to much of Mediterranean Europe and also Ireland, and the springtail *Tetracanthella arctica*, which is common in Iceland, Spitzbergen and Greenland and is found also in the Pyrenees Mountains between France and Spain and in the Tatra Mountains between Poland and the Czech Republic.

## Dispersal and vicariance

Disjunct populations, whether separated by thousands of kilometres or only a few kilometres, are isolated from one another either because they were founded following colonization events (dispersal), or because something has severed the connections between formerly continuous populations (**vicariance**). We have spent some time discussing dispersal in the previous chapter, so will touch only briefly on it here. Dispersal influences phylogeographic patterns through ongoing gene flow, which can have profound effects on population subdivision, $N_e$ and genetic diversity. Another way in which dispersal is important to phylogeography is through rare long-distance movements. These often entail the colonization of new habitats such as oceanic islands. Gigantic land tortoises in the past have colonized not just the Galápagos archipelago but also a number of other oceanic islands, including the Seychelles, Mauritius and Albemarle Island. They may have dispersed to these islands by riding on rafts of floating vegetation across hundreds or even thousands of kilometres of open ocean.

Vicariance is the term given to the splitting of formerly continuous populations by barriers such as rivers or mountains. The uplifting of the Isthmus of Panama, for example, was a vicariant event that caused the Atlantic and Pacific populations of numerous plant and animal species to become isolated from one another (Figure 5.9). Vicariance may also result if two populations become separated by an exaggerated intervening distance following the extinction of intermediate populations.

Examples of dispersal and vicariance as promoters of population differentiation are given in Table 5.3. There are two ways in which sequence data can help us to decide whether populations were separated by dispersal or vicariance. The first is to use an appropriate molecular clock to estimate the time since lineages diverged from one another and see if this coincides with the timing of a known vicariant event, such as the separation of continents following continental drift. When a molecular clock was applied to chloroplast sequences from species of the southern beech subgenus *Fuscospora* in Australasia and South America, it became apparent that some lineages diverged from each other at around the time that the two regions became separated, and therefore a vicariant event that occurred approximately 35 million years ago may explain the current distributions of these species (Knapp *et al.*, 2005).

**Figure 5.9** A red mangrove tree (*Rhizophora mangle*). This is an unusually salt-tolerant tree that grows along coastlines. Uplifting of the Isthmus of Panama approximately 3 million years ago was a vicariant event that caused red mangrove populations along the Atlantic and Pacific coasts to become isolated from one another (Nunez-Farfan *et al.*, 2002). The bird on this mangrove tree is a brown pelican (*Pelecanus occidentalis*). Author's photograph

A second approach for differentiating between dispersal and vicariance is to look at the branching order of gene genealogies; by comparing the evolutionary relationships of populations to their geographical distribution, we can gain some insight into the relative importance of past dispersal versus vicariant events (Figure 5.10). This method was used to investigate which force promoted the speciation of Queensland spiny mountain crayfish (genus *Euastacus*) in the upland rainforests of Eastern Australia (Ponniah and Hughes, 2004). Each of these rainforests, which are separated by lowlands, is home to a unique species of *Euastacus*, and two competing hypotheses could explain their current distribution.

**Table 5.3** Some examples in which either vicariance or dispersal has been identified as the most likely explanation for population differentiation and, in most cases, speciation

| Species | Rationale | Reference |
| --- | --- | --- |
| **Vicariance** | | |
| Sonoran Desert cactophilic flies (*Drosophila pachea*) | Genetic differentiation between, but not within, the continental and peninsular populations (barrier is Sea of Cortez) | Hurtado *et al.* (2004) |
| Marine gastropods (*Tegula viridula* and *T. verrucosa*) | Sister species located either side of the Isthmus of Panama | Vermeij (1978) |
| Sand gobies (genera *Pomatoschistus*, *Gobiusculus*, *Knipowitschia*, and *Economidichthys*) | Rapid evolution dating to the salinity crisis (end of the Miocene) in the Mediterranean Sea | Huyse, Van Houdt and Volckaert (2004) |
| **Dispersal** | | |
| Mouse-sized opposums (*Marmosops* spp.) in Guiana Region | Genetic data suggest recent origin of populations, rapid population growth, and dispersal from small ancestral population | Steiner and Catzeflis (2004) |
| Two frogs in the genera *Mantidactylus* and *Boophis* (species not yet described) | Recently discovered in Mayotte, an island in the Comoro archipelago (Indian Ocean) | Vences *et al.* (2003) |
| Freshwater invertebrates (*Daphnia laevis*, *Cristatella mucedo*) | Genetic lineages roughly follow waterfowl migratory routes | Taylor, Finston and Hebert (1998); Freeland, Noble and Okamura (2000) |

The first hypothesis states that a widespread ancestor was subdivided into populations by 'simultaneous vicariance' such as habitat fragmentation, after which time each population would have followed its own evolutionary path. Alternatively, a dispersal hypothesis states that colonization of each rainforest occurred in a northwards stepping-stone manner.

Because spiny mountain crayfish are known to have originated in the south, Ponniah and Hughes (2004) assumed that populations originally followed a north–south pattern of isolation by distance. From this they reasoned that if a single vicariant event had occurred, and all populations were split simultaneously, a pair of neighbouring populations in the south should now show a similar level of genetic differentiation to a pair of neighbouring populations in the north. Alternatively, if a stepping-stone dispersal pattern had occurred then southern populations should show greater genetic differentiation than northern populations

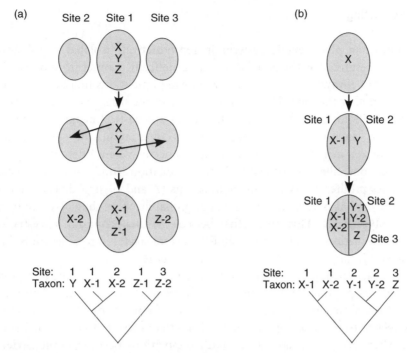

**Figure 5.10** The phylogenetic relationships of populations or species are expected to vary, depending on whether they arose following dispersal (a) or vicariance (b). Under a dispersal scenario, sites 2 and 3 are colonized by species (or populations) X and Z. If populations in sites 2 and 3 remain reproductively isolated from the populations in site 1, the descendants of the original populations eventually will evolve into pairs of related species (X-1 and X-2, Z-1 and Z-2), a pattern that is reflected in the phylogenetic tree. Under a vicariance scenario, site 1 first is split into sites 1 and 2, which leads to the evolution of species X-1 and Y from the ancestral species X. After site 2 is split into sites 2 and 3, the descendants of species Y in site 3 evolve into species Z. Meanwhile, speciation is also occurring within sites 1 and 2, leading to closely related species pairs (X-1 and X-2, Y-1 and Y-2). Note that in the vicariance phylogenetic tree those species from the same site are most closely related to one another, whereas the nearest neighbours in the dispersal phylogenetic tree are from different sites. Adapted from Futuyma (1998)

because they would have had a longer time to evolve population-specific haplotypes. The two hypotheses were tested using mitochondrial sequence data, which provided a genealogy consistent with the former scenario. The authors therefore concluded that vicariance was a more plausible explanation than dispersal for the current distribution of *Euastacus*. However, it is important to note that past events in this and other studies may be obscured by factors that cannot be controlled for easily, including unknown historical population sizes, the amount of time that has passed since populations diverged, and the fact that vicariance and dispersal may not be mutually exclusive. We will pursue this further later in the chapter, but first will look at how the genealogical relationships of two reproductively isolated populations are likely to change over time.

## Lineage sorting

The contrasting phylogenetic patterns in reproductively isolated populations in Figure 5.10 assume that the populations are genetically distinct from one another, but this is not always the case because when two populations first become isolated from one another they may both harbour copies of the same ancestral alleles. Over time, they will go through a process known as stochastic **lineage sorting** (Avise *et al.*, 1983), which must occur before alleles become population-specific. Lineage sorting is driven primarily by genetic drift, and occurs when differential reproduction causes some alleles to be lost from the population simply by chance, whereas other alleles proliferate. When two populations (A and B) first diverge, and little lineage sorting has occurred, there is a high probability that these two populations will be **polyphyletic**. This means that because of their common ancestry some alleles in population A will be more similar to some alleles in population B than to other alleles in population A, and vice versa (Figure 5.11).

After lineage sorting has progressed for a time, populations will be **paraphyletic** if the alleles in population A are more closely related to one another than they are to any of the alleles in population B, but some of population B's alleles are more closely related to some of population A's alleles than they are to each other (or vice versa). After more time has elapsed both populations become **monophyletic**, a situation that is also known as **reciprocal monophyly**. When this stage has been reached, all alleles within populations are genetically more similar to each other

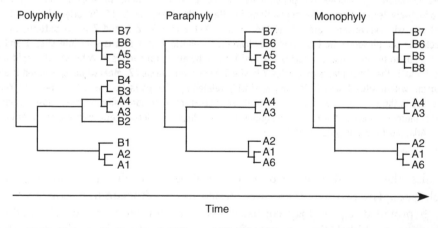

Time

**Figure 5.11** Progression from polyphyly to monophyly in two recently separated, reproductively isolated populations that are undergoing lineage sorting. Letters A and B refer to the populations in which the different alleles were found. After the populations are separated they are polyphyletic, because some of the alleles in population A are most closely related to some of the alleles in population B, and vice versa. Over time, alleles are both gained (following mutation) and lost (following selection or drift), leading to an intermediate stage in which population A is paraphyletic with respect to population B. Eventually the populations become monophyletic, which occurs when all A alleles are genealogically most similar to one another and all B alleles are genealogically most similar to one another than they are to the alleles that are found in other populations (Figure 5.11). It is only at this point that populations are genealogically distinct from one another.

The time that it takes for a pair of unconnected populations to reach the stage of reciprocal monophyly is directly proportional to the sizes of the populations in question. It will also depend on which genome is represented by the molecular markers that are being used. For mitochondrial and plastid DNA – which, as we know, are haploid and in most cases uniparentally inherited – time to monophyly is approximately $N_e$ generations. In diploid species, unless there are unusual circumstances such as a biased sex ratio, time to monophyly is usually around four times longer for nuclear than mitochondrial genes because of the proportionately larger effective population size of nuclear genes ($4N_e$ generations; Pamilo and Nei, 1988). Lineage sorting is even slower in polyploid genomes because they will have a correspondingly larger number of alleles at each locus.

The potentially confounding effects that lineage sorting has on the phylogenetic reconstructions of closely related populations or species was illustrated by a study of *Solanum pimpinellifolium*, a wild relative of the cultivated tomato *S. lycopersicum* (Caicedo and Schaal, 2004). Samples were taken from 16 populations along the northern coast of Peru and sequenced at a nuclear gene called fruit vacuolar invertase (*Vac*). One allele was identified as a recombinant and removed from the genealogical analyses. A maximum parsimony phylogeny was uninformative because it yielded five equally parsimonious trees, whereas a parsimony network revealed an unambiguous genealogical relationship among alleles. Perhaps the most surprising result was a lack of geographical structuring, which was unexpected because gene flow in this species is generally low and therefore some population differentiation was anticipated. The most likely explanation for these findings was the retention of ancestral polymorphism at the *Vac* locus, i.e. there has been insufficient time for lineage sorting to result in monophyletic (genetically distinct) populations.

Differential rates of lineage sorting provide one reason for disagreement between the nuclear and mitochondrial gene genealogies that are used in phylogeographic studies (Table 5.4). Because different genes 'sort' at different rates, the potential for discrepant genealogical relationships based on nuclear and mitochondrial genes will remain until populations have become reciprocally monophyletic with respect to all genes. Although widespread, monophyly is far from universal. A review of 584 studies compared the mitochondrial haplotype distributions of 2319 animal species (mammals, birds, reptiles, amphibians, fishes and invertebrates) and found that 23.1 per cent were either paraphyletic or polyphyletic (Funk and Omland, 2003). Other reasons for discordance between nuclear and mitochondrial phylogeographic inferences include recombination, sex-biased dispersal and hybridization (Table 5.4). The latter is a widespread phenomenon that has often obscured the evolutionary histories of populations and species. In the following section we will therefore look in more detail at how past hybridization can influence – and sometimes confound – our understanding of phylogeography.

**Table 5.4**  Some examples showing how nuclear and mitochondrial data can generate conflicting results in phylogeographic analyses. There are a number of reasons for such discrepancies, including different dispersal patterns of males and females, introgression of some but not all genomic regions following past hybridization events, and different rates of genetic drift (lineage sorting) in nuclear and cytoplasmic genomes

| Species | Mitochondrial phylogeography | Nuclear phylogeography | Possible explanation | Reference |
|---|---|---|---|---|
| Humpback whales (*Megaptera novaeangliae*) | Differentiation between Hawaii and California populations | Hawaii and California populations show little genetic differentiation | Different rates of genetic drift in the two genomes; differential dispersal of males and females | Palumbi and Baker (1994) |
| Rainbow trout (*Oncorhynchus mykiss*) | Population differentiation twice as high compared with that based on nuclear DNA | Populations showed relatively low genetic differentiation. Lack of concordance between mitochondrial and nuclear phylogenies | Differential introgression of mitochondrial and nuclear genes following past hybridization | Bagley and Gall (1998) |
| Green turtle (*Chelonia mydas*) | Significant population subdivision on regional and global scales | Population subdivision between, but not within, ocean basins | Male-mediated gene flow; female nest site philopatry | Bowen *et al.* (1992); Karl, Bowen and Avise (1992) |
| Eider duck (*Somateria mollissima*) | Substantial differentiation among both local colonies and distant geographical regions | Little differentiation among colonies; isolation by distance at a regional scale | Male-mediated gene flow; female breeding site philopatry | Tiedemann *et al.* (2004) |

# Hybridization

Once thought of as an anomaly, interspecific hybridization is now known to be widespread in many different taxonomic groups. We do not know how many species hybridize in nature, although the total number includes at least 23 675 species of plants (Knoblach, 1972), 3759 species of fish (Schwartz, 1972, 1981), and approximately 10 per cent of bird species (Grant and Grant, 1992b). Probably even more widespread is the hybridization that occurs within species following the interbreeding of 'individuals from two populations, or groups of populations, which are distinguishable on the basis of one or more heritable characters' (Harrison, 1990). Many of the examples in this section will refer to *inter*specific hybridization, but while reading this you should keep in mind that much of the theory is equally applicable to hybridization between individuals from *intra*specific populations.

If the two parental forms are morphologically distinct, then hybrids often can be identified from their intermediate phenotypes. Traditional breeding experiments and controlled crosses can also be valuable techniques for investigating hybridization. In wild populations, however, an increasing number of hybrids have been identified in recent years from genotypic data. As we saw in Chapter 2, this is possible because hybridization leads to the introgression of alleles across species boundaries, a process that often leads to cytonuclear disequilibrium. This introgression, combined with the preponderance of hybrids in virtually every taxonomic group, means that hybridization has become a relatively common explanation for unexpected distributions of alleles. Substantial introgression can occur even when hybridization is infrequent. Hybrids between native red deer (*Cervus elaphus*) and introduced Japanese sika deer (*C. nippon*) on the island of Argyll in Scotland were identified from mtDNA sequences and microsatellite alleles (Goodman *et al.*, 1999). Hybridization between the two species is infrequent and therefore introgression of alleles is rare at any one locus; however, where the two species come into contact with each other, up to 40 per cent of deer carry alleles that apparently have been transferred from one species to the other.

Patterns of introgression and cytonuclear disequilibrium can be complicated further by **Haldane's rule** (Haldane, 1922), which states that hybrids are less likely to be viable in the heterogametic sex, i.e. the one that has two different sex chromosomes (e.g. male mammals that are XY or female birds that are ZW; Chapter 2). Evidence for Haldane's rule has been found in a number of taxonomic groups, including mammals, birds and insects. In one example, the effects of Haldane's rule on cytonuclear disequilibrium were illustrated by a study of two species of swallowtail butterflies, *Papilio machaon* and *P. hospiton*, on the islands of Sardinia and Corsica. In Lepidoptera, females are heterogametic. If, as predicted by Haldane's rule, female butterfly hybrids are unfit, mtDNA will not be transferred between species because it is transmitted maternally. On the other hand, nuclear genes may move between species via male hybrid butterflies, which are homogametic and therefore more likely to be viable. This was indeed the pattern in the two *Papilio* species: hybridization has resulted in the introgression of alleles at some nuclear allozyme loci but there is no evidence of mitochondrial introgression (Cianchi *et al.*, 2003). Haldane's rule can therefore explain how nuclear introgression can occur in the absence of mitochondrial introgression.

*Hybrid zones*

It should be evident by now that alleles can be shared between species either as a result of shared ancestral polymorphism (incomplete lineage sorting) or hybridization (introgression of alleles), but how easy is it for researchers to differentiate between these two possibilities? This will depend on both the history and the geographical distributions of the two species. If they are recently diverged species that live entirely in sympatry, then it may be impossible to distinguish between the two scenarios solely on the basis of molecular data. In other situations, however, it may be possible to identify the source of shared alleles by looking at the distribution of these alleles across the species' geographical ranges. If two putatively hybridizing species share alleles when they are living **sympatrically**, but not when they are living **allopatrically**, then hybridization is a likely explanation for these shared alleles. On the other hand, if shared alleles occur in both sympatric and allopatric populations, incomplete lineage sorting may be a more plausible explanation. Additional clues about the evolutionary histories of species may therefore be found in **hybrid zones**, the areas of contact in which species meet and interbreed (see also Box 5.1).

Hybrid zones are very common; some estimates suggest that they may subdivide as many as one-half to one-third of species (Hewitt, 1989). The geographical extent of these zones varies dramatically, with some that are only a few metres wide and others that span many kilometres. Hybrid zones are often linear, such as the one represented by a narrow line throughout central Europe that subdivides the two house mouse species *Mus musculus* and *M. domesticus* (Hunt and Selander, 1973). There are also mosaic hybrid zones, which are patchy in their distribution. The field crickets *Gryllus pennsylvanicus* and *G. firmus* in eastern USA interbreed throughout a well-studied mosaic hybrid zone (Harrison, 1986; Rand and Harrison, 1989). Initial studies suggested that the two species were largely allopatric, with *G. firmus* populations near the coast giving way to *G. pennsylvanicus* populations further inland, and a roughly linear hybrid zone at their juncture. This interpretation was later modified when *G. pennsylvanicus* populations were found within 10 km of the coast, and hybrid populations were found approximately 100 km inland. We know now that the hybrid zone between these two species is in fact a complex mosaic that spans a relatively broad geographical area.

There are two ways in which hybrid zones can be maintained. The first is based on the premise that hybrid zones are independent of their environment and are simply the by-product of two populations coming into contact with each other. Under this model, hybrids are less fit than either of the parental populations and the hybrid zone is maintained by a balance between dispersal of parental genotypes and selection against hybrids, creating a so-called **tension zone** (Barton and Hewitt, 1985). One example of this occurs in the zones of hybridization between the smooth-shelled marine mussels *Mytilus edulis* and *M. galloprovincialis*. Both species have a planktonic larval stage and therefore can disperse over relatively long distances. Despite this potential for frequent contact, hybridization is restricted to discrete hybrid tension zones in which the viability of hybrids, relative to either parental form, is substantially reduced at the larval stage (Bierne *et al.*, 2002).

Hybrid zones can also be maintained in regions where hybrids have greater fitness than non-hybrids, a phenomenon that is known as **bounded hybrid superiority** (Moore, 1977). When hybrids are favoured under certain conditions, parental forms and hybrids will often be distributed along an environmental gradient that alternately favours either form. One example of this was found in a study that used AFLP markers to compare the genotypes of four hybridizing western North American oak species (*Quercus wislizenii, Q. parvula, Q. agrifolia* and *Q. kelloggii*) with climatic variables to see if there were particular conditions that seemed to favour hybrids (Dodd and Afzal-Rafii, 2004). Hybridization between these species is not restricted to bounded hybrid zones, in part because pollen dispersal can lead to allopatric hybridization. In the *Quercus* study, hybrids were sometimes separated by at least 300 km from one of the parental species, suggesting a mosaic hybrid structure across a broad spatial scale. Parental and hybrid forms were each associated with a particular set of environmental variables, including temperature, precipitation, vapour pressure deficits and solar radiation. These associations imply that 'pure' species and hybrids are being selected for under different environmental conditions.

---

### Box 5.1   Hybridization and speciation

One reason why hybrid zones have been the focus of so many studies over the past few decades is that they provide fascinating arenas of evolution and, in some cases, speciation. Reticulated evolution occurs when hybridization results in the formation of new species. This may occur either within hybrid zones or following seemingly random hybridization events. Recall from Chapter 3 that hybridization can lead to speciation through the creation of genetic diversity (see Box 3.5). A simpler mechanism of hybrid speciation occurs when two parental species interbreed and a new species results. This may or may not entail an increase in ploidy (Figure 5.12). In plants, polyploidy is extremely common; in fact, the frequency of allopolyploidy (Chapter 1) has been interpreted as evidence that the majority of plant species are either of direct hybrid origin or descended from a hybrid species (Grant, 1981). Less commonly, new species arise following homoploid hybrid speciation, which occurs when the hybrid has the same ploidy as its parental species. An example of this has been found in *Iris nelsonii*, which, according to data from allozymes, cpDNA and RAPDs, was derived from hybridization between three species (*I. fulva*, *I. hexagona* and *I. brevicaulis*) (Arnold, 1993).

In vertebrates, bisexual allopolyploid hybrid species are not uncommon in fishes and frogs. In addition, virtually all of the 70 unisexual vertebrate taxa that have been identified so far, which include species of fish, lizards and salamanders, arose following the hybridization of parental forms. More recently, hybrid speciation has been documented in marine invertebrates. Within the soft coral genus *Alcyonium*, *A. hibernicum* contains two groups of closely related sequences from the nuclear ribosomal internal transcribed spacer (ITS) region. One of these groups matched sequences that were also found in *A. coralloides*, whereas sequences in the other group matched those found in an, as yet, undescribed species that is currently designated *A.* sp. M2 (McFadden and Hutchinson, 2004). The most likely explanation for this pattern is that *A. hibernicum* arose following hybridization between *A. coralloides* and *A.* sp. M2. The conclusion that this species has a hybrid origin is further supported by its mode of reproduction, which is either parthenogenesis or obligate selfing, two mechanisms that are common to other well-studied unisexuals of hybrid origin.

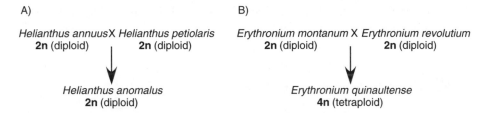

**Figure 5.12**   Hybrid speciation in plants. (A) Homoploid speciation – in this case in sunflowers – occurs when the newly formed species has the same ploidy level as its progenitors (Rieseberg *et al.*, 1993). (B) Polyploid speciation in lilies occurred when two hybridizing diploid species gave rise to a tetraploid species (Allen, 2001)

## Comparative Phylogeography

Phylogeographic studies that are based on only one or a few closely related species predominate in the literature, but the progressive accumulation of data means that we can now also start looking for geographical trends by comparing the genetic distributions of multiple species across a common geographical area. The advantages of this comparative approach are twofold. First, it may improve our understanding of the ways in which historical events have directly influenced the evolution of populations and species. If, for example, we suspect that a geological process such as the formation of a river was a vicariant event that separated multiple terrestrial populations, then we would expect to find similar levels of divergence between sister species from either side of the river. In this case, a consistent pattern could strengthen our argument that vicariance and not dispersal has promoted population divergence and speciation. Second, identifying common threads in the history of multiple taxa can be important from the perspective of conservation biology; to continue with our river example, we may find that the populations of many species on opposing banks are genetically distinct, and therefore a greater proportion of existing biodiversity would be preserved if some habitat was retained on both sides of the river as opposed to a longer stretch on just one side of the river. But how much agreement are we likely to find among a group of ecologically diverse taxa? A river will obviously have different impacts on species that can float, swim, fly or be wind-borne compared with those with a very low probability of crossing the water. Patterns of concordance may also depend on the spatial scales across which they are expected to occur, which is why we shall look first at some examples of comparative phylogeography at regional scales, and then extend this to see if any patterns remain at the continental level.

## Regional concordance

One of the earliest and best-known examples in which molecular data have identified concordant regional phylogeographical patterns in multiple species was reported by Avise in 1992 (Avise, 1992). He compared the genetic distributions of 19 freshwater, coastal and marine species (15 fish, one bird, one reptile and one mollusk) inhabiting southeastern USA. Twelve of these species, representing both freshwater and marine taxa, showed comparable genetic divisions between the Gulf of Mexico and the Atlantic populations, suggesting that vicariance has similarly influenced the population histories of multiple taxa in this region. Marine taxa are divided by the Florida peninsula, which extends into subtropical waters thereby effecting a barrier to species that live in the temperate waters along the Atlantic coast and the Gulf of Mexico. At the same time, many freshwater taxa are divided by the eastern and western river drainage systems that enter the south Atlantic and the Gulf, respectively. This study has become embedded in the literature as a classical example of comparative phylogeography and is nicely reviewed by Avise (2000).

Since then, as molecular phylogeographic studies have accumulated, other regional examples of concordance have emerged from various geographical regions around the world. In the wet tropical rainforests of northeastern Australia, mitochondrial DNA sequences from several bird, reptile and frog species reflect the division of species into southern and northern clades. The genetic break, which occurs in an area known as the Black Mountain Corridor, must represent a fairly ancient split because conspecific populations from either side of this corridor have sequences that diverge by up to 14.4 per cent (Joseph, Moritz and Hugall, 1995; Schneider, Cunningham and Moritz, 1998). The concordance in this region suggests vicariant population differentiation either side of the corridor, most likely a result of the rainforest repeatedly contracting into separate refugia north and south of the Black Mountain Corridor during the climatic oscillations of the Quaternary period.

Considerable interest has been directed towards the possible role that vicariant mechanisms may have played in creating the extraordinarily high levels of biodiversity in the lowland forests of the Amazon Basin. If vicariance has been particularly common in this part of the world, populations could have regularly become isolated from one another and this in turn could lead to a relatively high speciation rate. One possible mechanism for vicariant speciation in the Amazon Basin is outlined in the riverine barrier hypothesis that dates back to Alfred Russel Wallace, a contemporary of Darwin who made invaluable contributions to the development of evolutionary theory. Wallace spent a number of years in South America, during which time he observed that rivers often seem to create boundaries to species communities (Wallace, 1876). Many years later, the riverine barrier hypothesis was supported by molecular studies that provided evidence for very low gene flow between the populations of some forest understorey birds and saddle-back tamarins (*Saguinus fuscicollis*) either side of the Rio Juruá, a major tributary of the Amazon River in Brazil (Capparella, 1992; Peres, Patton and Da Silva, 1996).

In contrast, studies of frogs and small mammals have provided little support for the riverine barrier hypothesis (Da Silva and Patton, 1998; Gascon *et al.*, 2000). In these species, the phylogeographic divisions between populations tend to occur not on either side of the river but, instead, between upstream and downstream populations. These phylogeographic boundaries in the central section of the river are longstanding; conspecific populations from the two areas have sequences that diverge by as much as 13 per cent (Da Silva and Patton, 1998; Lougheed *et al.*, 1999). The age and pervasiveness of this division have led to an alternative hypothesis for Amazonian vicariance known as the ridge hypothesis. This is based on evidence for major ridges, or arches, that originated 5–10 million years ago when the Andes were forming, and that define a series of basins in western lowland Amazonia. These ridges have been obscured over the years by an accumulation of sediment but one of them, the Iquitos ridge, is an underlying geological structure that runs perpendicular to the Rio Juruá. In several species, the break between major phylogeographic groups occurs around the site of the Iquitos ridge.

The dart-poison frog (*Epipedobates femoralis*; Figure 5.13) is one species that supports the ridge hypothesis. Populations sampled from either bank of the Rio Juruá were not monophyletic and had a maximum sequence divergence of 6.13 per

**Figure 5.13** A dart-poison frog (*Epipedobates femoralis*). Photograph provided by Claude Gascon and reproduced with permission

cent. In contrast, populations from upstream and downstream, i.e. from either side of the Iquitos Arch, did form monophyletic groups and, at up to 12.36 per cent divergence, showed substantial differentiation from one another (Lougheed *et al.*, 1999; Figure 5.14). As data from a growing number of species become available, we must conclude that population differentiation and biodiversity around the Rio Juruá have been facilitated by both the river itself and the Iquitos Arch. This example illustrates how phylogeographic patterns in any particular region will, in all likelihood, have been influenced by a number of different factors. In the next

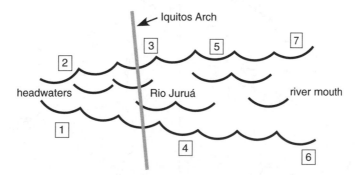

**Figure 5.14** Representation of the Rio Juruá in the Amazon Basin, showing the orientation of the Iquitos Arch. Numbers represent the dart-poison frog populations that were sampled by Lougheed *et al.* (1999). If the riverine barrier hypothesis was correct, we would expect higher genetic similarity among populations on the same side of the river compared with those on opposite sides (populations 2, 3, 5 and 7 versus populations 1, 4 and 6). In fact, the greatest genetic differentiation was seen between populations on either side of the Iquitos Arch (populations 1 and 2 versus populations 3–7), thereby lending support to the ridge hypothesis. Adapted from Lougheed *et al.* (1999)

section we will see examples of the even greater complexities that surround comparative phylogeography at a continental scale.

## Continental concordance

Although a degree of phylogeographic concordance can sometimes be found at regional scales, the picture becomes more complicated when we extend the geographical area of comparison. In many parts of the world, broad-scale phylogeographic patterns have been influenced by the Pleistocene glaciations (Hewitt, 2004). Over the past 700 000 years, major climatic oscillations caused large areas of land to be covered intermittently by vast sheets of ice that receded as the temperatures increased, and then spread out once more when temperatures began to drop. During the late Pleistocene, climatic variation occurred particularly rapidly, with temperatures sometimes changing by 10–12 per cent within a 10-year period. In North America, the most recent glaciation period reached its maximum ice coverage between 18 000 and 23 000 years ago, with the ice sheets reaching their maximum retreat between 8 000 and 15 000 years ago. Similar time scales of glaciation and deglaciation occurred in Europe, where ice sheets were less extensive but nevertheless covered a substantial area (Figure 5.15).

Preserved pollen, beetle fragments and fossil records have illustrated how dramatically species' distributions fluctuated throughout glacial–interglacial cycles. When large expanses of land were covered by ice sheets, many species could survive only in glacial refugia, which were located in areas free of ice. In Europe these tended to be located south of the ice front, although there is some

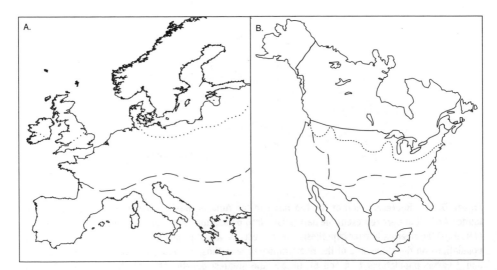

**Figure 5.15** Maps of (A) western Europe and (B) North America, showing the approximate southernmost extent of ice (dotted lines) and tundra (dashed lines) during the last Ice Age

evidence for several more northerly refugia in which temperate or cold-tolerant species, including trees (Willis, Rudner and Sumegi 2000) and forest mammals (Deffontaine et al., 2005), may have persisted. The survival of species in isolated refugia, alternating with their more widespread distribution during the interglacial periods when the ice retreated, has had a profound effect on their current genetic distributions. During glacial periods, populations were kept at relatively small sizes within refugia, and therefore a combination of genetic drift and new mutations promoted substantial levels of genetic divergence between these isolated populations. Dispersal away from refugia during interglacial periods often brought these previously separated populations into contact with one another, at which point hybridization between lineages would often occur. Numerous phylogeographic studies have provided us with considerable information on postglacial recolonization routes and the current distribution of genetic diversity in formerly-glaciated versus refugial areas.

## European postglacial recolonization routes

During interglacial periods, as the ice began to retreat, species dispersed out of refugia and recolonized the newly available habitats further north. In Europe, the three main refugia were in Iberia (Portugal and Spain), Italy, and the Balkans (see also Box 5.2). Fossil and molecular data have been used to reconstruct the dispersal routes of species spreading northwards out of these refugia as the ice sheets receded. By comparing the genetic similarity of populations in formerly glaciated regions to those in or near glacial refugial sites, we can sometimes get a reasonably clear picture of which refugial populations the current non-refugial populations are descended from. This in turn allows us to retrace the colonization routes that were followed by individuals emanating from each refugium. In Europe, dispersal was often impeded by major mountain ranges, including the Cantabrians, Pyrenees, Alps and Transylvanians, all of which tend to run in an east–west orientation. The Pyrenees and Alps, for example, presented a barrier to the meadow grasshopper (*Chorthippus parallelus*) travelling north from Iberia, although other species such as the hedgehog (*Erinaceus europaeus*) apparently were able to cross the mountains (Hewitt, 1999).

  Differences in ecology, particularly with respect to dispersal abilities, inevitably will mean that species followed a variety of colonization routes. Nevertheless, there is some degree of concordance in the routes that were followed, even between some markedly different taxa. For example, the route taken from a refuge in Iberia towards southern Scandinavian seems to be remarkably similar for the brown bear (*Ursus arctos*) and white oaks (*Quercus* spp.) (Taberlet and Bouvet, 1994; Dumolin-Lapègue et al., 1997); similarly, the common beech (*Fagus sylvatica*) and the meadow grasshopper (*Chorthippus parallelus*) apparently followed a comparable route from their refugia in the Balkans to the southeast of France

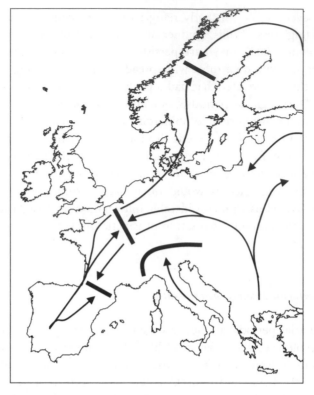

**Figure 5.16**   The main routes of postglacial colonization (arrows) that were followed by a wide range of taxa as individuals moved away from their refugia in western Europe. The thick lines represent the zones where populations emanating from different refugia tend to come into contact with one another. Hybridization between conspecific populations has often occurred at these so-called contact zones. Redrawn from Taberlet *et al.* (1998)

(Cooper, Ibrahim and Hewitt, 1995; Demesure, Comps and Petit, 1996). Figure 5.16 shows what are believed to be the main routes of postglacial dispersal for species moving away from European refugia.

The unravelling of postglacial dispersal routes has helped researchers to identify contact zones where populations from separate refugia have met. Hybridization between species or genetic lineages frequently occurs at these zones, which means that they provide another aspect of phylogeographic concordance. We therefore find a degree of concordance at the continental level in Europe with respect to both recolonization routes and zones of hybridization. Nevertheless, despite some similarities, overall phylogeographic patterns across Europe are broadly dissimilar. A review of cpDNA from 22 European tree and shrub species found significant differences in the distribution of genetic variation within and among populations that reflected the different history of each species (Lascoux *et al.*, 2004). Another study surveyed ten taxa, including mammals, amphibians, arthropods and plants, and this too revealed markedly different phylogeographies for each species. Differences included variable numbers of genetic lineages, discordant distributions of lineages, and varying levels of genetic divergence (Taberlet *et al.*, 1998). The author of that study did note that expectations of substantial concordance among such an ecologically diverse group of taxa may be unrealistic (Taberlet, 1998). Further evidence of continental concordance may emerge as more data become available, but so far it seems that relatively large spatial scales lead to highly variable phylogeographic patterns. There is, however, one other aspect of phylogeography that has been remarkably consistent across a wide range of taxa in several continents, and that is the distribution of genetic diversity.

---

### Box 5.2  Marine glacial refugia

Most of the research into the effects that glacial–interglacial cycles have had on the current distributions of species and their genetic lineages has been conducted on terrestrial plants and animals, although some recent studies have obtained evidence of marine glacial refugia. Both on land and in the sea, populations at the sites of former glacial refugia are expected to show higher levels of genetic diversity than populations in formerly glaciated areas, largely because the former often contain numerous genotypes that did not disperse northwards following glacial retreat. In the red seaweed *Palmaria palmata*, haplotype and nucleotide diversity estimates were higher in the English Channel compared with a number of other sites along the European and North American coasts, a finding that identifies the English Channel as a likely marine glacial refugium (Provan, Wattier and Maggs, 2005). Similar patterns of genetic diversity in the brown seaweeds *Fucus serratus* and *Ascophyllum nodosum* add further support to the identity of this refugium (Stam, Olsen and Coyer, 2001; Coyer *et al.*, 2003). These initially may seem paradoxical because the English Channel was mostly dry land during the last glacial maximum, but a trench known as the Hurd Deep was maintained at this site throughout the glacial–interglacial cycles. A marine lake was formed in this trench, and it is here that an unknown number of marine species could have persisted until the ice sheets retreated (Lericolais, Auffret and Bourillet, 2003).

---

### Distribution of genetic diversity

We know from previous chapters that the genetic diversity of populations is influenced by many factors, including population size, gene flow, mating system and natural selection, and therefore we may expect little agreement between species in the way that genetic diversity is distributed across broad spatial scales. Alternatively, it is possible that widespread environmental phenomena such as glaciation had similar impacts on the distribution of genetic diversity in multiple species, for example the postglacial colonization of more northerly sites was typically associated with founder effects and therefore we may expect to find relatively high levels of genetic diversity in or near glacial refugia and relatively low levels in formerly glaciated regions. This would depend to some extent on dispersal patterns. Founder effects should be particularly pronounced in species that often disperse over short distances but also periodically achieve long-distance dispersal. In these species there will be a time lag between the arrival of occasional long-distance dispersers and most other representatives of the gene pool, and this delay will allow the early arrivals to establish themselves and colonize neighbouring sites so that when the slower dispersers arrive they will struggle to compete with established populations (Ibrahim, Nichols and Hewitt, 1996). In contrast, species that show more constant dispersal rates should maintain a greater proportion of genetic diversity throughout their range. This should be true of both slow and rapid dispersers, as long as multiple genotypes disperse and colonize new sites *en masse*.

We therefore have the potential for postglacial dispersal with and without associated founder effects, and a review of the literature provides examples of both scenarios. A number of highly dispersive freshwater invertebrate species have relatively low levels of genetic diversity at northerly latitudes, which may be attributed to generally low levels of gene flow combined with occasional long-distance dispersal. Furthermore, because these species can reproduce asexually, populations can undergo rapid growth before receiving additional immigrants, and therefore a founder effect may persist for thousands of generations (Boileau, Hebert and Schwartz, 1992; Freeland, Rimmer and Okamura, 2004). At the other extreme, several species of Haircap mosses (genus *Polytrichum*) provide an example of how the rapid simultaneous dispersal of multiple genotypes led to postglacial dispersal without founder effects. During the ice ages, Haircap mosses were confined to southern refugia, but extensive spore dispersal allowed them to quickly colonize more northerly latitudes once the ice had retreated. This rapid dispersal has resulted in very low levels of genetic differentiation among populations, and no decrease in genetic variation with increasing latitude (Van der Velde and Bijlsma, 2003).

Given these two contrasting scenarios, we must now ask whether one predominates over the other, in other words whether there is any evidence for concordant distributions of genetic diversity along a latitudinal gradient from southerly refugia to northerly postglacial sites. A survey of the literature suggests that many northerly populations have indeed retained a postglacial founder effect that has reduced the genetic diversity of populations at high latitudes. In one review, data from 42 species of freshwater and anadromous fishes in North America yielded a significant inverse relationship between nucleotide diversity and latitude over formerly glaciated regions. Similar patterns of relatively low genetic diversity in formerly glaciated versus refugial areas have been found in many other taxa, including North American voles (*Microtus longicaudus*; Conroy and Cook, 2000), dragonflies (*Anax junius*; Freeland *et al.*, 2003), mosquitoes (*Wyeomyia smithii*; Armbruster, Bradshaw and Holzapfel, 1998), herbs (*Asclepias exaltata*; Broyles, 1998), and black spruce (*Picea mariana*; Gamache *et al.*, 2003), and European rock ferns (*Asplenium* spp.; Vogel *et al.*, 1999), grasshoppers (*Chorthippus parallelus*; Hewitt, 1999), woodmice (*Apodemus sylvaticus*; Michaux *et al.*, 2003), bryophytes (*Leucodon sciuroides*; Cronberg, 2000) and butterflies (*Polyommatus coridon*; Schmitt and Seitz, 2002). Thus it would appear that, although there are exceptions to the rule, there is a degree of phylogeographic concordance in the form of an inverse correlation between genetic diversity and latitude.

## Introduced Species

In this final section we shall look at how phylogeography can be used in the ongoing fight against introduced species (also known as invasive or alien species). This is a particularly topical application because, although the colonization of new habitats is sometimes a natural process, in recent years the scale of introductions has accelerated dramatically as a result of human activities that have introduced countless species into habitats that they would not otherwise have encountered. Examples of deliberate human introductions include agricultural plants and animals, pets, decorative plants for gardening, and animals for fishing or hunting. Unintentional introductions include parasites and pests associated with the deliberately introduced species, species transported in the ballast water of ocean-going ships, organisms that hitch-hike aboard shipped goods, plus species such as rats and cockroaches that often accompany human settlements.

Introduced species are a major problem because they often outcompete or prey upon native species (Figure 5.17), and in fact are such a threat to biodiversity that they have been ranked as the second greatest problem in global conservation today, exceeded only by habitat loss. A recent survey of extinct animal species revealed that, of those for which an extinction cause could be identified, 54 per cent ($n = 91$) of species had been eradicated at least partially by invasive species (Clavero and García-Berthou, 2005). In the Hawaiian archipelago, for example, introduced species such as the black rat have contributed to an alarming reduction in Hawaiian biodiversity over the past 200 years: nearly 40 per cent of birds unique to Hawaii are now extinct, 40 per cent of native plants either have been designated as endangered or are candidates for such classification, and approximately 900 species (71 per cent) of around 1263 historically described species of Hawaiian land snails are extinct, partly because of increased predation, disease and

**Figure 5.17**   A North American red squirrel (*Tamiasciurus hudsonicus*). The interactions between red and grey squirrels provide a good example of how introduced species can adversely affect native species. In North America, native red and grey squirrels have a long, shared history and co-exist in many areas. In Britain, however, the only native squirrel is the European red squirrel (*Sciurus vulgaris*). Grey squirrels were introduced onto this island in 1876. Since that time the red squirrel has been declining, and one important reason for this is competition from the grey squirrel. In addition, grey squirrels carry the parapox virus, which is often lethal to European red squirrels. As a result, although habitat loss is also implicated, invasive grey squirrels are playing an important role in the decline of native red squirrels in Britain (Yalden, 1999). Photograph provided by Kelvin Conrad and reproduced with permission

competition from alien species. This loss of biodiversity is expected to continue, because each year more and more species are being introduced; by 1995 an estimated 2600 insect species and 861 plant species had been introduced into Hawaii (Howarth, Nishida and Asquith, 1995).

There are also enormous economic costs associated with invasive species. In the USA, for example, billions of dollars are spent every year on controlling the damage inflicted by invaders. This includes keeping waterways clear of such plants as Sri Lankan hydrilla (*Hydrilla verticillata*) and Central American water hyacinth (*Eichhornia crassipes*), controlling or eradicating outbreaks of European and Asian gypsy moths (*Lymantria dispar*) in forests, and managing populations of the virulent Asian tiger mosquito (*Aedes albopictus*). The tremendous biological and economic costs of invasive species around the world highlight the importance of detecting and controlling invaders and possibly preventing future outbreaks.

One way in which phylogeographic analyses can help us to control invasive species is through the identification of cryptic invaders. Differentiating between invasive taxa and local populations on the basis of morphology may be problematic, particularly if there is substantial morphological variation within the native populations. When this occurs, molecular comparisons between endemic populations, putative invaders and possible source populations can be invaluable. This was the situation found in the gastropod *Melanoides tuberculata* in Lake Malawi in East Africa (Genner *et al.*, 2004). *Melanoides tuberculata* is native to most of tropical Africa, Asia and Oceania and is also an introduced species in much of the tropical and subtropical New World. When a non-native morph of *M. tuberculata* was found within Lake Malawi, researchers were unable to determine from shell morphology whether this had been introduced from a neighbouring population or from another continent. The latter is potentially more serious, because the introduction of a foreign genotype is more likely to have disruptive effects on an ecosystem.

Mitochondrial sequences from 38 *M. tuberculata* individuals sampled from 20 populations in Africa (including Lake Malawi), Israel, Sri Lanka and South-East Asia were used to reconstruct the phylogenetic relationships of the various populations. Samples from Lake Malawi fell into two distinct groups. Sequence divergence *within* each of these groups was 0 – 0.86 per cent, whereas sequence divergence *between* the two groups was 13.28–14.31 per cent. One group contained the native Lake Malawi samples plus all other African populations and samples from Israel and Sri Lanka. The other group contained the non-native Lake Malawi samples plus the South-East Asian samples. These molecular data strongly suggest that the non-native morph of *M. tuberculata* did not come from an allopatric African population, but instead was introduced from South-East Asia. It is not known how this introduction could have occurred, although one possibility is via discarded ornamental aquarium contents.

A cryptic invasion of genotypes also has been blamed for the explosive increase in the distribution and density of the common reed (*Phragmites australis*) in the USA over the past century (Saltonstall, 2002). Chloroplast sequences were obtained from 345 plants around the world, including herbarium samples that were collected before and after the range expansion of 1910. These revealed 27 haplotypes, one of which occurs at a high frequency throughout Europe and continental Asia. This same haplotype was very rare in the USA before 1910 but is now the most widespread haplotype in that country. It is likely that this invasive haplotype was first introduced to North America in the early 19th century, probably through several ports along the Atlantic coast. Since then it has dramatically expanded its range and now dominates many North American wetland areas at the expense of native *P. australis* genotypes and numerous other species.

Phylogeography may also help us to retrace the intercontinental dispersal routes of invasive species by using techniques similar to those that allowed researchers to identify postglacial recolonization routes. This was the approach taken in a study of the amphipod *Echinogammarus ischnus* (Cristescu *et al.*, 2004). This species is endemic to the Ponto-Caspian region, an area that encompasses the Black, Azov, Caspian and Aral Seas. Sometime during the past century, a single genotype of *E. ischnus* spread from the Black Sea to Western Europe and also to the Great Lakes of North America. By comparing DNA sequences from multiple sites, the authors of this study concluded that the most likely sequence of events involved *E. ischnus* first colonizing Western Europe and then invading the Great Lakes. Interestingly, the colonization route of *E. ischnus* appears very similar to that of *Cercopagis pengoi*, another Ponto-Caspian crustacean that recently invaded the Great Lakes (MacIsaac *et al.*, 1999). The two species have markedly different life histories and dispersal capacities, and their common colonization route suggests that invasions from the Ponto-Caspian region into Western Europe and North America may be more common than was previously believed.

## Overview

Studies of phylogeography confirm the importance of historical processes in shaping the current distributions of genes and species, and therefore we must incorporate history into the theoretical underpinnings of molecular ecology. In the final three chapters we will build on the theory of molecular markers and population genetics that was presented in Chapters 1–5 by looking at some specific applications of molecular ecology. We will start this in the next chapter by asking how molecular data have contributed to our understanding of behavioural ecology.

## Chapter Summary

- Phylogeography seeks to identify which historical processes have most influenced the current distributions of species and their genetic lineages. Phylogeography therefore embraces aspects of both time (evolutionary relationships) and space (geographical distributions).

- Mitochondrial sequences traditionally have been the marker of choice in phylogeography, although chloroplast and nuclear markers that provide either DNA sequences or allele frequencies are becoming increasingly popular.

- Molecular clocks may be used to estimate the amount of time that has passed since populations or species have diverged, although they should be interpreted with caution unless specifically calibrated for the taxonomic group and genetic region that is being studied.

- Traditional phylogenetic analyses based on distance, parsimony or maximum likelihood methods are often used in phylogeographic studies, although the resulting bifurcating trees may not be appropriate for reconstructing the evolutionary relationships of recently diverged lineages.

- The coalescent, which is central to phylogeographic theory, is based on estimates of the time to the most recent common ancestor (MRCA). Time to MRCA can be estimated from current population size although it may be obscured by fluctuating population size, natural selection, and non-random mating.

- Networks are often used in phylogeography because they represent multi-furcating trees that can accommodate low levels of sequence divergence, hybridization, and the co-existence of ancestors with their descendant lineages.

- Nested clade phylogeographic analysis (NCPA) can help us to link geographical features and historical events to networks. An alternative, newly emerging approach for testing specific phylogeographic hypotheses within a statistical framework is known as statistical phylogeography.

- Populations are physically separated from one another as a result of either vicariance or dispersal. Once separated, populations are initially polyphyletic with respect to one another. Over time, lineage sorting will render them first paraphyletic, and then reciprocally monophyletic.

- Discordance between phylogeographic data from nuclear versus organelle genes may result from hybridization, sex-biased gene flow, recombination, or different rates of lineage sorting.

- Hybrid zones may be maintained by dispersal of parental genotypes coupled with selection against hybrids (tension zones), or by greater fitness in hybrids versus parental genotypes in particular environmental conditions (bounded hybrid superiority).

- Comparative phylogeography has revealed some examples of regional concordance in different species. Concordance at the continental scale is often low, although there is some agreement between species with respect to postglacial dispersal routes, hybrid zones, and the distribution of genetic diversity.

- Phylogeography can provide valuable information about the identity and source of invasive genotypes and can also help us to retrace colonization routes, all of which can help in the ongoing fight against introduced species.

## Useful Websites and Software

- *Phragmites australis* in the USA: http://www.nwcb.wa.gov/weed_info/Phragmts.html

- TCS – a computer program to estimate gene genealogies using networks: http://darwin.uvigo.es/software/tcs.html

- Phylip – a package of programs for inferring phylogenies: http://evolution.genetics.washington.edu/phylip.html

- Mesquite – a modular system for evolutionary analysis. Software for evolutionary biology with modules that include the reconstruction of bifurcating trees and coalescence simulations: http://mesquiteproject.org/mesquite/mesquite.html

- DnaSP – a software package that estimates several measures of DNA sequence variation within and between populations, quantifies recombination and gene flow, and uses coalescent-based models to estimate the confidence intervals of some test statistics: http://www.ub.es/dnasp/DnaSP32Inf.html

- GEODIS – a tool to perform the nested clade phylogeographic analysis (NCPA): http://darwin.uvigo.es/software/geodis.html

## Further Reading

### Books

Avise, J.C. 2000. *Phylogeography: The History and Formation of Species.* Harvard University Press, Cambridge, Massachussets.
Nei, N. and Kumar, S. 2000. *Molecular Evolution and Phylogenetics.* Oxford University Press, New York.
Ridley, M. 2004. *Evolution* (3rd edn). Blackwell, Oxford.

# Review articles

Bromham, L. and Penny, D. 2003. The modern molecular clock. *Nature Reviews Genetics* **4**: 216–224.

Hare, M.P. 2001. Prospects for nuclear gene phylogeography. *Trends in Ecology and Evolution* **16**: 700–706.

Hewitt, G.M. 2000. The genetic legacy of the Quaternary ice ages. *Nature* **405**: 907–913.

Knowles, L.L. 2004. The burgeoning field of statistical phylogeography. *Journal of Evolutionary Biology* **17**: 1–10.

Posada, D. and Crandall, K.A. 2001. Intraspecific gene genealogies: trees grafting into networks. *Trends in Ecology and Evolution* **16**: 37–45.

Rosenberg, N.A. and Nordborg, M. 2002. Genealogical trees, coalescent theory and the analysis of genetic polymorphisms. *Nature Reviews Genetics* **3**: 380–390.

Zhang, D-X. and Hewitt, G.M. 2003. Nuclear DNA analyses in genetic studies of populations: practice, problems and prospects. *Molecular Ecology* **12L**: 563–584.

# Review Questions

**5.1.** A 750 bp long region of the cytochrome *b* mitochondrial gene was sequenced from two geminate fish species living either side of the Isthmus of Panama. When the sequences were aligned, 23 of the bases did not match. Calculate the molecular clock for cytochrome *b* in this genus. What assumptions must you make before you apply this clock to other studies that are based on cytochrome *b* sequence data?

**5.2.** How many generations ago should all copies of a) a nuclear (diploid) gene, and b) a mitochondrial gene, have coalesced in a population with $N_e = 200$? What are some of the assumptions that we make when calculating this probability?

**5.3.** Assuming that haplotype 1 is central to the network, construct a parsimony network from the six haplotypes that are identified by the following sequences (variable sites are written in bold):

```
Haplotype 1: GATAGTAGGGGGTTAGATACTAAATACGTAGACTAGCTAT
Haplotype 2: GATAGTAGGGATTAGATACTAAATACGTAGACTAGCTAT
Haplotype 3: GATAGTAGGGATTAGATACTAAATATGTAGACTAGCTAT
Haplotype 4: GATAGTAGGGGGTTAGAAACTAAATACGTAGACTAGCTAT
Haplotype 5: AATAGTAGGGGGTTAGATACTAAATACGTAGACTAGCTAT
Haplotype 6: AATAGTAGGGGGTTAGATACTAAATACGTAGACTAACTAT
Haplotype 7: GGTAGTAGGGGGTTAGATACTAAATACGTAGACTAGCTGT
```

Although you have no information on the frequencies of these haplotypes, does coalescent theory allow you to identify the most likely ancestral haplotype in this network?

**5.4.** Briefly describe two processes that can explain the sharing of alleles between two populations or species.

**5.5.** Two closely related species live mostly in allopatry, with species A occurring in western Canada and species B occurring in eastern Canada. A narrow hybrid zone can be found where they come into contact with one another, but the majority of hybrids are males.

(i)    Are these species more likely to be birds or mammals? Why?

(ii)   Is this more likely to be a tension zone, or to show bounded hybrid superiority? Why?

(iii)  If hybrid males backcross into both parental populations, would you expect to find similar levels of introgression of mitochondrial and nuclear DNA? Why?

# 6

# Molecular Approaches to Behavioural Ecology

## Using Molecules to Study Behaviour

Behavioural ecology is a branch of biology that seeks to understand how an animal's response to a particular situation or stimulus is influenced by its ecology and evolutionary history. Areas of research in behavioural ecology are varied, and include mate choice, brood parasitism, cooperative breeding, foraging behaviour, dispersal, territoriality, and the manipulation of offspring sex ratios. As with other fields of ecological research, the study of behavioural ecology was traditionally based on either laboratory or field work. Laboratory work has made many important contributions because it allows us to manipulate organisms under controlled conditions and observe them at close quarters. At the same time, laboratory-based research is limited because many species cannot be kept in captivity; of those that can, observations often must be interpreted in context because captive conditions can never exactly mimic those in the wild. Observations and experiments involving wild populations have also been a valuable source of information, although again there are limitations, for example it may not be possible to identify individuals or to follow and observe them for prolonged periods.

In recent years, molecular data have often been used to supplement the more traditional approaches, particularly when studying individuals in the wild. From small amounts of blood, hair, feathers or other biological samples we can generate genotypes that can tell us the genetic relationships among individuals, or can identify which individual a particular sample originated from. In this chapter we shall concentrate first on how calculations of the relatedness of individuals based on molecular data have greatly enhanced our understanding of mating systems and kin selection. We shall then look at some of the applications of sex-linked markers, before moving on to an overview of how gene flow estimates and individual

*Molecular Ecology*   Joanna Freeland
© 2005 John Wiley & Sons, Ltd.

genotypes have helped us to understand a number of behaviours that are associated with dispersal, foraging and migration.

# Mating Systems

When we talk about mating systems in behavioural ecology we are not referring to different types of sexual and asexual reproduction, which are described in earlier chapters as modes of reproduction; instead, we are interested in the social constructs that surround reproduction, such as the formation of pair bonds. Over the past 20 years a tremendous number of studies have used molecular data to quantify some of the fitness costs and benefits associated with different types of mating behaviour, and these have collectively provided a number of surprising results. A direct consequence of this work is that we now differentiate between social mating systems, which are inferred from observations of how individuals interact with one another, and genetic mating systems, which reflect the biological relationships between parents and offspring. Molecular genetic data have played an important role in helping us to understand the extent to which social and genetic mating systems can differ from one another.

## Monogamy, polygamy and promiscuity

There are five basic types of animal mating systems (Table 6.1). **Monogamy** involves a pair-bond between one male and one female, whereas in **polygamy**, which includes **polygyny**, **polyandry** and **polygynandry**, social bonds involve multiple males and/or females. **Promiscuity** refers to the practice of mating in the absence of any social ties. Note that many species will adopt two or more different mating systems, and the examples used throughout this text are not meant to imply that a particular species engages only in the mating system under discussion.

Social monogamy is actually very rare in most taxonomic groups, one notable exception being an estimated 90 per cent of bird species. Because it is generally so uncommon, behavioural ecologists have long been interested in why any species should choose social monogamy. In a number of species, including the California mouse (*Peromyscus californicus*) (Gubernick and Teferi, 2000), black-winged stilts (*Himantopus himantopus*) (Cuervo, 2003) and largemouth bass (*Micropterus salmoides*) (DeWoody *et al.*, 2000), offspring survival is substantially higher when both parents are looking after their young. This is known as biparental care and is generally more common in birds than in mammals because both male and female birds can incubate eggs and bring food to nestlings, whereas gestation and lactation in mammals mean that much of the parental care is performed by females. Biparental care, therefore, may at least partially explain why social monogamy is so common in birds.

**Table 6.1** The five basic types of animal mating systems

| Mating system | No. of males | No. of females | Examples[a] |
|---|---|---|---|
| Monogamy | 1 | 1 | Prairie vole (*Microtus ochrogaster*) |
| | | | Hammerhead shark (*Sphyrna tiburo*) |
| | | | Polynesian megapodes (*Megapodius pritchardii*) |
| Polygyny | 1 | Multiple | Red-winged blackbirds (*Agelaius phoeniceus*) |
| | | | Fanged frog (*Limnonectes kuhlii*) |
| | | | Spotted-winged fruit bat (*Balionycteris maculata*) |
| Polyandry | Multiple | 1 | Galápagos hawk (*Buteo galapagoensis*) |
| | | | Gulf pipefish (*Syngnathus scovelli*) |
| Polygynandry | Multiple | Multiple | Variegated pupfish (*Cyprinodon variegatus*) |
| | | | Smith's longspur (*Calcarius pictus*) |
| | | | Water strider (*Aquarius remigis*) |
| | | | Jamaican fruit-eating bat (*Artibeus jamaicensis*) |
| Promiscuity | Multiple | Multiple | Soay sheep (*Ovis aries*) |
| | | | Long-tailed manakins (*Chiroxiphia linearis*) |

[a]Examples refer to social mating systems, which in some cases may differ from genetic mating systems.

If offspring can survive without paternal care, and if a male can make himself attractive to multiple mates, then polygyny may result. In many species this occurs when resources such as food are distributed patchily, because males can then defend high quality territories that will each attract multiple females. In Gunnison's prairie dogs (*Cynomys gunnisoni*) for example, monogamy prevails when resources are uniformly distributed whereas polygyny or polygynandry is often found when resources are distributed in patches that are guarded by one or several males (Travis, Slobodchikoff and Keim, 1995).

Very occasionally, the sexual roles of males and females are reversed and females, which in these cases tend to be larger and more colourful than males, will compete for and defend territories to which they attract multiple males. The males will then perform most of the parental care. This mating system is known as polyandry, of which the American jacana (*Jacana spinosa*) is a well-studied example. In this species the female defends large territories on a pond or lake, and in each territory several males will each defend their own floating nest and incubate the eggs that the female lays there. The most likely explanation for this unusual mating system is the habitat in which it occurs (Emlen, Wrege and Webster, 1998). Suitable nest sites are scarce and predation is high. If female jacanas laid only one clutch at a time, then very few of her offspring would survive and the fitness of both males and females would be low. If, however, females simultaneously lay multiple clutches and a proportion of these survive, the female will increase her fitness. Although males appear to be disadvantaged by this mating system, they may have little choice in the matter when there is such strong competition for suitable nest sites.

Polygynandry refers to the situation in which two or more males within a group are bonded socially with two or more females. This differs from promiscuity, a system in which any female can mate with any male without any social ties being formed. Differentiating between polygynandry and promiscuity may require a detailed study of a particular social group, and in fact the two terms are sometimes used interchangeably. Promiscuity is very common in mammals, occurring in at least 133 mammalian species (Wolff and Macdonald, 2004). It has also been documented in birds such as sage grouse (*Centrocercus urophasianus*) (Wiley, 1973) and in a number of fish species including guppies (*Poecilia reticulata*) (Endler, 1983). Promiscuity can have high fitness benefits to males if they can fertilize multiple females. Females may also benefit from promiscuous mating, as illustrated by field experiments on a number of species, including adders (*Vipera berus*) (Madsen *et al.*, 1992) and crickets (*Gryllus bimaculatus*) (Tregenza and Wedell, 1998), that have shown increased offspring survival when females mated with multiple males. This may result from one or more of a number of factors, including genetically variable offspring, increased parental investment and a reduced risk of male infanticide.

## Parentage analysis

The above characterization of mating systems was originally based on field and laboratory observations and experiments, and has been modified substantially in recent years. Key to our improved understanding of mating systems has been the application of molecular genetic data to parentage analyses, an approach that has allowed us to identify the genetic relationships of offspring and their putative parents. From these data it has become increasingly apparent that a social mating system can be very different from a genetic mating system. However, before we look at the findings that have come from parentage studies, we need to understand how we can determine whether or not a putative parent is in fact an offspring's genetic parent.

In studies of behavioural ecology we may wish to identify both of an offspring's genetic parents. In many cases we will be confident about the identity of the mother because in species that require parental care of young, she is unlikely to feed or care for offspring that she did not produce. Biological fathers, on the other hand, may be harder to identify because they may offer no parental care (most mammals) or may unknowingly care for young that are not their own (many birds). If we have genotypic data from an offspring and its putative parents then the simplest form of parentage analysis is exclusion. If an offspring's genotype at a single locus is *AA* then it must have received an *A* allele from each parent. If the mother's genotype is *AB* then we have no reason to believe that she is not a biological parent. If, however, the putative father's genotype is *BC* then we know that he cannot possibly be the genetic father of this chick. By using multiple loci we may be able to use this approach to exclude all males from the population except one, in which case we would conclude that the single non-excluded male is the genetic father.

If the number of candidate fathers in a population is small, and a sufficiently large number of polymorphic loci are used, then identifying the true father based on exclusions may be possible. It is often the case, however, that there are multiple males that we cannot exclude, in which case an alternative approach must be used to assign the true father. These assignments are often done using maximum likelihood calculations (Marshall *et al.*, 1998). Likelihood ratios for each non-excluded male can be calculated by dividing the likelihood that he *is* the father by the likelihood that he *is not* the father. These likelihoods are based on both the expected degree of allele sharing between parents and offspring, and the frequencies of these alleles within the population. Likelihood ratios are calculated separately for each locus, and then the overall likelihood that a given male is the biological father is obtained by multiplying all likelihood values. This approach assumes that the loci behave independently from one another, i.e. they are in linkage equilibrium. The male with the highest likelihood ratio will generally be considered as the biological father, provided that his likelihood is sufficiently high.

Successful identification of parents depends in part on the molecular markers that are used. The likelihood of assigning the correct parent will often be directly proportional to the number and variability of the loci that are being genotyped, although there is also a risk that by using too many hypervariable markers we increase the chance of revealing a mutation that occurred between generations, in which case we may inappropriately exclude a biological parent (Ibarguchi *et al.*, 2004). In general, the most useful markers for likelihood analysis in parentage assignments are microsatellites; dominant markers such as AFLPs also can be used, although many more loci are needed. In one study, researchers compared the performance of markers in assigning parentage within a stand of white oak trees (*Quercus petraea, Q. robur*) in northwestern France (Gerber *et al.*, 2000). They found that fewer than ten microsatellite loci were sufficient for parentage studies, whereas 100–200 AFLP loci had to be used before parents could be assigned with comparable confidence. Of course, successful parentage analysis also depends on an adequate sampling regime. It is often not possible to sample every candidate parent from a population, particularly if dispersal is high, but the likelihood of finding the correct parent increases if a large proportion of breeding adults is included in the analysis.

Not surprisingly, assigning parentage is easiest when the identity of one parent is known, although it can also be done when neither parent is known. The authors of

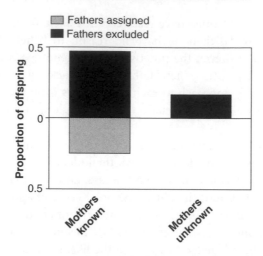

**Figure 6.1** Proportion of bottlenose dolphin offspring from which fathers could be excluded, and also to which fathers could be assigned, when the mothers were known and unknown. Data from Krützen *et al*. (2004)

a study of bottlenose dolphins (*Tursiops* sp.) in Shark Bay, Australia, attempted to identify the fathers of 34 offspring with known mothers and 30 offspring for which neither parent was known. They tried initially to identify the fathers through exclusions and then, in the cases where multiple males remained unexcluded, they attempted to assign the correct father using likelihood ratios. In the group for which the mothers were known, exclusions allowed them to identify the fathers of 16 juveniles, and assignments subsequently identified a further 11 fathers at the 95 per cent confidence level. In the group for which neither parent was known, only five fathers could be identified through exclusions, and no further identification of fathers was made possible by assignments (Figure 6.1; Krützen *et al.*, 2004).

*Extra-pair fertilizations*

Parentage studies occasionally tell us that individuals are less promiscuous than was previously believed. Both male and female Arctic ground squirrels (*Spermophilus parryii plesius*), for example, often copulate with multiple mates, but molecular genetic data have shown that more than 90 per cent of the pups whose mothers mated with more than one male were fathered by her first mate (Lacey, Wieczorek and Tucker, 1997; Figure 6.2). Far more common, however, is the finding that males and females are *more* promiscuous than their social mating systems would suggest. **Extra-pair fertilizations** (EPFs) occur when individuals choose mates that are not their social partners, a trend that has been documented in a wide range of taxa and in every type of mating system that involves pair-bonds. Table 6.2 provides just a few examples of studies that have uncovered EPFs.

**Figure 6.2** An Arctic ground squirrel (*Spermophilus parryii plesius*). Both males and females of this species typically mate with multiple partners and therefore, like the majority of mammals, its mating system is promiscuous. However, parentage studies have shown that most litters have only one genetic father (Lacey, Wieczorek and Tucker, 1997). This is therefore an unusual example of an animal whose genetic mating system is less promiscuous than its social mating system. Author's photograph

We can gain some idea of how pervasive this phenomenon is from the fact that fewer than 25 per cent of the socially monogamous bird species that had been studied up to 2002 were found to be genetically monogamous (Griffith, Owens and Thuman, 2002; see also Figure 6.3).

There are several evolutionary repercussions associated with EPFs. For one thing, their preponderance means that although it may be relatively easy to quantify a female's fitness based on the number of young that she produces, a male's fitness may be unrelated to the number of offspring that he rears. If there is a possibility that a male did not father all of the offspring produced by his mate then, in species that engage in biparental care, he is faced with a conundrum. Providing offspring and guarding them from predators is costly and is therefore worthwhile from an evolutionary perspective only if it increases a male's fitness. This clearly would not be the case if he were defending unrelated young. At the same time, males may risk losing all of their reproductive success if they neglect a brood that at least partially

**Table 6.2**  Some of the frequencies of extra-pair fertilizations (EPFs) that have been found in monogamous and polygamous species following molecular genetic parentage analyses. There are also species that very rarely engage in EPFs, and therefore the proportion of extra-pair young in all mating systems that involve pair-bonds ranges from essentially zero to more than half

| Species | Frequency of extra-pair fertilizations | Reference |
|---|---|---|
| **Social monogamy** | | |
| Reed bunting (*Emberiza schoeniclus*) | 55% of young | Dixon *et al.* (1994) |
| Common swift (*Apus apus*) | 4.5% of young | Martins, Blakey and Wright (2002) |
| Australian lizard (*Egernia stokesii*) | 11% of young | Gardner, Bull and Cooper (2002) |
| Island fox (*Urocyon littoralis*) | 25% of young | Roemer *et al.* (2001) |
| Hammerhead shark (*Sphyrna tiburo*) | 18.2% of litters | Chapman *et al.* (2004) |
| **Social polygyny** | | |
| Gunnison's prairie dog (*Cynomys gunnisoni*) | 61% of young | Travis, Slobodchikoff and Keim (1995) |
| Dusky warbler (*Phylloscopus fuscatus*) | 45% of young | Forstmeier (2003) |
| Rock sparrow (*Petronia petronia*) | 50.5% of young | Pilastro *et al.* (2002) |
| **Social polyandry** | | |
| Wattled jacana (*Jacana jacana*) | 29% of young | Emlen, Wrege and Webster (1998) |
| Red phalarope (*Phalaropus fulicarius*) | 6.5% of young | Dale *et al.* (1999) |

comprises their genetic offspring, and therefore paternal care often appears to be unconditional. In some cases, however, males appear to hedge their bets and provide parental care in proportion to their confidence in paternity. This was the strategy followed by males in a population of socially monogamous reed buntings (*Emberiza schoeniclus*) that raise two broods each year. A comparison of EPF

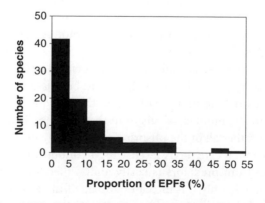

**Figure 6.3**  Proportion of EPF offspring within the broods of 95 socially monogamous or polygynous bird species. Adapted from Griffith, Owens and Thuman (2002) and references therein

frequencies, along with observational data, showed that of the two broods, the males provided most food to the one in which they had the highest confidence of paternity (Dixon *et al.*, 1994). Similar adjustments of male parental care in response to levels of genetic paternity have been found in a number of other taxonomic groups including bluegill sunfish (*Lepomis macrochirus*; Neff, 2003) and dung beetles (*Onthophagus Taurus*; Hunt and Simmons, 2002).

Another important consequence of EPFs is that, even in socially monogamous species, males do not have to form pair-bonds in order to achieve reproductive success. Genetic data have revealed successful fertilizations by floater (also known as sneaker) males, i.e. males who are not pair-bonded. Tree swallows (*Tachycineta bicolor*; Figure 6.4) typically engage in a high frequency of EPFs (around 55 per cent Conrad *et al.*, 2001), and in one study at least 8 per cent of these were accomplished by unmated males (Kempenaers *et al.*, 2001).

The potential reproductive success of unmated males has been further demonstrated by species that embrace a variety of reproductive strategies, such as the bluegill sunfish (*Lepomis macrochirus*). In bluegill populations in eastern Canada, parental males mature when they are around 7 years old, at which time they construct nests and attract females. They then defend the nest site, eggs and hatchlings against any intruders until the young are old enough to leave the nest. Sneaker males, on the other hand, may be only 2 years old and they attempt to

**Figure 6.4** A female tree swallow (*Tachycineta bicolor*) tending to her nest at the Queen's University Biological Station in Ontario, Canada. Researchers have been studying tree swallows here since 1975. Photograph provided by P.G. Bentz and reproduced with permission

fertilize eggs by darting into a nest and quickly releasing sperm while the resident male is spawning with a female, in the hope that they too will fertilize some of the eggs. A third strategy is followed by satellite males, which are usually aged 4–5 years and use colour and behaviour to mimic females. This disguise sometimes enables them to deposit sperm in the nest while the unsuspecting resident male is busy with a spawning female. Molecular studies have shown that the parental males achieve an average of 79 per cent of fertilizations, with the remaining 21 per cent achieved by sneaker or satellite males. Because about 80 per cent of the males in the studied population were parental males, the overall fitness of each of the three male strategies may be similar, although estimates of lifetime reproductive success are needed before this suggestion can be confirmed (Philipp and Gross, 1994; Neff, 2001; Avise *et al.*, 2002).

When weighing the fitness costs and benefits that are associated with alternative reproductive tactics we must also consider the degree to which males are cuckolded. Different rates of EPFs have been found in species that engage in both monogamy and polygyny. Comparisons of EPFs in willow ptarmigan (*Lagopus lagopus*; Figure 6.5) and house wrens (*Troglodytes aedon*), for example, have shown that the benefits to males of attracting multiple mates are often counteracted by an increased level of cuckoldry in polygynous males compared with monogamous males (Freeland *et al.*, 1995; Poirier, Whittinghan and Dunn,

**Figure 6.5** A male willow ptarmigan (*Lagopus lagopus*) in the sub-Arctic tundra of northwest Canada defends his territory at the start of the breeding season. Author's photograph

2004). In other words, although polygynous males appear to have a greater number of offspring, an increased frequency of EPFs in their broods means that they may not have fathered any more chicks than the monogamous males. If there is no increase in fitness associated with the additional costs that are incurred by polygynous males, who must guard relatively large territories, multiple mates and numerous offspring, then social monogamy should prevail.

## 6.1 Conspecific brood parasitism

Although extra-pair fertilizations provide the main explanation for the differences that we often find between genetic and social mating systems, genetic evidence has shown that in some bird populations a resident female may not be the biological mother of the young that are in her nest. This is because of a behaviour known as **conspecific brood parasitism** (CBP), which occurs when females lay their eggs in the nests of other conspecific birds. In species that require biparental care, the reproductive success of both the resident male and the resident female will suffer because they will end up rearing a bird that is not their own (Rothstein, 1990); the parasitic female, on the other hand, will benefit from an increase in fitness. A related behaviour known as **quasi-parasitism** (QP) occurs when a parasitic female lays her egg in the nest of the biological father with whom she achieved the EPF. In this case, it is only the resident female whose fitness is likely to suffer, because she will be the only one rearing an unrelated chick. Although much less common than EPFs, brood parasitism has been documented at low frequencies in a number of socially monogamous bird species, including European starlings (*Sturnus vulgaris*; Sandell and Diemer, 1999) and white-throated sparrows (*Zonotrichia albicollis*; Tuttle, 2003).

Conspecific brood parasitism also occurs in some socially monogamous fish species, such as the largemouth bass (*Micropterus salmoides*) that engages in biparental care for up to a month after eggs hatch. In one study, genetic monogamy was the norm in this species, although in 4/26 offspring cohorts there was evidence that some of the eggs had been deposited by an extra-pair female (DeWoody *et al.*, 2000). Parentage studies have also revealed brood parasitism in polygamous species, such as the polygynandrous Australian magpie (*Gymnorhina tibicen*) that lives in groups that strongly defend their territories from outsiders. Despite this territorial nature, one study found that an astounding 82 per cent of young had been fathered by males from outside the group and that 10 per cent of young were the result of CBP by females from outside the territory (Hughes *et al.*, 2003).

**Figure 6.6**   A blue-footed booby (*Sula nebouxii*) on Isla San Cristóbal in the Galápagos Archipelago. This is a socially monogamous bird that engages in relatively high levels of extra-pair fertilizations. The colourful feet are used in courtship displays, and males prefer females with particularly bright feet (Torres and Velando, 2005). Author's photograph

## Mate choice

Many studies have now used molecular data to conduct parentage analyses, and perhaps the most general conclusion that we can reach is that even in socially monogamous species both males and females will often mate with multiple partners. However, not all individuals are equally successful at attracting mates, and this leads us to the question of what makes a mate particularly attractive to a member of the opposite sex. Mate choice may be exercised by both males and females. Female blue-footed boobies (*Sula nebouxii*; Figure 6.6), for example, experienced a greater degree of intra- and extra-pair courtship if their feet were particularly colourful, suggesting that this is a trait that promotes male mate choice (Torres and Velando, 2005). Generally speaking, however, females are choosier than males because usually they invest more in eggs than males do in sperm. Understanding why individuals choose particular mates – both social and extra-pair – and not others is necessary before we can understand the evolution of mating systems.

Studies on mate choice, which have been accumulating rapidly in recent years, have been based on a combination of field, experimental and molecular work. In this section we will concentrate on two hypotheses that may explain mate choice, and that have benefited particularly from molecular data: the good genes hypothesis and the genetic compatibility hypothesis. While reading this section, bear in mind that forced copulations, mate guarding and intrasexual competition may mean that females do not always mate with their male of choice. Nevertheless, the relative ease with which we can determine the genetic parentage of offspring has provided us with some interesting data on why females (and sometimes males) choose particular mates with which to copulate.

The good genes hypothesis states that mates will be chosen on the basis of some characteristic that will always confer high fitness values on offspring. In Atlantic salmon, for example, individuals with an MHC *e* allele have the highest survivorship in populations that are infected by *Aeromonas salmonicida* bacteria, and therefore must be regarded as good gene donors (Lohm *et al.*, 2002). The good gene hypothesis can provide a plausible explanation for EPFs if a female's extra-pair male has one or more beneficial genes that are lacking in her social partner. Female great reed warblers (*Acrocephalus arundinaceus*), for example, obtained EPFs from neighbouring males that had larger song repertoires than the female's social mate. Because the survival of offspring was positively correlated with the size of their genetic father's song repertoire, females appeared to be selecting males with good genes (Hasselquist, Bensch and von Schantz, 1996).

The genetic compatibility hypothesis is based on the idea that a particular paternal allele will increase the fitness of offspring only when it is partnered with specific maternal alleles. In other words, genes are not universally good but, instead, each is more compatible with some genotypes than with others. Under this hypothesis, an individual will choose his or her mate on the basis of their combined genotypes. Female mice (*Mus musculus*) and female sand lizards (*Lacerta agilis*), for example, tend to choose mates whose MHC loci are as dissimilar to theirs as possible (Jordan and Bruford, 1998; Olsson *et al.*, 2003), a tactic that may be designed either to increase heterozygosity at the MHC in particular or to decrease inbreeding in general. Under some circumstances the genetic compatibility hypothesis seems to be the most plausible explanation for EPFs, for example one study found that in three different species of shorebird, the females were more likely to engage in EPFs when they were socially partnered with genetically similar males (Blomquist *et al.*, 2002). Interestingly, this study also found that males were more likely to fertilize quasi-parasitic females (Box 6.1) when they had a genetically similar social mate. The most likely explanation here seems to be inbreeding avoidance.

### Post-copulatory mate choice

In females, mate choice is not limited to pre-copulatory behaviour. After copulation, cryptic female mate choice may occur through the selection of sperm genotypes. In the flour beetle (*Callosobruchus maculatus*), unrelated sperm had a higher fertilization success rate than related sperm, suggesting cryptic female choice that was being driven by genetic compatibility in an attempt to decrease inbreeding and maximize the genetic diversity of offspring (Wilson *et al.*, 1997). In the marsupial *Antechinus agilis*, fertilization success was inversely correlated with the number of alleles that were shared by copulating males and females, once again suggesting post-copulatory mate choice based on genetic compatibility (Kraaijeveld-Smit *et al.*, 2002). There is also evidence to suggest that in mice, sperm are at least partially selected on the basis of their MHC haplotypes (Rülicke *et al.*, 1998). Similarly, although invertebrates lack MHC, fertilization in the colonial tunicate *Botryllus* is influenced by a polymorphic histocompatibility locus that controls allorecognition (Scofield *et al.*, 1982).

Somewhat surprisingly, even when fertilization is external it may be influenced by female choice. This is true of the ascidian *Ciona intestinalis*, in which external fertilization is partially regulated by maternal cells. Broods that were of mixed male parentage showed a relatively high proportion of fertilizations by males that were distantly related to the female compared with more closely related males (Olsson *et al.*, 1996). Finally, post-copulatory mate choice may sometimes be based on good genes, the quality of which may vary depending on environmental conditions. Female yellow dung flies (*Scathophaga stercoraria*) have three spermathecae (sperm-storage organs) in which they can partition sperm. In one study, the genotypes of offspring varied depending on whether the eggs were laid in the sun or in the shade, and this suggested that the females of this species use the egg-laying environment as a cue for choosing different sperm genotypes (Ward, 1998).

So far in our discussion of mating systems we have been looking at how parentage analyses based on molecular data have highlighted some of the differences between social and genetic mating systems (see also Box 6.2), and have also provided insight into several aspects of mate choice. Ultimately, parentage analysis has enabled us to quantify more accurately the fitness of individuals. However, not all reproductive success is achieved through the direct production of offspring, and in the following section we will take a look at how fitness can be enhanced through social breeding.

## 6.2   Extra-pair fertilizations and $N_e$

We know from Chapter 3 that variation in reproductive success (VRS) can influence the effective size of a population ($N_e$). In species such as the elephant seal, in which a few males with harems achieve most of the reproductive success, we expect to find a high male VRS and hence a low $N_e/N_c$, but how do EPFs affect the VRS, and hence the $N_e$, of other species with less extreme mating systems? In theory, EPFs may either decrease VRS by enabling unpaired males to reproduce, or increase VRS by allowing a handful of males to father a disproportionately high number of offspring.

Representatives of the endangered hihi bird (*Notiomystis cincta*) were translocated to several islands off the coast of New Zealand in an attempt to establish new populations. Because these were small populations there was a concern that genetic diversity would be low, and researchers therefore investigated the possibility that $N_e$ would be reduced further by VRS. The hihi is predominantly socially monogamous, although will sometimes form polygamous units. Parentage analysis of 56 clutches from one island over the course of 4 years revealed that 46 per cent of all chicks were fathered by extra-pair males. From one year to the next, the effects of EPFs on VRS were varied; in some years EPFs increased VRS but in other years they decreased it (Figure 6.7). However, although fluctuations in VRS were fairly pronounced, mortality rates were high, which meant that the net effect of VRS was to cause relatively modest fluctuations in the $N_e/N_c$ ratio from one year to the next, ranging from a 4 per cent decrease to an 8 per cent increase (Castro *et al.*, 2004). These results are similar to those of another study that found an EPF-driven decrease in $N_e/N_c$ of approximately 2 per cent in purple martins (*Progne subis*) and 8 per cent in blue tits (*Parus caeruleus*), two other socially monogamous bird species. In contrast, two socially breeding bird species, strip-backed wrens (*Campylorhynchus nuchalis*) and Arabian babblers (*Turdoides squamiceps*), had estimated increases in $N_e/N_c$ of 5 and 15 per cent respectively, that were attributable to EPFs (Parker and Waite, 1997).

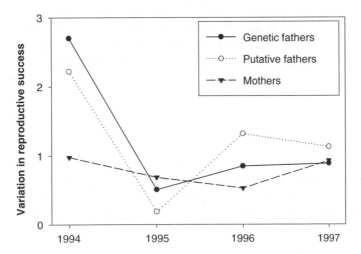

**Figure 6.7** Effects of EPFs on the variation in reproductive success (VRS) in hihi birds, with reproductive success calculated as the number of young that fledged from each nest. The VRS of putative fathers (i.e. *without* the effects of EPFs) may be either higher or lower than that of genetic fathers (i.e. *with* the effects of EPFs). The VRS of mothers is included for comparison. Adapted from Castro *et al.* (2004)

## Social breeding

In some species, helpers may assist breeding adults to raise their young, and this creates a system that is known as social breeding. There are several categories of social breeding, the most developed of which is found in **eusocial** species. These are characterized by a division of labour that results in numerous workers assisting relatively few reproductive nest mates to raise their offspring. In most cases these workers will never reproduce themselves, often because they are sterile. Most eusocial species are insects, including termites, ants and some species of wasps, bees, aphids and thrips. Eusociality in other orders is very rare, with two notable exceptions being the snapping shrimp (*Synalpheus regalis*) and several species of naked mole rat (*Heterocephalus glaber* and *Cryptomys* spp.). Less stringent forms of sociality involve helpers that may reproduce in later years, and can be found in diverse taxa including about 3 per cent of bird species (e.g. the white-throated magpie-jay, *Calocitta formosa*), a number of mammalian species (e.g. meerkats, *Suricata suricatta*) and multiple fish species (e.g. cichlids, *Neolamprologus brichardi*).

From an evolutionary perspective, scientists have long debated why individuals should invest time and effort in raising young that were clearly not their own. One common explanation for this behaviour is **kin selection**, which refers to the indirect benefits that an individual can accrue by helping its relatives (and therefore some of its genes) to reproduce. Kin selection is based on the concept of **inclusive fitness**, which is a fitness value that reflects the extent to which an individual's genetic material is transferred from one generation to the next, either through its own offspring or through the offspring of its relatives.

Kin selection was first proposed by Hamilton (1964), who suggested that an altruistic trait such as helping at the nest will be favoured if the benefits ($b$) of this trait, weighted by the relationship ($r$) between the helpers and the recipients, exceed the costs ($c$) to the helper, because under these conditions an individual's alleles will proliferate more rapidly under kin selection compared with personal reproduction. This can be expressed as:

$$rb > c \qquad\qquad (6.1)$$

If helping at the nest meant that an individual would die before he had produced any offspring, the cost to his fitness would be one ($c = 1$). If he helped to raise full-siblings, then the relatedness between the helper and the chicks would be 0.5 ($r = 0.5$; Box 6.3). If kin selection was the driving force, this altruistic behaviour would be favoured only if it meant that more than two full-siblings would survive, because $(0.5)(2) = 1$, but $(0.5)(3) > 1$. Hamilton is said to have worked out this rule in the pub one night, when he claimed that he would lay down his life for more than two siblings or eight cousins, a statement that can be understood in light of the relatedness values that are given in Table 6.3.

**Table 6.3** Some coefficients of relatedness in diploid species. Two individuals that have a relatedness coefficient of 0.5 will have 50 per cent of their alleles in common

| Coefficient of relatedness ($r$) | Examples |
| --- | --- |
| 1.0 | Identical twins |
| 0.50 | Parents and offspring |
| | Full-siblings (both parents in common) |
| 0.25 | Grandparents and grandchildren |
| | Aunts/uncles and nieces/nephews |
| | Half-siblings (one parent in common) |
| 0.125 | Cousins |
| | Great grandparents and great grandchildren |

## 6.3 Estimating relatedness from molecular data

The genetic relationships between individuals are usually referred to as $r$, the coefficient of relatedness, some examples of which are given in Table 6.3. Relatedness refers to the proportion of alleles that two relatives are expected to share, i.e. the probability that an allele found in an individual will also be present in that individual's parent, sibling, cousin, and so on. In a sexual diploid species, the coefficient of relatedness between parents and offspring is 0.5 because an offspring will inherit half of its DNA from each parent and will therefore share 50 per cent of its alleles with its mother and 50 per cent with its father. After another generation has passed, the new offspring once again has a 50 per cent probability of inheriting an allele from one of its parents, and the likelihood that it has inherited a particular allele from one of its grandparents is $(0.5)(0.5) = 0.25$, therefore $r = 0.25$ between grandchildren and grandparents.

The examples shown in Table 6.3 are straightforward but in ecological studies we are more likely to be interested in the relatedness between two individuals for whom we have no prior information, and we cannot estimate this from the total proportion of their shared alleles. We therefore need other methods to estimate the $r$ values of individuals whose relationships are unknown. One approach is to use the frequencies of alleles in individuals and populations to determine whether or not alleles are more likely to be shared because of common descent or because of chance. The more closely related two individuals are to each other, the more likely they are to share alleles because of common descent. If, however, they share only alleles that occur at high frequencies in the population, we may conclude that these alleles are shared simply as a result of chance.

We already know how to estimate population allele frequencies, and the frequency of an allele in a diploid individual must be either 1.0 (homozygote), 0.5 (heterozygote) or 0 (allele absent). Based on this information, the relatedness of one individual to one or more other individuals can be calculated from allele frequency data as:

$$\Sigma(p_y - p)/\Sigma(p_x - p) \qquad (6.2)$$

where for each allele $p$ is the frequency within the population, $p_x$ is the frequency within the focal individual, and $p_y$ is the frequency within the individual whose relationship to the focal individual we wish to know. Only those alleles that are found in the focal individual ($x$) are included in the equation (Queller and Goodnight, 1989). This method is incorporated into the software program 'Relatedness' (see useful websites and software at end of chapter). Note that this equation can generate either positive or negative numbers, with negative values resulting from very low levels of relatedness.

We shall work through this equation using a relatively straightforward example in which we are interested in whether a focal individual (individual $x$) within a cooperatively breeding group of birds is related to a single female whose brood he is helping to raise. In this example, genotypes are given as the sizes of the amplified microsatellite alleles. The focal individual is homozygous at microsatellite locus 1 (120, 120) and heterozygous at microsatellite locus 2 (116, 118). The potential relative is heterozygous at locus 1 (120, 122) and homozygous at locus 2 (118, 118). When calculating relatedness, we consider only the three alleles that are found in the focal individual (120, 116 and 118). The frequencies used in this calculation are:

| Allele | $p_x$ | $p_y$ | $p$ |
|--------|-------|-------|------|
| 120    | 1.0   | 0.5   | 0.65 |
| 116    | 0.5   | 0     | 0.20 |
| 118    | 0.5   | 1.0   | 0.35 |

Relatedness is therefore calculated as:

$$[(0.5 - 0.65) + (0 - 0.20) + (1 - 0.35)]/[(1 - 0.65) + (0.5 - 0.20) \\ + (0.5 - 0.35)]$$

$$= 0.30/0.80$$

$$= 0.375$$

This suggests that the two birds are quite closely related to each other, although in practice we would interpret this finding with caution because it is based on only two loci, and more data – possibly from up to 30–40 microsatellite loci or >100 SNP loci – are needed before relatedness coefficients can be calculated with a high degree of confidence (Blouin *et al.*, 1996; Glaubitz, Rhodes and DeWoody, 2003).

Genetic data have enabled us to calculate the relatedness of breeders and their helpers with relative ease (Box 6.3), and these relatedness values have helped biologists to determine whether or not kin selection is a plausible explanation for social breeding. One species in which this seems to be the case is the bell miner (*Manorina melanocephala*), which breeds within discrete social units that consist of a single breeding pair plus up to 20 helpers. One study found that the majority of these helpers (67 per cent) were closely related ($r>0.25$) to the breeding pair (Figure 6.8; Conrad *et al.*, 1998). Kin selection may also explain cooperative breeding in the eusocial Damaraland mole-rat (*Cryptomys damarensis*), in which the mean colony relatedness was found to be $r = 0.46$ (Burland *et al.*, 2002). In some cases the overall relatedness between helpers and offspring may be reduced by EPFs, for example the moderately high level of EPFs (19 per cent of 207 offspring) in western bluebirds (*Sialia mexicana*) meant that the mean relatedness between chicks and the males that were helping their parents to raise these young was 0.41 (Dickinson and Akre, 1998). This was lower than the relatedness value of 0.5 that is expected if the helpers and chicks were all full-siblings, although the reduction from 0.5 to 0.41 does not necessarily preclude kin selection as a driving force.

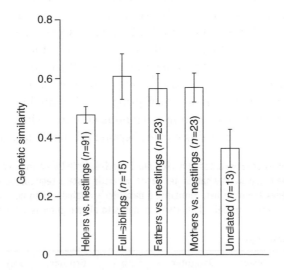

**Figure 6.8** Genetic similarity between different groups of bell miners (±CI), based on the proportion of shared genetic markers. These data show that helpers at the nest are related to the nestlings. Redrawn from Conrad *et al.* (1998)

On the other hand, genetic data have shown that fairy-wren helpers (*Malurus cyaneus*) often assist in the rearing of young to which they are unrelated (Dunn, Cockburn and Mulder, 1995), and male white-browed scrubwrens (*Sericornis frontalis*) that are unrelated to the breeding female actually are more likely to help raise her young (Magrath and Whittingham, 1997). Social breeding clearly cannot be explained by kin selection in these species and therefore other factors must be taken into account. These may include gaining experience in parental care, increasing the likelihood of being allowed to remain in the colony, or improving the chance of future survival or reproduction. Ecological constraints may also favour social breeding if there is a limited supply of food, nest sites or other resource, and this may explain why socially breeding bird species are relatively common in the environmentally harsh arid and semi-arid regions of Africa and Australia where high quality habitat is in short supply.

### Social insects

The previous examples were based on the relatedness values between diploid individuals, but no discussion of social breeding would be complete without reference to social insects, many of which are haplodiploid. This means that males develop from unfertilized eggs and therefore are haploid ($n$), having only one set of chromosomes that come from the female parent. In contrast, females, which can be either sterile workers or reproductive queens, develop from fertilized eggs and so inherit one set of chromosomes from their mother and one set from their father, which makes them diploid ($2n$). The relatedness between haplodiploid family members is not the same as that between diploids (Table 6.4). An important difference is that, unlike sexually reproducing diploid species, haplodiploid females are more closely related to their full-sisters ($r = 0.75$) than to their offspring ($r = 0.5$), and therefore female workers can increase their fitness by rearing sisters instead of producing their own young provided that the number of sisters is not less than two-thirds of the number of offspring that they might otherwise produce. This of course will be true only in **monogynous colonies** (single queen) in which the queen is inseminated by a single male, because it is only under these conditions

**Table 6.4** Coefficients of relatedness in haplodiploid species. Note that a mother's relatedness to her son is 0.5 because he received only half of her genes, whereas a son's relatedness to his mother is 1.0 because all of his genes are from her. Similarly, a daughter's relatedness to her father is 0.5 because half of her genes are from him, but a father's relatedness to his daughter is 1.0 because he is haploid and therefore she has all of his genes. There is no relatedness between fathers and sons because males result from unfertilized eggs

|  | Mother | Sister | Daughter | Father | Brother | Son | Niece/nephew |
|---|---|---|---|---|---|---|---|
| Female | 0.5 | 0.75 | 0.5 | 0.5 | 0.25 | 0.5 | 0.375 |
| Male | 1 | 0.5 | 1 | 0 | 0.5 | 0 | 0.25 |

**Table 6.5** Some examples showing the average relatedness values within monogynous (one queen) and polygynous (multiple queens) colonies. In monogynous colonies a relatively high proportion of workers have at least one parent in common, and therefore overall relatedness tends to be relatively high compared with polygynous colonies

| Species | Average relatedness | Type of colony | Reference |
|---|---|---|---|
| Crab spider (*Diaea ergandros*) | 0.44 | Monogynous | Evans and Goodisman (2002) |
| Giant hornet (*Vespa mandarinia*) | 0.738 | Monogynous | Takahashi *et al.* (2004) |
| Carpenter ant (*Camponotus ocreatus*) | 0.65 | Monogynous | Goodisman and Hahn (2004) |
| Argentine ant (*Linepithema humile*) | 0.007 | Polygynous | Krieger and Keller (2000) |
| Greenhead ant (*Rhytidoponera metallica*) | 0.082 | Polygynous | Chapuisat and Crozier (2001) |
| Honey bee (*Apis mellifera*) | 0.25--0.34 | Polygynous | Laidlaw and Page (1984) |

that workers will be full-sisters. In colonies of the slavemaker ant (*Protomognathus americanus*), for example, workers are usually full-sisters with a relatedness of 0.75 and therefore will benefit by helping to raise more sisters (Foitzik and Herbers, 2001). In situations such as this, kin selection can explain why workers forego reproduction.

The situation is more complex in monogynous colonies when the queen has multiple mates, and also in **polygynous colonies** (multiple queens), because in these situations the relatedness of workers can range from almost 0 to around 0.75 (see Table 6.5). In polygynous colonies, worker relatedness depends not just on the number of queens but also on how closely the queens are related to one another (Ross, 2001, and references therein). Since the helpers in polygynous colonies often share few genes with the offspring, an explanation other than inclusive fitness is needed to explain this type of social breeding. Ecological factors may provide at least part of the answer, one possibility being that multiple queens are needed to ensure that enough eggs will be laid to support a colony that is large enough for long-term survival. However, this cannot explain the prevalence of helpers in colonies in which a single queen mates with multiple males, because each time a new male inseminates the queen a new set of half-siblings will be introduced into the colony and the overall within-colony relatedness will be reduced. One possible explanation in these cases is the need to increase genetic diversity within the colony.

## Manipulation of Sex Ratio

Another aspect of behavioural ecology that has benefited from molecular data and that, like mating systems, is linked to reproductive behaviour, is the way in which parents manipulate the sex ratio of their offspring. In 1930 the biologist and statistician R.A. Fisher wrote an influential book on evolutionary genetics in which he addressed, among many other things, the importance of sex ratios (Fisher, 1930). Fisher maintained that a sex ratio should remain stable if the production of males and females provides equal fitness, per unit of effort, for the individuals that are controlling sex ratios. If, on the other hand, greater fitness can be obtained by producing an excess of one sex, then either males or females will be favoured, at least until the time when there is no longer an advantage to biasing the sex ratio.

Research into adaptive sex ratios really got under way in the 1970s after Trivers and Willard (1973) wrote a seminal paper in which they resurrected the argument that parents may manipulate the sex ratio of their offspring for adaptive reasons. Over the years considerable support for this has come from a wide range of taxonomic groups, but until recently investigations were mainly limited to species in which males and females were easily distinguished on the basis of external morphology. In many species we are now able to use sex-specific markers to identify the sexes of morphologically indistinguishable adults and juveniles. In addition, by genotyping tissue from eggs we can sometimes use molecular data to calculate the **primary sex ratio** (that found in eggs) of many species. This allows us to compare the primary and **secondary sex ratio** (that found in hatchlings) of a population, which is sometimes a necessary distinction to make before we can determine whether or not a secondary sex ratio has been influenced by disproportionate egg mortality in either males or females, as opposed to adaptive parental behaviour.

## Adaptive sex ratios

The use of molecular data to obtain sex ratios has been particularly widespread in studies of birds. It is almost impossible to sex the adults of many bird species or the newly hatched chicks of virtually all bird species on the basis of external phenotypic characters, but they can be sexed from their genotypes. Recall from Chapter 2 that female birds are the heterogametic sex (ZW) whereas males are homogametic (ZZ). A chromo-helicase-DNA-binding (CHD) gene is located on each of the W and Z sex chromosomes of most bird species (CHD-W and CHD-Z, respectively). A pair of primers has been characterized that will anneal to a conserved region and amplify both of the CHD genes in numerous species (Griffiths *et al.*, 1998). A variable non-coding region that is a different length in each gene means that the size of the product will depend on whether it was the CHD-W gene or the CHD-Z gene that was amplified. As a result, a single band (CHD-Z only) will result from the PCR of male genomic DNA, whereas two bands (CHD-Z and CHD-W) will result from amplified female genomic DNA (Figure 6.9).

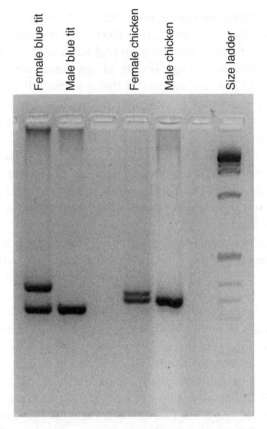

**Figure 6.9** A portion of CHD genes was amplified from male and female blue tits and chickens using primers P2 and P8 (Griffiths *et al.*, 1998). Note that in both species two bands were generated from the female samples but only one band from the male samples. Photograph provided by Kate Orr and reproduced with permission

These avian sex markers can be used on tissue that has been taken from eggs, although more accurate results are obtained from newly hatched nestlings. In a number of studies, the sex ratios determined from molecular data have added support to the theory of adaptive parental manipulation. Female blue tits (*Parus caeruleus*) produce more sons when mated to males that have a higher survival rate, a characteristic that females can gauge on the basis of the male's ultraviolet plumage ornamentation (Sheldon *et al.*, 1999). In kakapo (*Strigops habroptilus*) and house wren (*Troglodytes aedon*) populations, females were produced in excess when conditions were not conducive to the growth of particularly large and healthy offspring (Albrecht, 2000; Clout, Elliott and Robertson, 2002), presumably because weaker males are less likely to obtain mates than weaker females, particularly in polygynous species.

Many more examples of biased sex ratios have been found in birds (see Komdeur and Pen, 2002; Pike and Petrie, 2003). As yet there is no single theory that can explain this adaptive behaviour, in part because the reasons seem to vary both within and between species. Timing of egg production, parental quality, environmental conditions, and helpers at the nest may all influence sex ratios. Furthermore, the mechanisms for sex ratio manipulation remain unclear. Non-random segregation of sex chromosomes, selective resorbtion of yolk, selective ovulation, sex-specific fertilization, and sex-specific inhibition of zygote formation are just some of the mechanisms that have been proposed (Pike and Petrie, 2003, and references therein). So far, molecular data have helped to demonstrate the existence of sex ratio allocation in birds, but there is considerable work to be done before we understand the adaptive reasons and the mechanisms for producing an excess of males or females.

Birds are not the only taxonomic group in which molecular markers have been used to identify the sex of morphologically similar juveniles. Neither embryos nor tadpoles can be sexed in amphibians on the basis of external phenotypes, but a sex-linked gene, ADP/ATP translocase, has been used to differentiate between the homogametic and heterogametic forms of the Japanese frog *Rana rugosa*. In this species, interpretation of molecular data depends on which form is being studied, because in different forms the heterogametic sex is either the male (XX/XY) or the female (ZZ/ZW), and in some forms both sexes are homogametic (Miura *et al.*, 1998). In one study, embryos from two populations of the ZZ/ZW form were genotyped by PCR-RFLP analysis of ADP/ATP translocase (Sakisaka *et al.*, 2000). These data showed a significant bias towards male offspring at the start of the breeding season and a female-biased sex ratio towards the end of the breeding season. The authors of this study suggested that this switch could be explained by the relatively fast development of males which typically metamorphose into adults by the autumn, whereas the more slowly developing female tadpoles often hibernate throughout the winter.

## Sex ratio conflicts

In social insects, sex allocation is complicated further by the relatedness between haplodiploids. Table 6.4 shows us that a reproductive female (queen) shares the same level of relatedness ($r = 0.5$) with both her sons and her daughters, and therefore her ideal sex ratio is 1:1. Female workers, on the other hand, who do not produce offspring, share a higher degree of relatedness with their sisters ($r = 0.75$) than their brothers ($r = 0.25$), and therefore their ideal sex ratio in the colony is 3 : 1 in favour of females. This leads to a conflict over sex allocation between workers and queens, particularly in monogynous colonies in which the queen has a single mate, because then the offspring will all be full-siblings of the workers. Because workers outnumber the queens and are also the ones that rear the larvae, they should have the upper hand in this conflict and we therefore may expect that sex ratios should approach the workers' ideal. This has prove to be the case in a number of monogynous species, for example in the ant species *Colobopsis nipponicus* and *Leptothorax tuberum* the proportion of females in numerous colonies was found to be around 0.75 (Hasegawa, 1994; Pearson, Raybould and Clarke, 1995).

Workers may control sex ratios in a colony either by killing male larvae or by controlling the proportion of females that develop into reproductive adults (potential queens) versus sterile workers. Since adult males and females can be easily identified in social insect colonies, their sex ratios can be obtained without the aid of molecular markers, but the mechanisms of sex ratio manipulation cannot be understood fully without using molecular data to compare primary and secondary sex ratios. In a study of the ant *Leptothorax acervorum*, researchers used microsatellite markers to genotype eggs, and from these data they learned that the sex ratio did not change between eggs and adults. They therefore concluded that workers were obtaining their optimal sex ratio by manipulating the proportion of females that developed into sterile workers and not by killing male larvae (Hammond, Bruford and Bourke, 2002). The situation is different in fire ants (*Solenopsis invicta*), which often have sex ratios that are intermediate to the ratios that should be favoured by workers and by queens. Once again, microsatellite data were used to genotype eggs and obtain a primary sex ratio, and these data revealed that queens were biasing the sex ratio of their eggs in favour of males, thereby forcing workers to raise a higher proportion of males than that dictated by their optimal sex ratio (Passera *et al.*, 2001; see also Box 6.4).

## 6.4 Biased sex ratios in adult populations

Many surveys of wild populations have revealed an excess of either male or female adults. The reasons for this are not always well understood, although molecular sex probes have allowed researchers to test a number of possible explanations. Populations of western sandpipers (*Calidris mauri*) in the northern part of their range show a male to female ratio of around 3:1. Predation by peregrine falcons is high in these populations, and researchers wanted to know whether or not the shortage of females could be attributed to disproportionately high predation rates. They removed feathers from the remains of individuals that had been recently preyed upon and from these they amplified the CHD genes. This told them that around 24 per cent of the carcasses tested were female. Because the proportion of females was comparable in living and dead birds, sex-biased predation was not a plausible explanation for the male-biased sex ratios in these populations (Nebel, Cloutier and Thompson, 2004).

In contrast to western sandpipers, females often predominate in populations of the knobbed whelk (*Busycon carica*), sometimes outnumbering males by as much as 10:1. A polymorphic microsatellite locus has been identified as a sex-specific marker in this species, because it is usually heterozygous (two alleles) in females and **hemizygous** (only one allele) in males. Researchers used this marker to genotype the embryos from two knobbed whelk broods, and found that 383/768 embryos (49.9 per cent) were female. As a result, the female-biased sex ratios in knobbed whelk populations cannot be attributed to a skewed primary sex ratio, but instead must be explained by some other factor such as a relatively high mortality of juvenile or adult males, or the use of different habitats by males, and females (Avise, Power and Walker, 2004).

## Sex-Biased Dispersal

We will look now at a behaviour that can be maintained for a variety of reasons, most of which are related to reproduction, and that is **sex-biased dispersal**. This occurs in many species when one sex is more likely than the other to disperse between populations. In mammals, females are usually **philopatric**, meaning that they tend to remain at, or return to, their natal site for breeding, whereas males often disperse from their birthplace and never return. The opposite is true in birds, with males more likely to be philopatric and females more likely to disperse, although exceptions to the rule can be found in both groups.

One possible explanation for sex-biased dispersal is described by the resource–competition hypothesis (Greenwood, 1980), which predicts that the sex remaining at its natal site will be the one that benefits most from home-ground familiarity. In birds this will often be males, who can benefit from site familiarity when acquiring and defending territories. In mammals, females may benefit most from knowledge of a particular area because they may be able to produce more young if they are familiar with local food resources. A second possible explanation is summarized by the local mate competition hypothesis (Perrin and Mazalov, 1999), which proposes that individuals disperse so that they will not have to compete with their relatives for mates, thereby increasing their inclusive fitness. Alternatively, sex-biased dispersal may be explained by the inbreeding avoidance hypothesis (Pusey, 1987), which is based on the idea that the sex that incurs the greatest cost from inbreeding is more likely to disperse.

Deciding which hypothesis provides the most appropriate explanation for sex-biased dispersal in a particular species can be difficult, in part because relevant hypotheses may not be mutually exclusive. In addition, reasons for dispersal may change over time, depending on a variety of factors such as environmental conditions or the density of a local population. The biggest contribution of molecular ecology to this area of research has been through the quantification of sex-biased dispersal, and we shall look now at four ways in which molecular data can be used to contrast the dispersal patterns of males and females.

## Nuclear versus mitochondrial markers

One method for measuring sex-biased dispersal, which we have already referred to in earlier chapters, is the comparison of population differentiation estimates that are based on biparentally (autosomal nuclear) versus uniparentally (mtDNA, Y-chromosome) inherited markers. When males disperse and females are philopatric, mitochondrial markers should show higher levels of population differentiation than autosomal nuclear markers; conversely, at least in mammalian species, if females disperse and males are philopatric then Y-chromosome data should show higher levels of differentiation than either autosomal or mitochondrial DNA.

This method has been used to compare the dispersal of males and females from the nurseries of several bat species. Female bats often form maternity colonies in which they raise their young, and therefore they tend to be highly philopatric despite being proficient fliers that could easily move between sites. In Bechstein's bat (*Myotis bechsteinii*), colonies are closed to non-native females and yet one study found that overall relatedness within colonies was only 0.02 (Kerth, Safi and Konig, 2002b). In the absence of female-mediated gene flow, these low levels of relatedness must mean that males are regularly dispersing among colonies, a suggestion that was supported by a comparison of mitochondrial and nuclear differentiation among multiple colonies. Genetic differentiation based on nuclear alleles was much lower ($F_{ST} = 0.003-0.031$) than that based on mtDNA ($F_{ST} = 0.658-0.961$), a pattern that reflects extremely rare female dispersal in conjunction with widespread male dispersal, with the latter preventing the colonies from becoming inbred (Kerth, Mayer and Petit, 2002a).

In the previous example of Bechstein's bats, the genetic data were supplemented by field observations and an understanding of the species' ecology, and therefore the conclusions were well-supported. However, studies of sex-biased dispersal that are based solely on comparisons between mitochondrial and nuclear DNA should be interpreted with caution for two reasons. First, the different mutation rates in the two sets of markers can influence the levels of observed genetic differentiation. Second, as we know from Chapter 2, the effective population size of mtDNA is expected to be approximately a quarter of that of nuclear DNA, although this is true only if mating is random, which often is not the case. In a strongly polygynous mating system, for example, many more females than males will reproduce each breeding season, and as a result the effective population size of maternally inherited genes can be larger than that of biparentally inherited genes (Chesser and Baker, 1996). It can be difficult, therefore, to anticipate how the population sizes of different genomes will influence observed levels of genetic differentiation.

## Relatedness

A second way to infer different levels of dispersal between the sexes is to compare male–male and female–female relatedness within populations, because the sex that does not disperse should show higher levels of relatedness than the sex that does disperse. The simplest way to test for this is to compare the relatedness between all male–male pairs and all female–female pairs within each population to see if overall values are higher in one sex than the other. Estimates of relatedness were used to compare female and male dispersal in the Australian lizard *Egernia stokesii*. The breeding partners of this genetically monogamous species live within aggregates that include offspring and other relatives. Although both sexes show some degree of philopatry, females within groups had higher overall levels of relatedness to one another ($r = 0.1380$) compared with the relatedness between males ($r = 0.0433$), and this was taken as evidence for male-biased dispersal (Gardner *et al.*, 2001).

## $F_{ST}$ Values

A third way to compare the dispersal patterns of males and females is to calculate genetic differentiation (typically an $F_{ST}$ value) separately for each sex based on data from the same biparentally inherited loci. The sex that disperses more should show lower levels of among-population differentiation than the philopatric sex. One drawback to this technique is that, because only biparentally inherited markers can be used, all of the relevant loci will be passed down to both male and female offspring once the dispersed adult reproduces. This means that, unless data are collected from dispersed individuals before they have reproduced, any sex-specific signals will be weakened. Discordant $F_{ST}$ values in males and females of the marine iguana *Amblyrhynchus cristatus* (Figure 6.10) on the Galápagos archipelago provided evidence of male-biased dispersal because the same nuclear loci, when analysed separately for each sex, revealed average $F_{ST}$ values of 0.14 in females and 0.04 in males. These data were supported by field observations of males actively dispersing between islands. Males also spent more time than females swimming off-shore to forage, a behaviour that could increase their likelihood of being passively dispersed by ocean currents (Rassmann *et al.*, 1997).

## Assignment tests

The fourth method for inferring sex-biased dispersal is based on the assignment tests that were introduced in Chapter 4. These tests can be used to compare the number of males and females that are assigned to a population other than the one

**Figure 6.10** A marine iguana (*Amblyrhynchus cristatus*) on Santa Cruz Island in the Galápagos archipelago. A comparison of $F_{ST}$ values that were calculated separately for each sex provided evidence of male-biased dispersal in this species (Rassmann *et al.*, 1997). Author's photograph

from which they were sampled, the rationale here being that members of the more dispersive sex will be misassigned most often. A slightly modified assignment test was used first by Favre *et al.* (1997) to infer sex-biased dispersal in the greater white-toothed shrew (*Crocidura russula*). For each individual they calculated an assignment index, which reflects the probability that a particular individual's genotype originated in the population from which it was sampled. Differences in population genetic diversity were then corrected for by subtracting assignment population means from the log-transformed assignment indexes, which provided corrected assignment indexes (*AIc*). These corrected indexes reflect the expected frequency of each individual genotype in the population from which it was sampled. Individuals with negative *AIc* values have rare genotypes (i.e. with a low expected frequency) and therefore are likely to be recent immigrants. A preponderance of negative *AIc* values in the female greater white-toothed shrew led Favre *et al.* (1997) to conclude that dispersal in this species is female-biased, an

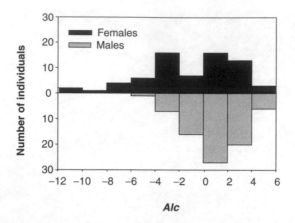

**Figure 6.11**  Corrected assignment indexes (*AIc*) in the white-toothed shrew. A higher proportion of females compared with males have negative *AIc* values, which means that females are more likely to disperse. Redrawn from Favre *et al*. (1997)

unusual pattern in mammals that may be explained by the fact that, unlike most mammals, this species is socially monogamous and therefore may be expected to show dispersal patterns that are more typically found in birds (Figure 6.11).

## Concordant results

Inferences of sex-biased dispsersal have been made in a wide range of taxa using all four genetic methods outlined above (Table 6.6). Because no method is infallible, conclusions may be more robust if multiple methods have provided concordant results. This was the case in a study of two hyrax species, the bush hyrax (*Heterohyrax brucei*) and the rock hyrax (*Procavia johnstoni*), in the Serengeti National Park. Both species live within a classic metapopulation structure characterized by distinct habitat patches consisting of rock outcrops that are separated by wide expanses of grass plains. Local populations are subject to extinction–recolonization events, and dispersal between populations is ongoing.

A recent genetic study generated several lines of evidence that all revealed female-biased dispersal in the bush hyrax, and equal dispersal in male and female rock hyrax. Part of this evidence came from microsatellite-based $F_{ST}$ values, which were higher in males than females in the bush hyrax but comparable in both sexes in the rock hyrax. In addition, assignment tests identified a greater proportion of immigrant females than immigrant males in the bush hyrax but not in the rock hyrax (Figure 6.12), and relatedness among females within sites was relatively high in the rock hyrax but not in the bush hyrax. These results were somewhat surprising, because earlier mark–recapture studies had concluded that dispersal in both species was male-biased. This discrepancy may be explained if

**Table 6.6** Examples of sex-biased dispersal in a variety of taxa. Note that although examples of female-biased dispersal in mammals and male-biased dispersal in birds are exceptions to the rule, some have been included here to show that they can occur

| Species | Dispersing sex | Method of analysis[a] | Reference |
|---|---|---|---|
| **Mammals** | | | |
| Eastern grey kangaroo (*Macropus giganteus*) | Male | MI | Zenger, Eldridge and Cooper (2003) |
| Bechstein's bat (*Myotis bechsteinii*) | Male | MI | Kerth, Nayer and Petit (2002a); Kerth, Safi and Konig (2002b) |
| Dall's porpoise (*Phocoenoides dalli*) | Male | MI | Escorza-Trevino and Dizon (2000) |
| Wolverine (*Gulo gulo*) | Male | GD, A | Cegelski, Waits and Anderson (2003) |
| Greater white-toothed shrew (*Crocidura russula*) | Female | GD | Favre *et al.* (1997) |
| Yellow-spotted rock hyrax (*Heterohyrax brucei*) | Female | GD, A, R | Gerlach and Hoeck (2001) |
| River otter (*Lontra canadensis*) | Male | GD | Blundell *et al.* (2002) |
| **Birds** | | | |
| Great reed warbler (*Acrocephalus arundinaceus*) | Female | A | Hansson, Bensch and Hasselquist (2003) |
| Red grouse (*Lagopus lagopus scoticus*) | Female | MI | Piertney *et al.* (2000) |
| Yellow warbler (*Dendroica petechia*) | Male | MI | Gibbs, Dawson and Hobson (2000) |
| Red-billed quelea (*Quelea quelea*) | Male | A | Dallimer *et al.* (2002) |
| **Fish** | | | |
| Brown trout (*Salmo trutta*) | Male | A | Bekkevold, Hansen and Mensberg (2004) |
| Lake Malawi cichlids (*Pseudotropheus zebra* and *P. callainos*) | Male | R | Knight *et al.* (1999) |
| Lake Tanganyika cichlid (*Eretmodus cyanostictus*) | Female | R | Taylor *et al.* (2003) |
| **Reptiles** | | | |
| Australian lizard (*Egernia stokesii*) | Male | R | Gardner *et al.* (2001) |
| Marine iguana (*Amblyrhynchus cristatus*) | Male | GD, MI | Rassmann *et al.* (1997) |

**Table 6.6** *(Continued)*

**Amphibians**

| | | | |
|---|---|---|---|
| Bullfrog (*Rana catesbeiana*) | Female | GD, A | Austin *et al.* (2000) |
| Tungara frog (*Physalaemus pustulosus*) | Male | A | Lampert *et al.* (2003) |

**Invertebrates**

| | | | |
|---|---|---|---|
| Tick (*Ixodes ricinus*) | Male | R | de Meeus *et al.* (2002) |
| Ponerine ant (*Diacamma cyaneiventre*) | Male | MI | Douhms, Cabrera and Peeters (2002) |
| Narrow headed ant (*Formica exsecta*) | Male | A | Sundström, Keller and Chapuisat (2003) |

[a]A, assignment tests; GD, genetic differentiation in males and females based on data from the same loci; R, relatedness; MI, marker inheritance, which refers to genetic differentiation based on biparentally versus uniparentally inherited markers.

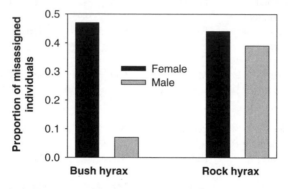

**Figure 6.12** Male and female dispersal in bush and rock hyraxes on the Serengeti, as revealed by assignment tests. The bush hyrax shows evidence of female-biased dispersal, whereas dispersal of males and females in the rock hyrax is comparable. Data from Gerlach and Hoeck (2001)

bush hyrax males disperse more often than females but either die *en route* or are less successful than females at reproducing in their new populations, in which case male dispersal would not equate with male-mediated gene flow (Gerlach and Hoeck, 2001).

# Foraging

In the final section of this chapter we shall move away from reproduction and look at how molecular data have provided insight into the foraging behaviours of animals. We will do this by looking first at how molecular techniques can be used to identify prey remains and individual food preferences. Knowledge of prey items may help us to answer ecological questions about food webs and resource partitioning and can also be an important tool in conservation biology. Second, we shall look at how genetic markers can be used to track individuals as they move around in search of food, covering distances of only one or two kilometres in some species, and hundreds of kilometres in others.

## Identifying prey

A survey of the literature suggests that the molecular identification of prey remains is becoming an increasingly popular tool. If we know what animals are feeding on, we will have a better understanding of ecology at both the species and community levels because feeding is relevant to predator–prey and host–parasite interactions, food chains, intra-and interspecific competition, and niche partitioning. Identifying prey is sometimes possible through observations of predator–prey interactions, although this approach is often impractical for small, secretive or wideranging species. Prey remains from gut contents or faeces can sometimes be identified from fragments of bone, carapace, seeds, feathers, scales or other resistant parts, although the capture, ingestion and digestion of prey often leaves characters too mutilated for identification. Furthermore, soft prey items such as slugs or invertebrate eggs are unlikely to leave any identifiable remains. Molecular identification of prey remains therefore presents a useful alternative because it does not require morphological features to be preserved.

Provided that appropriate primers are available, prey items can often be identified by amplifying DNA from composite samples such as faeces, gut contents or even the entire predator, and then matching the characterized DNA to sequences or allele sizes from existing DNA databases. Deep-sea marine invertebrates provide a good example of animals whose feeding habits cannot be observed easily and whose prey are unlikely to be recognizable on the basis of morphology once they have been partially digested. In a recent study, Blankenship and Yayanos (2005) used universal cytochrome $c$ oxidase I primers to amplify copies of this mitochondrial gene from the stomach contents of the deep-sea crustaceans *Scopelocheirus schellenbergi* and *Eurythenes gryllus* from the Tonga Trench. They found that both species fed on a wide range of prey, not all of which would have been carrion. This was an unexpected result because both species were previously believed to feed exclusively as scavengers, but the identity of some of their prey items means that they also may be predators.

Studies such as this are extremely useful because they represent a novel way of quantifying the prey items of relatively inaccessible species. However, although the technique is relatively straightforward, a note of caution is in order. Potential problems associated with degraded DNA, inhibitors and contamination of gut or faecal material are just some of the reasons why the molecular identification of

**Figure 6.13**   A coyote (*Canis latrans*). Food preferences in this omnivorous species can be identified from faecal genotypes. Author's photograph

prey can be a rather demanding task that requires meticulous attention to all aspects of sample collection, preservation and laboratory work.

### Individual food preferences

One interesting outcome of using molecular techniques to identify prey remains is that it can help us to understand which prey items individual predators are choosing. In one study, the food habits of coyotes (*Canis latrans*; Figure 6.13) in the California Santa Monica Mountains were investigated by researchers who collected and genotyped 115 coyote faeces (Fedriani and Kohn, 2001). Sequence data were used to identify which species the prey items belonged to, and microsatellite data were used to generate individual coyote genotypes that would tell the researchers which coyote had left each faecal sample. This is possible because faeces contain some cells from the gut of the animal that left them, and if variable markers such as microsatellites are used then it is possible to generate individual-specific genetic profiles that will link each sample to an individual animal (see Box 6.5). Sequence data were obtained from up to 11 faeces from each coyote, and from these it was possible to calculate the percentage of each animal's diet that was made up of small mammals, other vertebrates, invertebrates, fruit and rubbish. Although the majority of coyotes fulfilled expectations by taking small mammals as their primary food source, 18 per cent of the sampled coyotes had an alternative primary food source (Figure 6.14), which is a substantial portion of the population whose food preferences would have been overlooked if all prey items had been pooled.

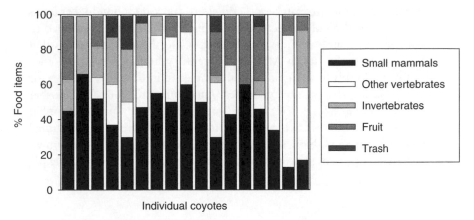

**Figure 6.14** A breakdown of individual coyote diets. Microsatellite data were used to assign faeces to individual coyotes, and mitochondrial sequence data were used to identify faecal prey remains. Adapted from Fedriani, and Kohn (2001)

### 6.5 Probability of identity

Matching samples such as faeces to individual animals is possible only if the molecular markers being used are sufficiently variable to generate a different genotype for each individual within a population. We can test for this by calculating the **probability of identity** (PI), which is the likelihood that two individuals chosen at random from the population will have the same genotype. A low probability of identity means that two or more individuals are unlikely to share the same genotype, in which case we would be confident that a sample has been linked to the correct individual. PI values are most commonly based on microsatellite loci, and before to calculate PI we need to know the frequency of each relevant allele within the population because high-frequency alleles are much more likely to be found in multiple individuals than are low-frequency alleles. One of the simplest ways to calculate the likelihood of two individuals in a population sharing the same genotype was developed for a study of black bears using the following equation (Paetkau and Strobeck, 1994):

$$PI = \sum_{i=1}^{n} (p_i^2)^2 + \sum_{i=1}^{n} \sum_{j=i+1}^{n} (2p_i p_j)^2 \qquad (6.3)$$

where $p_i$ and $p_j$ are the frequencies of the $i$th and $j$th alleles at each locus in a given population. The probability of identity is revisited in the context of wildlife forensics in Chapter 8.

> Probability of identity can be a useful aid to tracking individual animals that are rare or difficult to recapture. In one study, researchers wished to follow the movements of individual wolves (*Canis lupus*) from packs that had recolonized the Italian Alps after a century-long absence, and they were able to do this by matching faecal samples to individuals on the basis of multi-locus microsatellite genotypes (Lucchini *et al.*, 2002). In another study, a population of the highly endangered hairy-nosed wombat (*Ladiorhinus drefftii*) was censused accurately by determining individual genotypes from hairs that were collected at the entrances to their burrows, a method that presented no risk to the individual wombats (Sloane *et al.*, 2000).

## Biological control

A practical application of the molecular identification of prey was illustrated by a study of Linyphiid spiders. These are important aphid control agents in arable fields, although they supplement their diet with other invertebrates of higher nutritional value such as Collembola. Sequences from mitochondrial cytochrome oxidase I were used to determine which collembola species the spiders were consuming, and it turned out that a high percentage of spiders ate the collembolan *Isotoma anglicana* even though this was relatively scarce in the field. On the other hand the most common collembolan, *Lepidocyrtus cyaneus*, was seldom eaten by the spiders. The spiders therefore were clearly choosing their prey items on the basis of more than simply abundance. A practical outcome of this study is the knowledge that increasing the population of *I. anglicana* could indirectly help to keep the aphid population under control by boosting the lynyphiid spider population (Agusti *et al.*, 2003). Other examples of the molecular identification of prey from arthropod guts are given in Table 6.7.

## Predation and conservation

Predator–prey interactions can also be important in conservation biology. The coccinellid beetle *Halmus chalybeus* was introduced into the Hawaiian island Kaua'i in 1894 as a biological control of homopteran pests, but the recent invasion of these beetles into the Alaka'i swamp is now a cause for concern because many coccinellid beetles are generalist predators. In a recent study, DNA analysis was used to identify their various prey items, which turned out to include some species of considerable conservation value, notably endemic moths of the genera *Scotorythra* and *Eupithecia* (Sheppard *et al.*, 2004).

**Table 6.7**  Some of the prey items that have been identified from PCR amplification of arthropod gut contents using various molecular markers

| Predator | Prey | Molecular marker | Reference |
|---|---|---|---|
| Mirid bug (*Dicyphus tamaninii*) | Cotton bollworm (*Helicoverpa armigera*) | RAPDs | Agusti, De Vicente and Gabarra (1999) |
| Carabid beetles (*Pterostichus cupreus*) | Mosquitoes (*Culex quinquefasciatus*) | Esterase genes (nuclear) | Zaidi *et al.* (1999) |
| Ladybird beetles (*Hippodamia convergens*) and lacewings (*Chrysoperla plorabunda*) | Cereal aphid (*Rhopalosiphum maidis*) | Cytochrome oxidase II (mitochondrial) | Chen *et al.* (2000) |
| Lady beetle (*Coleomegilla maculata*) | European corn borer (*Ostrinia nubilalis*) | ITS 1 (internal transcribed spacer 1 of nuclear rDNA) | Hoogendoorn and Heimpel (2001) |
| Mite (*Anystis baccarum*) | Apple-grass aphid (*Rhopalosiphum insertum*) | NADH dehydrogenase 1 (mitochondrial) | Cuthbertson, Fleming and Murchie (2003) |
| Ground beetle (*Pterostichus melanarius*) | Slugs (*Arion intermedius* and *Deroceras reticulatum*), grass snail (*Vallonia pulchella*), aphids (*Aphis fabae, Rhopalosiphum padi*), beetles (*Sitona* sp.) | Cytochrome oxidase I and 12S rRNA (mitochondrial) | Harper *et al.* (2005) |

In the previous example an introduced species was jeopardizing native species through predation, but the feeding behaviour of introduced species may also threaten native species through competition for the same food sources. Biologists were concerned that a social wasp (*Vespula germanica*) that was introduced into Australia in the 1970s may be outcompeting a native wasp (*Polistes humilis*). They therefore sequenced a portion of the 16S rDNA mitochondrial gene from both species' prey items, and used these sequences to determine which invertebrate orders were being preyed upon by each species. They found that the native wasp *P. humilis* fed almost exclusively on Lepidoptera, whereas the introduced wasp *V. germanica* fed on a wide range of invertebrates including Lepidoptera, Hemiptera, Diptera, Orthoptera and Odonata. These results show that *P. humilis* is a specialist whereas *V. germanica* is a generalist; furthermore, comparisons with earlier studies revealed variations in the diet of *V. germanica*, suggesting that it is an opportunist that feeds on whatever prey is readily available. Generalism and opportunism are two characteristics that help invasive species to spread rapidly across their new terrain and give them the upper hand when competing with native species for food and other resources (Kasper *et al.*, 2004).

A final example of the relevance of feeding behaviour to conservation biology comes from a study in which microsatellite markers were used to identify which waterfowl chicks the glaucous gull (*Larus hyperboreus*) was eating in Alaska. Of particular concern was the possibility that the gulls were contributing to the decline of emperor geese (*Chen canagica*) and spectacled eiders (*Somateria fischeri*) by eating their young. An examination of the gut contents revealed no evidence of predation on the spectacled eider, although 26 per cent of all gulls examined had eaten emperor geese (Scribner and Bowman, 1998). Other species eaten were the white-fronted goose (*Anser albifrons*) and the cackling Canada goose (*Branta canadensis minima*). Glaucous gulls seem to feed preferentially on goslings, although they do not appear to discriminate between the different species of geese because they feed on them in proportion to their availability (Bowman, Stehn and Scribner, 2004).

## In search of food

In addition to characterizing prey remains, data from molecular markers have sometimes been used to identify the feeding grounds of individuals from different populations. In one study, the spatial foraging patterns of two species of bumble bees (*Bombus terrestris* and *B. pascuorum*) were analysed using genetic data (Chapman, Wang and Bourke, 2003). These species are common European bumble bees and both live in colonies headed by a single queen who mates only once; therefore, workers from each colony are full-siblings with a relatedness value of 0.75. When foraging, *B. pascuorum* specializes on flowers with deep corollae that accommodate the long-tongued workers, whereas the short-tongued workers of *B. terrestris* will visit a wider range of flower types. Workers in both species leave scent marks on flowers that they have visited, presumably to label these as having little remaining pollen, although whether this is for the benefit of themselves or their nest-mates was unknown prior to this study. The authors looked at the foraging patterns for both bumble bee species over three spatial scales in London. At the smallest scale, the workers that visited the same patches of flowers showed no relatedness to one another and therefore must have originated from multiple colonies. Since there should be little advantage to leaving scent marks for unrelated individuals, the marks are more likely to be for the benefit of the individual that is leaving them.

At the intermediate scale, the bumble bee visitors to entire sites were sampled (sites being defined as discrete areas containing numerous flower patches, such as parks or cemeteries). Because all the workers from each colony are full-sisters, the number of full-sisterhoods identified at a site represents the number of colonies that are visiting that site. An analysis of sibships revealed that an estimated 96 *B. terrestris* and 66 *B. pascuorum* colonies visited each site. Finally, over a larger spatial scale, total foraging distances were estimated for the workers of each species. The median foraging distance was estimated to be 0.62–2.8 km for *B. terrestris* and 0.51–2.3 km for *B. pascuorum*, distances that exceed previous estimates based on mark–recapture or radar-tracking studies. These surprisingly long-distance foraging trips will have obvious implications for gene flow among populations of plants that are pollinated by bees.

## Migration

The search for food is one of the most important reasons for migration, as illustrated by wolves (*Canis lupus*) in northwest Canada, which are migratory in the areas where their main prey is caribou because of their need to follow the seasonal movements of the herds. A microsatellite analysis of 491 wolves from nine regions in northwest Canada revealed an overall pattern of isolation by distance, but there was an unusually high level of genetic differentiation between wolf packs from either side of the Mackenzie River. Because this river is frozen for 6–8 months out of every year, it should not present an insurmountable physical barrier. The lack of dispersal across the river can, however, be explained by the behaviour of the resident caribou, which undertake an annual migration either side of the river in a north–south direction. In this case the migration patterns of the prey (caribou) seem to be directly influencing the migration patterns of the predators (wolves), because both travel in a north–south direction and neither undertake east–west migrations across the river (Carmichael *et al.*, 2001).

Migration in search of food may also result in temporary 'populations' that consist of individuals that were born hundreds of kilometres apart. This is true of juvenile green turtles (*Chelonia mydas*), which congregate in foraging grounds that are far away from their nesting grounds. One aggregation frequently forms off the coast of Barbados, even though there are no suitable nesting sites nearby. Mitochondrial haplotypes identified from control region sequences were compared with genetic data previously obtained from nesting beaches to determine where these foraging turtles originated. The group turned out to be of mixed stock, with 25 per cent originating in Ascension Island, 23 per cent from Aves Island/ Surinam, 19 per cent from Costa Rica, 18.5 per cent from Florida and 10.3 per cent from Mexico (Luke *et al.*, 2004). A foraging group of immature loggerhead sea turtles (*Caretta caretta*) off Hutchinson Island, Florida, turned out to be less cosmopolitan. Sixty-nine per cent of the members of this group came from south Florida, 20 per cent from Mexico, and 10 per cent from north Florida–North Carolina (Witzell *et al.*, 2002). Many sea turtle populations are declining, and these data highlight the importance of international cooperation to the future management of these species.

Some of the longest distances that are travelled in the search for food and other resources are found in migratory birds. Dunlin (*Calidris alpina*) breed in the Arctic and spend their winters in Europe, Africa and Asia. A comparison of mtDNA haplotypes from breeding, migrating and overwintering populations revealed that migration was occurring in a general north–south direction, with dunlin that breed in the western Palaearctic overwintering in the western part of their range (Portugal, Morocco), and populations that breed further east overwintering in the Middle East (Wennerberg, 2001).

In the case of dunlins it was possible to match haplotypes between breeding and overwintering grounds because populations were genetically structured, but

**Figure 6.15**  A Monarch butterfly larva (*Danaus plexippus*). Monarch butterflies migrate in a stepping-stone manner, with females laying their eggs on milkweed plants as they travel north towards their summer territory. Photograph provided by Kelvin Conrad and reproduced with permission

this approach has been less successful in some other species. Monarch butterflies (*Danaus plexippus*; Figure 6.15), for example, are well known for their migration between breeding sites across Canada and the USA and overwintering sites in California (western populations) and Mexico (eastern populations). Although the eastern and western populations do not interbreed, individuals from the two regions show almost no variation in mitochondrial DNA sequence and therefore haplotypes cannot be used to genetically track the Monarch migration routes (Brower and Boyce, 1991). Similarly, North American migratory songbirds tend to show little structuring of mtDNA haplotypes across broad spatial scales. Researchers studying the yellow-breasted chat (*Icteria virens*), the common yellowthroat (*Geothlypis trichas*) and the Nashville warbler (*Vermivora ruficapilla*), all of which are long-distance migrants, were able to do little more than assign individuals to either eastern or western breeding lineages on the basis of mtDNA haplotypes, and as a result they gained little information about which breeding populations different migrants had originated from (Lovette, Clegg and Smith, 2004).

An alternative and more successful approach for tracking migratory individuals has been based on a combination of microsatellite data and stable isotopes in Wilson's warbler (*Wilsonia pusilla*; Clegg *et al.*, 2003). The microsatellite data provided some evidence of genetic structure across the North American range of this species, although differentiation among western populations was negligible. The hydrogen isotope ratios from feathers collected on the breeding grounds provided information about the latitude at which the feathers were grown in the previous year. A combination of microsatellite (east–west differentiation) and isotope (north–south differentiation) data allowed researchers to conclude that birds from western Canada and the USA overwinter in Central America, birds that breed near the west coast overwinter in western Mexico, and birds that breed further inland and at higher elevations spend their winters in eastern Mexico. Tracking migration routes with these types of data is becoming an increasingly important conservation tool, since populations of many neotropical migrants are declining and conservation tactics must take into account the possibility of habitat loss at either the breeding or the overwintering ground.

## Overview

This chapter has provided a summary of the numerous ways in which molecular data can be used to improve our understanding of behavioural ecology, although it is important to stress that genetic data must generally be used in conjunction with field studies and observational data before firm conclusions about an animal's behavioural repertoire can be reached. In this and earlier chapters we have made a number of references to conservation biology and in the next chapter we shall build on this by taking a more in-depth look at what is undoubtedly one of the most important applications of molecular ecology: the contribution of molecular techniques to conservation genetics.

## Chapter Summary

- Behavioural ecology is concerned with how an animal's response to a particular stimulus or situation is influenced by its ecology and evolution. Molecular data have improved our understanding of numerous aspects of behaviour, including mating systems, sex ratio allocation and foraging behaviour.

- The basic types of animal mating systems are monogamy, polygamy (polyandry, polygyny and polygynandry), and promiscuity. These can be differentiated on the basis of whether or not social bonds are formed and, if so, how many males and females are included.

- Molecular data can identify extra-pair fertilizations (EPFs) through parentage analysis. Potential biological parents can be excluded if they are genetically incompatible with the offspring. If too few adults can be excluded to allow the identification of parents, a maximum likelihood analysis of genotypes can be used to assign the most likely biological parent.

- Molecular evidence for EPFs has shown that both sexes, whether monogamous or polygamous, will often copulate with partners with which they share no social bond. One outcome of this is that social monogamy seldom translates into genetic monogamy.

- Parentage analysis has revealed a lot of information about pre- and post-copulatory mate choice, and has provided support for both the good genes hypothesis and the genetic compatibility hypothesis.

- Kin selection can be a plausible explanation for social breeding if helpers are related to recipients. The relatedness between individuals can be estimated on the basis of within-individual and within-population allele frequencies.

- Sex-linked molecular markers tell us that various taxonomic groups, including birds, frogs and social insects, are able to adaptively manipulate the sex ratio of their offspring. In social insects this can lead to conflict between queens and workers because haplodiploidy results in relatedness asymmetries.

- Sex-biased dispersal is extremely widespread. In mammals, males are more likely to disperse than females, whereas the opposite is true in birds. This may be done to minimize local competition or to avoid inbreeding. Sex-biased dispersal can be inferred from relatedness values, assignment tests, comparisons of $F_{ST}$ from both males and females, or comparisons of $F_{ST}$ from uniparentally versus biparentally inherited markers.

- Foraging behaviour can be studied by genotyping prey remains from guts or faeces. If combined with individual-specific genotypes, this can provide detailed information on individual feeding preferences. Identifying prey can have practical applications in biological control programmes or conservation biology.

- Molecular data can also provide us with information on foraging patterns, e.g. they may tell us how far individuals travel and what routes they follow in their search for food and other resources.

## Useful Websites and Software

- Relatedness – software program for estimating genetic relatedness using data from co-dominant genes: http://www.gsoftnet.us/GSoft.html. See also http://es.rice.edu/projects/Bios321/relatedness.calc.html for an example that shows how to calculate the relatedness of wasps from the same and from different colonies.

- Kinship – software program for estimating maximum likelihood ratios for testing hypotheses of relatedness: http://gsoft.smu.edu/GSoft.html

- Cervus-software that conducts large-scale parentage analysis using co-dominant loci: http://helios.bto.ed.ac.uk/evolgen/cervus/cervus.html

- PrDM calculator for detecting multiple parentage in a sample of offspring: http://www.zoo.utoronto.ca/tpitcher/prdmsoftware

## Further Reading

### Books

Caro, T. 1998. *Behavioral Ecology and Conservation Biology.* Oxford University Press, New York.
Griffiths, R. 2000. Sex identification using DNA markers. In *Molecular Methods in Ecology* (Baker, A.J., ed.), pp. 295–321. Blackwell Scientific, Cambridge.
Hardy, I.C.W. 2002. *Sex Ratios: Concepts and Research Methods.* Cambridge University Press, Cambridge.

Krebs, J.R. and Davies, N.B. 1997. *Behavioural Ecology: an Evolutionary Approach* (4th edn). Blackwell Publishing, Oxford.

## Review articles

Avise, J.C., Jones, A.G., Walker, D. and DeWoody, J.A. 2002. Genetic mating systems and reproductive natural histories of fishes: Lessons for ecology and evolution. Annual Review of Genetics **36**: 19–45.

Blouin, M. 2003. DNA-based methods for pedigree reconstruction and kinship analysis in natural populations. *Trends in Ecology and Evolution* **18**: 503–511.

Fitzpatrick, M.J., Ben-Shahar, Y., Smid, H.M., Vet, L.E.M., Robinson, G.E. and Sokolowski, M.B. 2005. Candidate genes for behavioural ecology. *Trends in Ecology and Evolution* **20**: 96–104.

Griffith, S.C., Owens, I.P.F. and Thurman, K.A. 2002. Extra pair paternity in birds: a review of interspecific variation and adaptive function. *Molecular Ecology* **11**: 2195–2212.

Jones, A.G. and Ardren, W.R. 2003. Methods of parentage analysis in natural populations. *Molecular Ecology* **12**: 2511–2523.

Neff, B.D. and Pitcher, T.E. 2005. Genetic quality and sexual selection: an integrated framework for good genes and compatible genes. *Molecular Ecology* **14**: 19–38.

Pike, T.W. and Petrie, M. 2003. Potential mechanisms of avian sex manipulation. *Biological Reviews* **78**: 553–574.

## Review Questions

**6.1.**    Figure 6.16 is a picture of a microsatellite gel showing the genotypes of a chick, her mother and six different males from the population, any of which could be the chick's biological father. Does exclusion analysis allow you to identify the father?

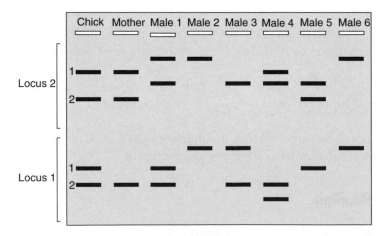

**Figure 6.16**

**6.2.**    Would you expect males and females to have a similar variation in reproductive success in a population that followed a polyandrous mating system? Why or why not?

**6.3.** Use the following microsatellite genotypes to calculate the relatedness between a female black-tailed prairie dog (focal individual) and the male that she has chosen to mate with:

|         | Female genotype | Male genotype |
|---------|-----------------|---------------|
| Locus 1 | 162, 162        | 162, 168      |
| Locus 2 | 124, 124        | 124, 126      |

The population allele frequencies are 0.45 for allele 162, 0.25 for allele 168, 0.50 for allele 124 and 0.15 for allele 126.

**6.4.** Would you expect conflict over sex allocation between queens and workers to be of a similar intensity in monogynous and polygynous colonies? Why or why not?

**6.5.** Do the following data suggest male-biased disperal, female-biased dispersal or neither?

(i) River otters: $F_{ST}$ calculated from nine microsatellites was 0.064 for males and 0.131 for females (Blundell *et al.*, 2002).

(ii) Eastern grey kangaroos: gene flow ($N_e m$) among populations, determined from the average genetic differentiation, was 22.61 for nuclear data and 2.73 for mitochondrial data (Zenger, Eldridge and Cooper, 2003).

(iii) The corrected assignment indexes ($AIc$) for red-billed queleas was −0.0121 for males and 0.0242 for females (Dallimer *et al.*, 2002).

(iv) The local relatedness values of the Lake Tanganyika cichlid *Eretmodus cyanostictus* ranged from −0.094 to 0.013 for males and from −0.0041 to −0.177 for females (Taylor *et al.*, 2003).

# 7

# Conservation Genetics

## The Need for Conservation

Biodiversity quite simply refers to all of the different life forms on our planet, and includes both species diversity and genetic diversity. There are many reasons why we value biodiversity, the most pragmatic being that ecosystems, which maintain life on our planet, cannot function without a variety of species. On a slightly less dramatic note, different species provide us with food (crops, livestock), fibres (wool, cotton), pharmaceuticals (25 % of medical prescriptions in the USA contain active ingredients from plants; Primack, 1998) and entertainment (countryside walks, ecotourism, zoos, gardening, fishing, birdwatching). From a less anthropocentric perspective, species may be considered worthwhile in their own right and not simply because they benefit humans, in which case there are important ethical considerations surrounding the predilection of one species, *Homo sapiens*, to drive numerous other species extinct.

We know from the fossil record that biodiversity has been increasing steadily over the past 600 million years, despite the fact that as many as 99 per cent of species that have ever lived are now extinct (Figure 7.1). Around 96 per cent of all extinctions have occurred at a fairly constant rate, creating what is known as the background extinction rate. This has been estimated from the fossil record as an average of 25 per cent of all living species becoming extinct every million years (Raup, 1994). The remaining 4 per cent or so of all extinctions occurred during five separate mass extinctions, which are identified from the fossil record as periods in which an estimated 75 per cent or more of all living species became extinct. The most recent, and also the most famous, mass extinction occurred in the late Cretaceous (65 million years ago) when approximately 85 per cent of all species, including the dinosaurs, were wiped out.

Many biologists predict that we are now entering a sixth mass extinction (Leakey and Lewin, 1995). Over the past 400 years or so, several hundred species are known

*Molecular Ecology*   Joanna Freeland
© 2005 John Wiley & Sons, Ltd.

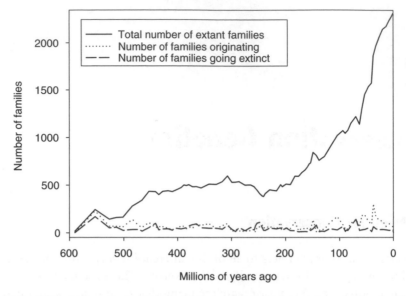

**Figure 7.1** Evidence from the fossil record tells us that the total number of living families has increased steadily over the past 600 million years. Numbers of originations and extinctions have fluctuated, but in most time intervals the former outnumbers the latter. Data from Benton (1993)

to have disappeared. Although this might sound like a lot, these recent extinctions actually represent a very small percentage of described taxa and therefore do not suggest anything close to a mass extinction (Table 7.1). Instead, it is the predicted rates of extinctions over the next century that are the main cause for concern. The best estimates of these are provided by the World Conservation Union (IUCN: International Union for Conservation of Nature and Natural Resources), which regularly compiles Red Lists on the numbers of species that are known to be at risk. Several categories are used (e.g. critically endangered, endangered, vulnerable) and these are based on a number of parameters, including current population size,

**Table 7.1** The numbers of species extinctions that have been recorded over the past 400 years (adapted from Primack, 1998). Note that the true numbers are undoubtedly higher than this because numerous undescribed species will also have gone extinct, e.g. a large number of plant and invertebrate species were probably wiped out during the destruction of tropical rainforests over the past few decades

| Taxonomic group | Number of extinctions | Percentage of taxonomic group |
| --- | --- | --- |
| Mammals | 85 | 2.1 |
| Birds | 113 | 1.3 |
| Reptiles | 21 | 0.3 |
| Amphibians | 2 | 0.05 |
| Fish | 23 | 0.1 |
| Invertebrates | 98 | 0.01 |
| Flowering plants | 384 | 0.2 |

number of mature adults, generation time, recent reductions or fluctuations in population size, and population fragmentation (see http://www.redlist.org/ for more details).

The Red List that was compiled by the IUCN in 2003 reported that 23 per cent of all mammal species and 12 per cent of all bird species are threatened. We know little about the total proportion of threatened species in other taxonomic groups simply because we lack the relevant information for most species. For example, 49 per cent of fishes that have been evaluated are classified as threatened, but because only around 5 per cent of all fish species have been adequately assessed, this value gives us limited insight into the status of fishes as a whole. Similarly, 72 per cent of evaluated insects have been placed in the threatened category, but <0.1 per cent of insect species have been investigated so far. Few data are available for most groups of plants, with the exception of conifers in which 93 per cent of species have been evaluated, and we know that 31 per cent of these are threatened. Clearly these data are far from complete, but if it turns out that similar proportions of *all* species in the various taxonomic groups are threatened then the fate of very many species hangs in the balance (Table 7.2). It is for this reason that many people believe that we are currently on the brink of a sixth mass extinction.

So why exactly are so many species threatened with extinction? In most cases, the answer to this is anthropogenic activity. Farming, logging, mining, damming and building have destroyed the habitats of countless species around the world.

**Table 7.2** Numbers and proportions of threatened species according to the IUCN 2003 Red List. Note that for most taxonomic groups only a very small proportion of species have been evaluated

| Taxonomic group | Number of described species | Number of evaluated species | Number of threatened species as % described | Number of threatened species as % evaluated |
|---|---|---|---|---|
| **Vertebrates** | | | | |
| Mammals | 4842 | 4789 | 23 | 24 |
| Birds | 9932 | 9932 | 12 | 12 |
| Reptiles | 8134 | 473 | 4 | 62 |
| Amphibians | 5578 | 401 | 3 | 39 |
| Fishes | 28 100 | 1532 | 3 | 49 |
| **Invertebrates** | | | | |
| Insects | 95 0000 | 768 | 0.06 | 72 |
| Molluscs | 70 000 | 2098 | 1 | 46 |
| Crustaceans | 40 000 | 461 | 1 | 89 |
| Others | 13 0200 | 55 | 0.02 | 55 |
| **Plants** | | | | |
| Mosses | 15000 | 93 | 0.5 | 86 |
| Ferns | 13025 | 180 | 1 | 62 |
| Gymnosperms | 980 | 907 | 31 | 34 |
| Dicotyledons | 199 350 | 7734 | 3 | 75 |
| Monocotyledons | 59 300 | 792 | 1 | 65 |

Many endemic species have suffered from human-mediated introductions of alien species, both deliberate and accidental. Hunting, fishing and trading have led to the overexploitation of many species, whereas countless others have suffered from industrial or agricultural pollution. Although these processes are diverse, a common outcome is a reduction in the sizes of wild populations. When this occurs, species begin to suffer from reduced genetic diversity and inbreeding, and this is where conservation genetics comes into play. In this chapter we will look at some of the most important aspects of conservation genetics by first examining how genetic data can be used to identify distinct species and populations as potential targets of conservation. In subsequent sections we shall build on some of the theory that was presented in earlier chapters by re-visiting genetic diversity, inbreeding, population sizes and relatedness, but this time paying particular attention to how they can be applied to some of the issues surrounding conservation biology.

## Taxonomy

Taxonomy is the science that enables us to quantify biodiversity, although its applications extend much further than this because without it our understanding of ecology and evolution would be greatly reduced. Taxonomy has therefore remained an important area of biological research since Linnaeus developed his extensive classification system in the 18th century. Over the years, organisms have been classified on the basis of a wide range of morphological, behavioural and genetic characters. In this section we will limit ourselves to a discussion on the importance of taxonomy to conservation biology, paying particularly attention to the contributions that have come from molecular data.

### Species concepts

Conservation strategies are often directed at individual species or at habitats that have been identified as species-rich and they therefore tend to assume that most individuals have been assigned correctly to a particular species. But is this necessarily the case? Although generally supportive of conservation initiatives, most biologists would argue that the identity of species is far from straightforward. Historically, researchers have often relied on the **biological species concept** (BSC), which defines species as '. . . groups of actually or potentially interbreeding natural populations, which are reproductively isolated from other such groups' (Mayr, 1942). Although conceptually straightforward, the BSC does have several short-coming, for example a literal interpretation does not allow for hybridization and few can agree on how this dilemma should be solved. In addition, the BSC cannot accommodate species that reproduce asexually or by self-fertilization.

More than 20 different species concepts can be found in the literature (Hey *et al.*, 2003). One alternative to the BSC that has been gaining support in recent years is the **phylogenetic species concept** (PSC). This defines species as groups of individuals that share at least one uniquely derived characteristic, and is often interpreted to mean that a species is the smallest identifiable monophyletic group of organisms within which there is a parental pattern of ancestry and descent (Cracraft, 1983). The PSC circumvents to some extent the problem of asexual reproduction, but it has been criticized for dividing organisms on the basis of characteristics that may have little biological relevance, and also for creating an overwhelmingly large number of species. Furthermore, two groups that are identified as separate species under the PSC may retain the potential to reproduce with one another. If reproduction between these two groups did occur, they would no longer be monophyletic and therefore would have to be reclassified as a single species.

The PSC tends to identify a greater number of species than the BSC. One review of 89 studies concluded that the PSC identified 48.7 percent more species than the BSC (Agapow *et al.*, 2004; see Figure 7.2). If the increasingly popular PSC replaces the BSC as the most widely accepted species concept, the number of endangered species will increase and the geographical range of many will decrease. This in turn would lead to a wide-scale re-evaluation of numerous conservation programmes, for example the location of high-profile biological hotspots, in which large numbers of endemic species can be found, may change depending on which concept is used to determine the number of species in a given region (Peterson and Navarro-Siguenza, 1999). Many biologists therefore advocate a less dramatic

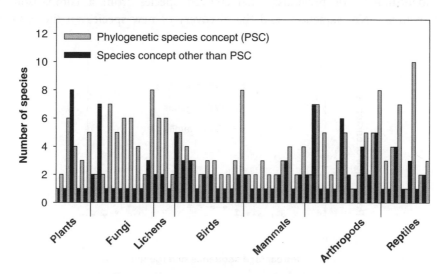

**Figure 7.2**  Some examples of how the number of species in different taxonomic groups varies, depending on whether or not the phylogenetic species concept (PSC) is used for classification purposes. Adapted from Agapow *et al.* (2004) and references therein

approach in which multiple species concepts are retained, provided that it is clear which concept is being employed at any given time; some situations will lend themselves to the BSC, others to the PSC, whereas others (e.g. those involving many unicellular or parasitic taxa) may lend themselves to another approach altogether (de Meeus, Durand and Renaud, 2003). This tactic has the advantage of being well balanced but suffers from the uncertainties that surround variable taxonomic criteria.

## Genetic barcodes

A more recently established approach to taxonomy seeks to identify species solely on the basis of a **genetic barcode** (also known as a DNA barcode) consisting of one or a few DNA sequences. For example, a 648 bp region of the mitchondrial cyctochrome *c* oxidase I gene (COI) is currently being developed as a barcode identifier in animals. In a comparison of 260 bird species, this gene region was found to be species-specific, and was also an average of 18 times more variable *between* species (7.05–7.93 per cent) than *within* species (0.27–0.43 per cent) (Figure 7.3; Hebert *et al.*, 2004b). This is one of the findings that led to an international collaboration known as the Consortium for the Barcoding of Life that is currently hosted by the Smithsonian's National Museum of Natural History in Washington, DC, and is promoting the eventual acquisition of genetic barcodes for all living species.

The use of genetic barcodes to identify species has two general applications: the identification of previously characterized species from a comparison of documented DNA sequences, and the discovery of new species on the basis of

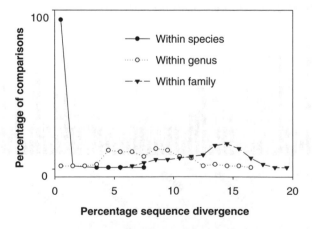

**Figure 7.3** The extent to which the mitochondrial cytochrome oxidase I gene varies among 260 species of North American birds. Comparisons are based on levels of sequence divergence within and among species, genera and families. Data from Hebert *et al.* (2004b)

novel DNA sequences. The former application is not particularly controversial and, as we saw in the previous chapter, the practice of identifying species or samples by matching up sequences is becoming increasingly widespread. Nevertheless, this approach does assume that sequences are species-specific and we know from Chapter 5 that both hybridization and incomplete lineage sorting mean that this will not always be the case. Because hybridization occurs between species within all major taxonomic groups, and an estimated one-quarter of all animal species have yet to reach the stage of reciprocal monophyly (Funk and Omland, 2003), DNA sequences sometimes will transcend the boundaries of putative species.

The second application of DNA barcodes, which is the identification of new species, is more controversial. This is partly because the range of intraspecific sequence divergence can be difficult to predict. Although Hebert *et al.* (2004b) found that avian intraspecific divergence was consistently <0.44 per cent and therefore lower than interspecific divergence, a study by Johnson and Cicero (2004) found that interspecific sequence divergences were 0–8.2 per cent in 39 comparisons of avian sister species. Inconsistencies such as these may be the exception rather than the rule, although data from a wider range of taxonomic groups are needed before we can reach this conclusion.

Before such data can be acquired, appropriate genetic regions first must be identified in these other taxonomic groups. Microbes, for example, transfer genes between putative species so often that sequence data from an estimated 6–9 genes will be required before closely allied species can be differentiated (Unwin and Maiden, 2003). In plants, hybridization and polyploidy can obscure evolutionary relationships, although proponents of genetic barcodes hope that a region of the chloroplast genome can be found that will reliably distinguish species. They also suggest that COI will be useful for identifying a number of protistan species, although anaerobic species lack mitochondria and therefore will require a different marker. In the meantime, DNA barcodes are becoming an increasingly acceptable tool for identifying species and may well become more widespread in the literature over the next few years (see also Box 7.1).

---

**Box 7.1  Defining species from molecular and ecological data**

Some of the potential problems associated with molecular taxonomy, such as incomplete lineage sorting or low sequence divergence between closely related species, sometimes can be overcome if molecular data are combined with ecological studies. The value of this combined approach was illustrated by a recent taxonomic re-evaluation of the neotropical skipper butterfly *Astraptes fulgerator*. For many years this was described as a single, variable, wide-ranging species that occurred in a variety of habitats distributed between the far southern USA and northern Argentina. The ecology of this species has been studied intensively throughout a long-term project in which the colour patterns and feeding preferences of >2500 wild-caught caterpillars were monitored. Once these had developed into adults, researchers recorded the sizes of the butterflies and their wing shapes, colours and patterns. Overall morphological similarity is high throughout the range because of recent shared ancestry, and also because selection has maintained mimicry of warning colouration against predators. Nevertheless, although morphological differences were subtle, the ecological data suggested that *A. fulgerator* was in fact a complex of at least six or seven species (Hebert *et al.*, 2004a, and references therein).

As a recent addition to the barcoding project, cytochrome oxidase I sequences were obtained from 465 *A. fulgerator* individuals. Morphological characters of caterpillars and adults, plus the identity of their food plants, were superimposed onto a neighbour-joining tree that was reconstructed from the COI sequence data. One group was paraphyletic and pseudogenes (nuclear copies of mitochondrial genes; Chapter 2) were amplified from several individuals, but for the most part the combined genetic and ecological data revealed ten distinct clusters suggesting that *A. fulgerator* is in fact a complex of at least ten distinct species. The sequence divergence between these ten species ranged from 0.32 to 6.58 per cent (Hebert *et al.*, 2004a).

Species are unlikely to be distinguished solely on the basis of sequence divergences as low as 0.32 %, which is why a combination of molecular and ecological data was necessary in this case before realistic species designations could be made. Although the initial investigations were lengthy, the authors suggest that future studies on *Astraptes* spp. can use the COI barcode as the sole identification tool, thereby bypassing the need for the relatively time-consuming acquisition of ecological and morphological data. In an ideal world, all species would be characterized on the basis of such comprehensive phenotypic and genotypic data, although in many cases this option will be logistically impossible.

## Subspecies

Possibly even more confusing than the species concept is the demarcation of subspecies. Although advocated by Linneaus, the classification of subspecies was seldom used until the mid-20th century. The adoption of subspecies around this time was particularly widespread in birds. Reclassification was usually based on morphological characteristics, and as a result the current classification of bird

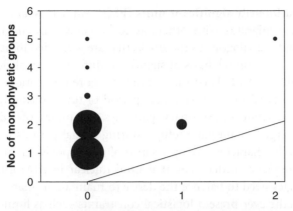

**Figure 7.4** Number of monophyletic mitochondrial lineages per species compared with the number of these lineages that currently match subspecies classifications. The size of each circle is proportional to the number of comparisons in each category. The diagonal line indicates where the circles would be located if the monophyletic mitochondrial lineages in each species were in complete agreement with designated subspecies. Because all circles are above this diagonal line, all species contain monophyletic groups that are not classified as subspecies. Adapted from Zink (2004) and references therein

subspecies does not agree with the distribution of monophyletic mitochondrial lineages. A review of the literature has shown that bird species contain on average around two monophyletic mtDNA lineages, but are subdivided into an average of 5.5 subspecies (Zink, 2004; Figure 7.4). The cactus wren (*Campylorhynchus brunneicapillus*), for example, has only two evolutionarily distinct mitochondrial lineages but has six named subspecies.

Discrepancies such as these may mean that conservation efforts are directed at genetically indistinct subspecies while distinct lineages receive less attention, and this has led Zink (2004) to call for the reclassification of subspecies. This is a somewhat controversial demand because there are a number of reasons why the morphology and genetics of recently diverged species may not agree, one of these being incomplete lineage sorting. Furthermore, as we learned in Chapter 4, quantitative trait variation may exceed the genetic differences that are revealed by neutral molecular markers. Subspecific status should therefore be revoked with caution because morphological differences, however slight, may reflect local adaptation even if neutral molecular markers show no differentiation.

## Conservation units

In an attempt to circumvent some of the problems that may be associated with taxonomy, conservation biologists sometimes concentrate on **management units**

(MU) and **evolutionarily significant units** (ESU). An MU is 'any population that exchanges so few migrants with others as to be genetically distinct from them' (Avise, 2000), and is analogous to the stocks that are identified in fisheries. Distinct MUs can be identified on the basis of significant differences in allele frequencies at multiple neutral loci. An ESU consists of one or more populations that have been reproductively isolated for a considerable period of time, during which they have been following separate evolutionary pathways. Examples of this may include lineages that diverged in alternate refugia during glacial periods (Chapter 5). The ESUs are typically characterized by reciprocal monophyly in mtDNA and significant allele frequency differences at neutral nuclear loci (Moritz, 1994). Conservation strategies need to balance the desire to maintain as many MUs and ESUs as possible with the ever-present logistical constraints such as limited finances and a shortage of suitable habitat.

The preservation of distinct MUs and ESUs is generally seen as desirable because each unit contributes to a species' genetic diversity. Conservation of hybrids, on the other hand, is a much more controversial issue. The US Endangered Species Act (ESA), for example, originally proposed that hybrids would not be protected. This clause has since been revoked, although a proposed replacement policy on 'intercrosses' (avoiding the sometimes pejorative term 'hybrids') has yet to be officially integrated into the ESA. This lack of resolution is partly attributable to the different categories of hybrids (Allendorf *et al.*, 2001). On the one hand, narrow hybrid zones that have been stable for many years are often adaptive (Chapter 5) and therefore may be considered ESUs. On the other hand, invasive species may threaten the genetic integrity of endemic species through hybridization, in which case the desirability of these hybrids becomes a matter for debate. In New Zealand, introduced mallard ducks (*Anas platyrhynchos*) have hybridized extensively with the native grey duck (*Anas superciliosa superciliosa*), and as a result there may no longer be any 'pure' grey duck populations remaining (Rhymer, Williams and Braun, 1994). In cases such as this, one option may be to eliminate populations of the invasive species and its hybrids; if this is unrealistic, the protection of hybrids may be the only way to preserve any of the threatened species' alleles.

Despite a number of unanswered questions regarding taxonomy and conservation, it is fair to say that molecular data provide us with an important window into the evolutionary history and genetic differentiation of species, and this may help us to make informed decisions about which populations constitute a conservation priority. There are some cases in which species boundaries have been altered solely on the basis of molecular data, for example in morphologically simple taxa such as the Cyanidiales, a group of asexual unicellular red algae (Ciniglia *et al.*, 2004), or in numerous other marine species for which ecological data are difficult to acquire. Substantial sequence differences between the ITS region of ribosomal DNA in Australian and South African populations of the marine green alga *Caulerpa filiformis*, for example, suggest that these are in fact two cryptic species (Pillmann *et al.*, 1997), whereas the negligible divergence between ribosomal genes of the seaweeds *Enteromorpha muscoides* and *E. clathrata* suggests that these should in fact be merged into a single species (Blomster, Maggs and Stanhope, 1999).

At other times the contributions of molecular data to taxonomy seem to have raised as many questions as they have answered. For example, all individuals are genetically unique but just how much genetic dissimilarity should we tolerate within a single MU? How much genetic divergence is required before ESUs are designated distinct species? How can molecular taxonomy accommodate incomplete lineage sorting and hybridization? Our inability to answer these questions to everyone's satisfaction does not mean that identifying the most appropriate units for conservation is an impossible task, although we need to remain aware of the limitations and assumptions that surround many taxonomic decisions. For the rest of this chapter we will, for the most part, be talking about species and populations as unambiguous entities, but we must keep in mind the possibility that species and population boundaries will be redrawn some time in the future.

## Population Size, Genetic Diversity and Inbreeding

Endangered species have, by definition, small or declining population sizes and are therefore sensitive to environmental perturbations simply because small populations lack a 'buffer' that helps them to survive periods of high mortality, for example following a disease outbreak or a temporary reduction in food supplies. Of equal or greater relevance to the long-term survival of small populations are their levels of genetic diversity. We know from Chapters 3 and 4 that the amount of genetic diversity within a population depends on the balance between mutation, gene flow, drift and selection (summarized in Figure 3.9). Although the effects of natural selection are variable, genetic diversity will be eroded by genetic drift and, in most cases, enhanced by gene flow. Because genetic drift acts more rapidly in small populations, we would expect overall genetic diversity to be roughly proportional to the size of a population, and this indeed appears to be the case. A review published by Frankham (1996) examined the relationship between population size and genetic diversity in 23 studies of plants and animals. Twenty-two of these species revealed a significant positive relationship between population size and genetic diversity when the latter was measured as $H_e$, $H_o$, allelic diversity ($A$) or proportion of polymorphic loci ($P$). Because endangered species typically have smaller population sizes than non-endangered species, they should also have relatively low levels of genetic diversity, and this too is generally the case (Table 7.3).

So what exactly are the dangers associated with genetically depauperate populations? For one thing, reduced levels of genetic diversity mean that populations may be unable to adapt to a changing environment. In addition, the fate of alleles in small populations is more likely to be determined by genetic drift than by

**Table 7.3** Mean heterozygosity values in endangered avian populations, calculated from allozyme data. Adapted from Haig and Avise (1996) and references therein

| Species | Number of populations surveyed | Mean heterozygosity |
|---|---|---|
| Wood stork (*Mycteria Americana*) | 15 | 0.093 |
| Trumpeter swan (*Cygnus buccinator*) | 3 | 0.010 |
| Hawaiian duck (*Anas wyvilliana*) | 2 | 0.035 |
| Laysan duck (*Anas laysanensis*) | 1 | 0.014 |
| Blue duck (*Hymenolaimus malacorhynchos*) | 5 | 0.002 |
| Lesser prairie chicken (*Tympanuchus pallidicinctus*) | 1 | 0.000 |
| Guam rail (*Rallus owstoni*) | 1 | 0.030 |
| Piping plover (*Charadrius melodus*) | 5 | 0.016 |
| Micronesian kingfisher (*Halcyon cinnamomina*) | 1 | 0.000 |
| Red-cockaded woodpecker (*Picoides borealis*) | 26 | 0.078 |

selection. All populations harbour deleterious alleles at low frequencies, and if drift is a stronger force than selection then these deleterious alleles are much more likely to reach fixation.

The accumulation of harmful mutations contributes to a population's **genetic load**, which is defined as the reduction in a population's mean fitness compared with the mean fitness that would be found in a theoretical population that has not accumulated deleterious alleles. Genetic load can be measured as the number of lethal equivalents, which is the number of deleterious genes whose cumulative effect is the equivalent of one lethal gene. Because a relatively high proportion of deleterious alleles will become fixed in a small population, genetic load tends to be inversely proportional to population size. If populations are small enough then the accumulation of lethal equivalents can lead to a substantial reduction in reproductive fitness, at which point the population will experience **mutational meltdown**, which means that it will continue to decline until it goes extinct. It is not clear just how often mutational meltdown occurs in the wild, although it will be accelerated by inbreeding, which poses the biggest short-term threat to small populations.

## Inbreeding depression

Inbreeding is more likely to occur in small populations simply because there is a greater chance that an individual will mate with a relative. In a diploid species, inbreeding increases the likelihood that an individual will have two alleles that are identical by descent at any given locus, and it therefore has the effect of increasing homozygosity at all loci. For this reason, the inbreeding coefficient $F$ is based on heterozygosity deficits (Equation 3.15). This relationship between inbreeding and heterozygosity also means that the rate at which heterozygosity is lost from a population following drift $(1/(2N_e)$; Chapter 3) is equal to the rate at which inbreeding accumulates, and this can be expressed as:

$$\Delta F = 1/(2N_e) \tag{7.1}$$

where $\Delta F$ equals the increment in inbreeding that will occur from one generation to the next (see also Box 7.2). In the absence of immigrants, inbreeding will therefore accumulate at a rate that is inversely proportional to population size (Figure 7.5).

---

**Box 7.2  Inbreeding and genetic diversity**

In outcrossing species, a small $N_e$, low genetic diversity and high inbreeding go hand in hand. Discussions may focus explicitly on only one or two of these topics, but it is important to remember that, over time, populations with small effective sizes will simultaneously experience an increase in inbreeding and a decrease in genetic diversity. This relationship can be shown by the following equation:

$$H_t/H_0 = [1 - 1/(2N_e)]^t = 1 - F \tag{7.2}$$

where $H_t$ and $H_0$ represent heterozygosity at generation $t$ and generation 0, respectively, and $F$ is the inbreeding coefficient (Frankham, Ballou and Briscoe, 2002). We were introduced to the first part of this equation in Chapter 3 (Equation 3.14) as a way to estimate the rate at which heterozygosity will be lost from a population. By expanding this equation to include the inbreeding coefficient, we can see how drift, which is strongly influenced by population size, will simultaneously reduce genetic diversity and promote inbreeding.

---

Inbreeding threatens the survival of small populations when it leads to a reduction in fitness, a phenomenon that is known as **inbreeding depression**. There are two ways in which this can occur. The first of these is known as **dominance**, so-called because the favourable alleles at a locus are usually dominant and the deleterious alleles have been maintained within the populations because they are recessive. The increased homozygosity that results from inbreeding means that deleterious alleles are more likely to occur as homozygotes; when this happens their effects cannot be masked by the dominant favourable allele, which results in inbreeding depression. The second phenomenon that can lead to inbreeding depression is known as overdominance, or heterozygote advantage, which means that individuals that are heterozygous at a particular locus have a higher

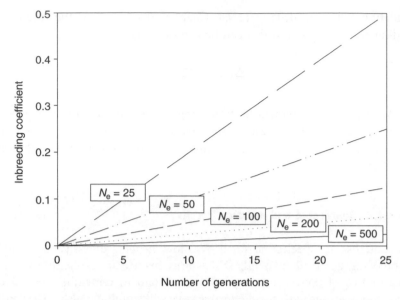

**Figure 7.5** The increase over time in the inbreeding coefficients (*F*) of five populations of different sizes, all of which were outbred completely at time zero (*F* = 0) and all of which are closed to immigrants. The rate at which inbreeding levels increase within a population is inversely proportional to its effective size

fitness than individuals that are homozygous for either allele. In Chapter 3 we were introduced to sickle cell anaemia, a classic example of overdominance in which heterozygotes benefit from a high resistance to malaria.

Inbreeding depression can be quantified using the following equation:

$$\delta = 1 - (X_I/X_O) \tag{7.3}$$

where $X_I$ is the fitness value of a particular trait in inbred offspring and $X_O$ is the fitness value of that same trait in outbred offspring. We can work through this equation by looking at the trait of survival in golden lion tamarins (*Leontopithecus rosalia*). In one study, the average survival of outbred offspring was found to be 0.829, whereas the average survival of inbred offspring was 0.474 (Dietz and Baker, 1993). The level of inbreeding depression revealed by this trait is therefore $\delta = 1 - (0.474/\ 0.829) = 1 - 0.572 = 0.428$. Some other examples of inbreeding depression are shown in Table 7.4.

As more and more studies of inbreeding accumulate in the literature, it is becoming apparent that inbreeding depression is actually far more widespread than was previously believed. The recent proliferation of inbreeding studies is partially attributable to the increasing accessibility of molecular data. In the past, inbreeding depression was typically inferred from lengthy investigations that sought to compare the overall fitness of inbred versus outbred individuals. Studies such as these have proved invaluable to our understanding of inbreeding in wild

**Table 7.4** Some examples of inbreeding depression in a variety of taxonomic groups. Adapted from Crnokrak and Roff (1999) and references therein

| Species | Trait | $X_O$ | $X_I$ | $\delta$ |
|---|---|---|---|---|
| Cooper's hawk (*Accipiter cooperii*) | Clutch size | 4 | 3.7 | 0.075 |
| Mexican jay (*Aphelocoma ultramarina*) | Nestling survival | 0.33 | 0.086 | 0.739 |
| Lion (*Panthera leo*) | Sperm mobility | 91 | 61 | 0.330 |
| Anubis baboon (*Papio anubis*) | % Offspring viability | 84.2 | 50 | 0.406 |
| Tree snail (*Arianta arbustorum*) | Number of clutches | 17 | 13.6 | 0.200 |
| Common adder (*Vipera berus*) | Brood size | 10 | 7 | 0.300 |
| Yellow trout lily (*Erythronium americanum*) | Seed production | 41.2 | 10.5 | 0.745 |
| Blue gilia (*Gilia achilleifolia*) | % Seedling establishment | 100 | 69 | 0.310 |

populations, but they have several drawbacks. In the first place, a lack of pedigree information, combined with the high frequency of extra-pair fertilizations in many wild populations, can make it difficult to determine whether individuals are outbred or inbred. Futhermore, not all species lend themselves to the intensity of study that is needed before data such as those presented in Table 7.4 can be accumulated; particularly problematic is the fact that field studies are typically limited to a short period of time – often a single breeding season – and the data collected during this period may not accurately reflect an individual's life-time fitness. Longer-term studies can sometimes be conducted under laboratory or captive conditions, but inbreeding depression may be greatly reduced in captive populations; according to one review, inbreeding depression is on average 6.9 times higher for mammals in the wild compared with mammals that are kept in captivity (Crnokrak and Roff, 1999). Some of these problems may be circumvented by a shortcut that uses molecular data and individual fitness components to look for heterozygosity fitness correlations.

### Heterozygosity fitness correlations

Heterozygosity fitness correlations (HFCs) are based on two principles: first, multilocus heterozygosity values can be used as a measure of inbreeding; and second, inbreeding depression leads to a reduction in fitness. If we combine these two principles, we reach the conclusion that inbreeding depression should be characterized by a correlation between low heterozygosity and reduced fitness. This is most commonly tested for by comparing observed heterozygosity values with one or more individual fitness components such as rate or percentage of seed germination, growth rate, time to reproduction, the number of flowers, fruits or seeds, sperm quality or volume, and longevity.

Although caution should be used when interpreting results that are based on a limited number of loci (Pemberton, 2004), a number of studies have suggested that

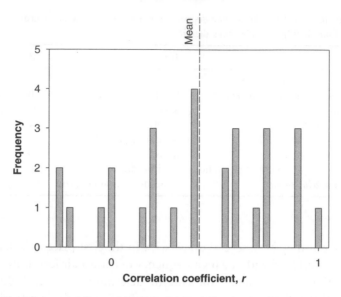

**Figure 7.6** Distribution of the correlation coefficients (*r*) from a meta-analysis of heterozygosity fitness correlations. The mean correlation between heterozygosity (or equivalent measure of genetic diversity) and fitness was 0.432. Adapted from Reed and Frankham (2003) and references therein

a correlation between heterozygosity and fitness is indeed a useful measure of inbreeding depression. In one study, a meta-analysis was conducted on 34 data sets, each based on a minimum of three populations. Each data set was examined to see if there was a correlation between the genetic diversity of a population and its overall fitness. Twenty-eight of the data sets showed a positive correlation between fitness and genetic diversity. The median correlation was 0.440 (Figure 7.6), which the author felt was substantial considering that the meta-analysis incorporated a diversity of species and a variety of methods for estimating both fitness and genetic diversity (Reed and Frankham, 2003). Some examples of inbreeding depression inferred from HFCs are given in Table 7.5.

A particularly detailed study of heterozygosity and fitness was conducted on a wild population of red deer (*Cervus elaphus*) on the Isle of Rum (Figure 7.7). This population has been monitored intensively since 1971, and these long-term records provide reliable measures of fitness based on the number of calves that each individual produced over its lifetime. Heterozygosity values were calculated using between six and nine microsatellite markers, and an analysis of the data showed that lifetime breeding success was positively correlated with heterozygosity in both males and females (Slate *et al.*, 2000). Another in-depth study was conducted on a large metapopulation of the Glanville fritillary butterfly (*Melitaea cinxia*) on the Åland islands in southwest Finland. This metapopulation consists of many small populations that breed in about 1600 meadows of different sizes and varying distances from one another. There is an average of 200 extinctions and 114

**Table 7.5** Species in which a positive correlation has been found between heterozygosity and fitness (HFCs). Inbreeding depression is inferred from a combination of reduced fitness and low heterozygosity values

| Species | Characteristic | Reference |
|---|---|---|
| Soay sheep (*Ovis aries*) | Higher parasite-mediated mortality in individuals with low heterozygosity | Coltman *et al.* (1999) |
| Pocket gopher (*Thomomys bottae*) | Metabolic cost of burrowing is higher in individuals with low heterozygosity | Hildner and Soulé (2004) |
| Butterfly blue (*Scabiosa columbaria*), a perennial plant | Populations were less able to compete with *Bromus* grass when heterozygosity was low | Pluess and Stocklin (2004) |
| Common mussel (*Mytilus edulis*) | Improved immune response in highly heterozygous individuals | Carissan-Lloyd, Pipe and Beaumont (2004) |
| American chestnut (*Castanea dentata*) | Higher growth rates in highly heterozygous individuals | Stilwell *et al.* (2003) |
| Dainty damselfly (*Coenagrion scitulum*) | Heterozygosity positively correlated with body size and mating success | Carchini *et al.* (2001) |
| Common toad (*Bufo bufo*) | Inverse correlation between heterozygosity and number of observed physical abnormalities in developing tadpoles | Hitchings and Beebee (1998) |

**Figure 7.7** Male red deer (*Cervus elaphus*) on the Isle of Rum. Lifetime breeding success in this population is positively correlated with heterozogosity. Photograph provided by Jon Slate and reproduced with permission

re-colonizations each year. In this study, females were sampled from 42 populations of varying size and isolation, and characterized at eight loci (seven allozymes
and one microsatellite). Heterozygosity was lower, and hence inbreeding was
higher, in small populations, a result that was not surprising because some of the
smallest populations consist entirely of full-siblings. In addition, these small
populations showed evidence of inbreeding depression in the form of reduced
egg hatching success, larval survival and female longevity, plus a longer pupal
period that increases the likelihood of pupae being parasitized. When 7/42
populations went extinct during the second year of the study, it became apparent
that the risk of a population becoming extinct was inversely proportional to its
overall heterozygosity and hence directly proportional to its level of inbreeding.
This study was the first to demonstrate that inbreeding in the wild can cause
populations to become extinct (Saccheri *et al.*, 1998).

### Purging

Inbreeding does not automatically lead to inbreeding depression. One way in
which inbreeding depression can be avoided is through the **purging** of deleterious
alleles. As we know, inbreeding increases the homozygosity of recessive deleterious
alleles, which means that the associated deleterious traits are more likely to be
expressed. These traits then will be selected against, which can lead to elimination
(purging) of the deleterious alleles from the population. This process is particularly effective against alleles that are lethal in the homozygous state. However,
purging is unlikely to have much effect on alleles that are only mildly deleterious,
which instead may become fixed within a small population following genetic drift.
Furthermore, purging cannot ameliorate the effects of inbreeding that are
associated with overdominance, and can be counteracted by the introduction of
novel deleterious alleles following mutation. The effectiveness of purging therefore
remains a subject of debate, in part because purging is more often inferred than
unequivocally demonstrated. In the absence of empirical evidence, it is often
provided as a default explanation for the survival of species that have been through
extremely small bottlenecks, such as Père David's deer (*Elaphurus davidianus*).
This deer is undoubtedly one of the most inbred mammals in the world because
the global population was reduced to 13 individuals in the 19th century. Nevertheless, the deer are now thriving and show little evidence of inbreeding depression, possibly because purging eliminated lethal recessives during at least one of
the bottlenecks.

More precise evidence for purging was sought in a review of 28 experimental
studies of mammals, insects, molluscs and plants in which inbreeding depression
had been estimated over multiple generations of experimentally inbred strains.
Inbreeding depression initially increased over several generations, but after a while
the situation started to reverse in a number of species when fitness levels

rebounded. The most likely explanation for this reduction in inbreeding depression is the purging of deleterious alleles, although it is also possible that the increase in fitness reflected adaptation to laboratory conditions (Crnokrak and Barrett, 2002). In insects, there is some evidence that purging is responsible for the relatively low levels of inbreeding that have been found in haplodiploid compared with diploid species, because purging of genetic load can be accomplished relatively easily through the production of haploid males whose deleterious recessive alleles cannot be masked (Henter, 2003).

Because purging may be an effective way to reduce inbreeding depression, it has been suggested in the past that deliberate inbreeding can be used as a conservation management tool to rid small populations of deleterious alleles. However, many biologists believe that the risks associated with this outweigh the potential benefits, since it is extremely difficult to predict the efficacy of purging. This unpredictability has been illustrated by a number of studies, including an experiment in which inbreeding depression was monitored over ten generations of inbreeding in three subspecies of wild mice: *Peromyscus polionotus subgriseus, P. p. rhoadsi* and *P. p. leucocephalus*. Although comparable breeding programmes were set up for all three subspecies, the results were inconsistent. Over time, *P. p. rhoadsi* showed a reduction in inbreeding depression, *P. p. leucocephalus* showed an increase in inbreeding depression, and *P. p. subgriseus* showed no change. These differences may depend at least partially on whether or not populations had experienced previous bouts of inbreeding and purging in the wild, although this is unlikely to be the only relevant factor (Lacy and Ballou, 1998). Results such as these mean that many conservation biologists view deliberate attempts at purging to be a fairly desperate strategy for reducing inbreeding depression.

### Self-fertilization

So far we have been looking at inbreeding depression in species that reproduce solely by outcrossing. We will now turn our attention to **self-fertilization** (or selfing), which involves the fusion of gametes that have been produced by the same individual and is therefore the most extreme form of inbreeding. Around 40 % of all flowering plant species are capable of self-fertilization. We might expect selfing plants to exhibit high levels of inbreeding depression, but in fact they are often less prone to inbreeding depression than outcrossing species. This may be because they are more adept at purging deleterious alleles, although, as with obligately outcrossing species, purging seems to be more effective in some populations than in others.

In the eelgrass (*Zostera marina*), for example, selfing plants produce seeds more frequently and in larger numbers than outcrossing plants (Rhode and Duffy, 2004). In the wild daffodil *Narcissus longispathus*, on the other hand, inbreeding

**Figure 7.8** *Narcissus longispathus* (Amaryllidaceae), a rare self-compatible trumpet daffodil restricted to a few mountain ranges in southeastern Spain. Photograph by Spencer C.H. Barrett and reproduced with permission

depression can be pronounced (Figure 7.8). This is a herb that is endemic to a few mountain ranges in southeastern Spain and can reproduce by either self-fertilization or outcrossing. In one study, heterozygosity was found to be much higher in parental plants than in seedlings, a discrepancy that is taken as evidence for strong selection against inbred offspring (Barrett, Cole and Herrera, 2004). This is therefore an example of self-fertilization leading to inbreeding depression in the form of high seedling mortality. Despite these obvious drawbacks, the authors of this study suggest that self-fertilization is maintained in this species because it allows prolific reproduction during the founding of new populations, even if mates are unavailable.

Many hermaphrodite animals are capable of both outcrossing and self- fertilization, including a number of tapeworm, snail and ascidian species. The parasitic tapeworm *Schistocephalus solidus* has a complex life cycle, with a copepod (fresh-water zooplankton) as its first intermediate host, the three-spined stickleback (*Gasterosteus aculeatus*) as its second intermediate host, and one of several fish-eating bird species as its final host. Researchers who were interested in whether or not selfing led to inbreeding depression in this parasite used microsatellite data to compare the genotypes of adults and offspring to establish whether juveniles were the product of self-fertilization (one parent) or outcrossing (two parents). They then discovered that outcrossed parasites produced a significantly more intense infection than selfed parasites, and as a result they were more likely to progress in their life cycle to the point where they could reach their final host. Despite an advantage to outcrossing, this species nevertheless maintains an ability to self-fertilize, presumably for reproductive assurance because there is no guarantee that a tapeworm will be able to find a partner with which to outcross (Christen and Milinski, 2003).

### Inbreeding avoidance

A final testimony to the hazards associated with inbreeding are the lengths to which individuals will often go in order to avoid it. In Chapter 6 we were introduced to two important mechanisms of inbreeding avoidance. The first of these was sex-biased dispersal. If one sex is philopatric and the other disperses before reproducing, then the breeding males and females within a population or breeding group should not be related to each other. That is not to say that inbreeding avoidance is the only reason why sex-biased dispersal occurs, because other factors such as competition for territories or for mates may also come into play, but it is undoubtedly a driving force in some situations. One example of this was presented by a 9-year study of a savannah sparrow (*Passerculus sandwichensis*) population on an isolated archipelago in the Bay of Fundy, Canada. Both males and females were more likely to move to a different part of the island to breed if the parent of the opposite sex was still alive, presumably to reduce the risk of inbreeding (Wheelwright and Mauck, 1998).

Even when neither males nor females tend to disperse from their family groups, incestuous matings can often be avoided. A study that was published in 1995 reviewed data from a number of species that live in family groups and found that 18/19 avian species and 17/20 mammalian species showed a strong tendency to avoid mating with relatives (Emlen, 1995). This leads us to the second mechanism of inbreeding avoidance that was introduced in Chapter 6, and that is mate choice. If mates are chosen at least partially on the basis of inbreeding avoidance then species must have a basis for recognition. Some species will use phenotypic characters, for example the call of the American toad (*Bufo americanus*) is more similar in closely related individuals and can therefore be used as a cue to avoid inbreeding (Waldman, Rice and Honeycutt, 1992). Other species use olfactory

**Table 7.6** Some examples of endangered or critically endangered species that had $N_c$ <500 at their most recent assessment. These are not all necessarily doomed to extinction, but their risk is relatively high because of their low levels of genetic diversity and their often high levels of inbreeding. Sources: IUCN Red List 2003 and BirdLife International

| Species | Geographic range | $N_c$ |
|---|---|---|
| Baishanzu fir (*Abies beshanzuensis*) | Baishanzu Mountain, China | 5 |
| Greenflower Indian mallow (*Abutilon sandwicense*) | Oahu, Hawaii | 200–300 |
| Bastard quiver tree (*Aloe pillansii*) | Namibia, South Africa | <200 |
| Visayan wrinkled hornbill (*Aceros waldeni*) | Western Visayas, Philippines | 120–160 |
| Blue-eyed ground-dove (*Columbina cyanopis*) | Brazil | <250 |
| Anegada ground iguana (*Cyclura pinguis*) | Virgin Islands | <200 |
| Aruba Island rattlesnake (*Crotalus durissus unicolor*) | Caribbean | 350 |
| Javan rhinoceros (*Rhinoceros sondaicus*) | Java, Vietnam | <100 |
| Ethiopian wolf (*Canis simensis*) | Ethiopia | 400 |

cues, which may help them to identify genetically dissimilar mates, for example sand lizards prefer the odour of individuals that have distantly related MHC alleles (Olsson *et al.*, 2003).

Of course, inbreeding avoidance is impossible in very small populations, but remember that inbreeding does not necessarily lead to inbreeding depression. A question that often appears in conservation genetics is how small a population must be if it is to avoid inbreeding depression. The work of animal breeders suggests that populations with an effective size of 50 or more should usually be able to avoid inbreeding depression and retain reproductive fitness (Franklin, 1980). However, this estimate refers to the short-term avoidance of inbreeding depression, and other studies suggest that an effective size of between 500 and 1000 is necessary if populations are to maintain their long-term evolutionary potential (Franklin and Frankham, 1998). Note also that this is the *effective* population size, and if we accept that the average $N_e/N_c$ ratio is 0.1 (Chapter 3), the minimum population census size necessary for long-term survival will be closer to 5000. Somewhat alarmingly, many species have a total $N_c$ that is <500 (Table 7.6) and, although they are not all necessarily doomed to extinction, they are undoubtedly at a greater risk than large populations because of their relatively low levels of genetic diversity and their often high levels of inbreeding. In the next section we will look at how human management can help to reduce the risk of inbreeding through the introduction of novel genotypes.

## Translocations

Once we have identified which populations are most at risk from inbreeding depression, management strategies can be drawn up that will help to increase their chances of long-term survival. One of the most effective ways of slowing the decline of small, genetically depauperate populations is through the introduction of immigrants.

## Genetic rescue

When migrants are translocated from one population to another, they will often introduce new alleles into the recipient population. If this results in a reduction of inbreeding depression, it is known as **genetic rescue** (Thrall *et al.*, 1998). Genetic rescue will increase the growth rate of a population over multiple generations from the time when the novel genes were introduced. This is usually attributed to **heterosis**, which is elevated fitness in the offspring of genetically divergent individuals (sometimes known as **hybrid vigour**). Recall that inbreeding depression can be attributable to either dominance or overdominance. As we might expect from a process that effectively reverses inbreeding depression, heterosis can result from either the production of relatively fit heterozygous individuals or, more likely, the masking of deleterious alleles.

There are several success stories in which genetic restoration has dramatically improved the fitness of populations, possibly saving them from extinction. One extreme example of this occurred in the last remaining population of the Florida panther (*Felis concolor coryi*). In recent decades the effective size of this population has been around 25 and it is therefore not surprising that genetic variation, as revealed by microsatellite loci, is much lower than that found in populations of any other North American subspecies of *F. concolor* (Culver *et al.*, 2000). More importantly, the population became fixed for a number of deleterious traits, including a kinked tail, extremely poor quality semen, and unilaterally undescended testicles (Roelke, Martenson and O'Brien, 1993). In 1995 a plan was initiated to introduce Texas cougars (*Felis concolor stanleyana*), one of the Florida panther's closest relatives, in a bid to reverse the decline of the Florida population. Eight females were introduced, and by the year 2000 four of these were still alive and a minimum of 25 Florida panther × Texas cougar descendants were thriving. By that time the population numbered around 60–70 individuals. Of the animals with Texas ancestry that have so far been evaluated, only 7 % have a kinked tail and three out of five males had two descended testicles. Semen quality was evaluated in one male and was remarkably improved. Although it is early days yet, these preliminary results are highly encouraging and suggest that the genetic rescue of the Florida panther has been successful (Land and Lacy, 2000).

Another success story is an adder (*Vipera berus*) population in southern Sweden that is restricted to a coastal strip of grassy meadow and has been isolated from other adder populations for at least a century. Inbreeding depression began to manifest itself in the early 1980s in the form of a number of deleterious traits that led to a high proportion of deformed or stillborn offspring. The low survival rates meant that by 1992 there were only about ten males in the population. At this time, 20 adult males from a large, genetically diverse population further north were introduced and left for four breeding seasons, after which time the surviving immigrants were recaptured and returned to their natal population. In 1999 an assessment of MHC variability showed that genetic diversity within the population had increased substantially following this introduction (Madsen, *et al.*, 1999). The latest report was from 2003 when 39 adult males were collected, which is the largest number since the population was first monitored 23 years ago (Madsen, Ujuari and Olsson, 2004). Once again, the introduction of novel genes has ameliorated inbreeding depression and therefore increased the chance of a population's survival.

Perhaps the most striking example of genetic rescue comes from an isolated wolf pack that was discovered in southern Scandinavia more than 900 km from the nearest known wolf packs in Finland and Russia. In the 1980s there were fewer than ten individuals in this pack, but in 1991 the population started to grow exponentially and by 2001 it had expanded to between 90 and 100 individuals distributed among 10–11 packs. Researchers collected a total of 94 tissue or blood samples between 1984 and 2001 and genotyped them using mtDNA, Y-chromosome and nuclear (autosomal and X-chromosome) markers. All animals born before 1991 had the same mtDNA haplotype and the same Y-chromosome haplotype. These data, combined with very low allelic diversity at the nuclear loci, strongly suggest that the pack was founded by a single female and a single male. Between 1991 and 1992 the genetic diversity of this pack increased dramatically, when ten new alleles across 19 loci appeared in six sibling wolves that were born during that time. These siblings also carried a new Y-chromosome haplotype, showing that they had been fathered by an immigrant male who was new to the population. This single individual increased the heterozygosity of the pack from a mean of 0.49 in eight wolves that were born between 1985 and 1990 to a mean of 0.62 in 16 wolves that were born between 1991 and 1995. This increase in genetic diversity coincided with rapid population growth, and is testimony to the contribution that a single individual can make to the genetic rescue of a population (Vilà *et al.*, 2003a).

### Source populations

Although translocations are often successful, care must be taken when moving individuals across large geographical distances. If the extant source population at one site shows substantial genetic differences from the extinct or endangered population at the destination site, then there is the risk that the introduction will bring with it the hazards that are often associated with invasive species. This is why biologists screened several potential donor populations in China and Russia before identifying an appropriate source of Oriental white storks (*Ciconia boyciana*) for re-establishing the Japanese population, which became extinct in 1986 (Murata *et al.*, 2004). The lack of extant Japanese storks made genetic comparisons somewhat challenging, but researchers circumvented this problem by cutting small pieces of skin from 17 Japanese storks that had been stuffed and mounted on display in Toyooka City, Japan, and two nearby villages. They extracted DNA from these samples and identified mitochondrial haplotypes by sequencing a variable portion of the control region. These were compared to haplotypes from Chinese and Russian storks that had been previously used in captive breeding programmes. The maximum divergence between Japanese, Chinese and Russian storks was only 2.6 %, which is much lower than the levels of intraspecific control region sequence divergence that have been found in some other bird species; furthermore, one haplotype was found in both a Japanese and a Chinese stork, suggesting a relatively recent historic connection between the Japanese and continental populations. Finally, a maximum likelihood phylogenetic tree showed no distinction between the evolutionary lineages in Japan, China and Russia. The authors of this study therefore concluded that translocation of storks from the continent to Japan would be appropriate, at least on the basis of genetic compatibility.

Biologists may be more concerned about the genetic compatibility of source and destination populations if they are separated by large distances, although distance is not necessarily an accurate predictor of genetic differentiation. The Egyptian vulture (*Neophron percnopterus*) is an endangered species in Europe that is declining rapidly in many areas. Populations in the Canary and Balearic Islands are particularly vulnerable, and it has been suggested that they may benefit from future translocations. However, the island populations show significant levels of genetic differentiation from the continental European populations, and in fact the continental populations are genetically more similar to a subspecies that lives in India (*N. p. ginginianus*) than they are to the geographically closer island populations (Figure 7.9). It therefore seems appropriate to treat the mainland and island European populations as ESUs, and to avoid translocating birds from the mainland to the islands if possible (Kretzmann *et al.*, 2003).

### Founder effects

Successful translocation programmes must also ensure that the translocated individuals collectively harbour adequate levels of genetic diversity. This will depend partially on the number of immigrants that are introduced at any given time. Most extant populations of North American bison (*Bison bison*) were founded from a small number of individuals and, although genetic diversity is not universally low, it is correlated with the number of founders in each population and in some cases has left populations susceptible to inbreeding depression (Wilson and Strobeck, 1999). A relationship between inbreeding depression and number of founding individuals also emerged from a comparison

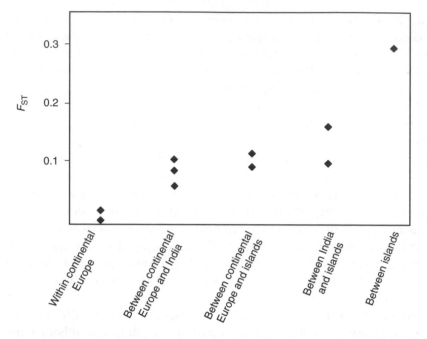

**Figure 7.9** Genetic differentiation ($F_{ST}$) between populations of the Egyptian vulture (*Neophron percnopterus*) from mainland Spain, the Canary and Balearic islands and the subspecies *N. p. ginginianus* from India. Because populations from continental Europe are more similar to the Indian subspecies than they are to populations on the Canary and Balearic islands, translocations from Spain to these islands may be inappropriate. Data from Kretzmann *et al.* (2003)

of 22 species of native New Zealand birds. Species that had not passed through population bottlenecks had an average hatching failure of 3.0 per cent, whereas species that had experienced bottlenecks of <150 individuals, some of which occurred during translocations, had an average hatching failure of 25.3 per cent (Briskie and Mackintosh, 2004). This finding has important practical implications because conservation programmes usually establish populations of endangered species from substantially fewer than 150 founding individuals.

It is not just the number of founders that influences the success of translocation programmes. Numerous attempts have been made to restore koala (*Phascolarctos cinereus*) populations to southeastern Australia following their disappearance from this region in the 1930s after they had been hunted extensively for their fur. Further north in Queensland the koalas remained common, and a well-managed translocation using individuals from Queensland populations should have minimized any accompanying founder effect. Instead, populations in mainland South Australia were restocked using individuals from Kangaroo Island, which is just off the south coast of Australia. This was unfortunate because the Kangaroo Island population was founded in the 1920s by a few individuals taken from a population on French Island that itself had been founded in the late 19th century by only two

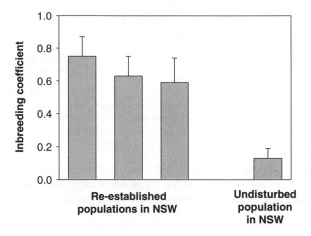

**Figure 7.10** Inbreeding coefficients in three translocated koala populations in New South Wales (NSW), Australia, compared with a single undisturbed population. Inbreeding depression in translocated populations has resulted in testicular aplasia in as many as 23.9 per cent of males. Adapted from Seymour *et al.* (2001)

or three individuals. The South Australia population was therefore restocked by individuals whose recent genetic history featured three substantial bottlenecks. It is therefore not surprising that populations in southern Australia have significantly lower levels of genetic diversity than those further north. They are also suffering from testicular aplasia, a unilateral or bilateral failure in testicular development that is evidence of inbreeding depression (Houlden *et al.*, 1996; Seymour *et al.*, 2001) (Figure 7.10).

## Outbreeding depression

Another potential problem that may affect the success of translocations is outbreeding depression. We know from the earlier section on genetic rescue that when two genetically divergent individuals reproduce heterosis may occur, which results in progeny whose fitness is higher than the average fitness of either parental population. This can happen through the masking of deleterious alleles or by overdominance associated with increased heterozygosity. In some cases, however, the fitness of the progeny will be lower than that of either parent, in which case **outbreeding depression** has occurred. Two genetic factors can lead to outbreeding depression. The first is the loss of locally adapted genotypes. If individuals from two populations that are each adapted to their natal environments hybridize, their offspring will contain a mixture of alleles that may not be well suited to either environment. When this occurs, outbreeding depression will be evident in the first generation of offspring.

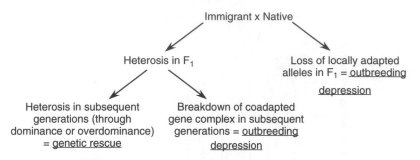

**Figure 7.11** Some outcomes that may result from matings between immigrant and native individuals. Note that genetic rescue assumes that the native population was experiencing some level of inbreeding depression prior to the mating between natives and immigrants

The second genetic cause of outbreeding depression is the loss of positive epistatic interactions. **Epistasis** refers to the interaction of genes from multiple loci and their collective influence on particular traits. The group of loci involved in an epistatic interaction is known as a **co-adapted gene complex**, and if this complex is broken up through recombination then outbreeding depression may result. Because the set of chromosomes from each parental lineage will remain intact in their offspring (the $F_1$ generation), outbreeding depression may not be evident immediately and in fact fitness may increase temporarily because the $F_1$ generation will have high heterozygosity values. By the $F_2$ generation, however, recombination will have disrupted adaptive gene combinations, causing a sudden reduction in fitness (Figure 7.11). This process was evident in the tidepool copepod *Tigriopus californicus* when broods representing both pure populations and interpopulation hybrids were raised under different conditions of temperature and salinity. As is typical following the disruption of co-adapted gene complexes, adverse affects were not evident in the $F_1$ of interpopulation hybrids, but the fitness of the $F_2$ generation was substantially reduced (Edmands and Deimler, 2004).

### Evidence for outbreeding depression

Although theoretically well established, the extent to which outbreeding depression threatens the survival of wild populations remains a matter of debate; some researchers maintain that its importance is overstated, whereas others believe that it is a widespread phenomenon that would be detected more often if appropriate studies were conducted. Outbreeding depression has been documented most commonly in plants because their relatively low dispersal rates, which are often combined with local adaptation to particular environmental conditions, mean that plants can show genetic differentiation over both small and large spatial scales. Under these conditions, outbreeding depression may result when individuals from either distant or neighbouring populations interbreed (see Box 7.3).

In wild populations of the herbaceous scarlet gilia (*Ipomopsis aggregata*), matings between parents separated by only 100 m produced offspring with a reduced lifetime fitness that was caused by outbreeding depression, although not surprisingly the decline in fitness depended in part on environmental hetero-geneity and the associated selection regimes (Waser, Price and Shaw, 2000). Effects at an even finer spatial scale were found in Nelson's larkspur (*Delphinium nelsonii*) when flowers were hand-pollinated using pollen from between 1 m and 30 m away. Progeny from intermediate crossing distances (3 m and 10 m) grew approximately twice as large as the progeny that resulted from crossings between either nearby or more distant plants. This was presumably because pollen from intermediate distances did not adversely affect the fitness of offspring, whereas pollen from plants that were 1 m or 30 m away led to inbreeding depression and outbreeding depression, respectively (Waser and Price, 1994).

The potential importance of outbreeding depression to conservation manage-ment was illustrated by a study of the Mediterranean perennial herb *Anchusa crispa*, which is endemic to sandy seashores on the islands of Corsica and Sardinia. Populations are small and patchy and have extremely low levels of heterozygosity. The performance of inbred (from the same patch) and outcrossed (from different patches in the same population) plants was evaluated in greenhouse experiments. Although the plants that were produced after two generations of selfing produced fewer seeds than outcrossed plants, they also produced more clusters of flowers and had significantly higher survival rates, and therefore the overall fitness of inbred plants was substantially higher than that of outcrossed plants. In this case, outbreeding depression seemed to be more of a hazard than inbreeding depression, and the authors of this study suggested that inbreeding may have purged deleterious alleles in this species, and furthermore may actually be favoured because of the adaptation conferred by homozygous alleles in different populations (Quilichini, Debussche and Thompson, 2001). Although this was a greenhouse experiment that needs to be verified by studies of wild populations, these preliminary results suggest that conservation of *A. crispa* would not be well served by the introduction of novel alleles into small, inbred populations.

## Box 7.3  Restoration genetics

Some of the issues associated with the regeneration of ecosystems fall under the heading of restoration genetics, an emerging field that combines restoration ecology with population genetics. A recent article by Hufford and Mazer (2003) highlighted some of the potential genetic pitfalls that can compromise the success of restoration programmes. We have dis-cussed some of these in earlier sections of this and other chapters but will now briefly revisit them in the context of a specific conservation management technique. First, extreme founder effects should be avoided by screening for genotype diversity. This may be particularly important in plants that are capable of clonal reproduction, because in these species a founding population could include multiple individuals that have the same genotype. Restored eelgrass (*Zostera marina*) populations in southern California and New England were found to have significantly lower levels of genetic diversity than undisturbed populations because the founders were genetically depauperate, and this in turn led to inbreeding depression within the transplanted populations (Williams, 2001). A second concern is the genetic swamping of local genotypes, either through hybridization or through aggressive growth such as that exhibited by the invasive genotype of the common reed *Phragmites australis* in the USA (Chapter 5).

Although less well-studied and therefore somewhat controversial, a third potential genetic problem associated with ecosystem restoration is outbreeding depression following the mating of genetically dissimilar individuals. Projects that aim to restore biodiversity in intensively managed farmland often use seed mixtures of wildflowers that are produced by commercial suppliers, and that may have originated many miles from the site of restoration. The potential consequences of using seeds from distant sources were investigated in a study of three arable weed species: common corncockle (*Agrostemma githago*), red poppy (*Papaver rhoeas*) and white campion (*Silene alba*) (Keller, Kollmann and Edwards, 2000). Swiss plants were crossed with plants that originated in England, Germany and Hungary, and the fitness of the hybrids was compared with that of the parental plants. Outbreeding depression was indicated by reduced biomass in both the $F_2$ generation of all red poppy crosses and in the $F_1$ generation that was generated by a cross between Swiss and German corncockles. Seed mass decreased in the $F_2$ generation of the white campion crosses and survival was reduced in both the $F_1$ and $F_2$ progeny of red poppies. Results such as these suggest that, whenever possible, habitats should be restored using seeds of relatively local origin. This should minimize the likelihood of both outbreeding depression and genetic swamping.

Outbreeding depression has been documented less frequently in animals, although some evidence has emerged from a number of recent studies. Hybrids between males and females from two different populations of pink salmon (*Oncorhynchus gorbuscha*) that are separated by around 1000 km showed a decreased survival in the $F_2$ generation relative to either parental population, which is consistent with an epistatic model of outbreeding depression (Gilk *et al.*, 2004). The male courtship song in offspring that were generated by interpopulation crosses of the fruitfly *Drosophila montana* had a frequency different to that found in either parental population, which led to reduced mating success and lowered fitness, once again suggesting outbreeding depression (Aspi, 2000).

As with plants, outbreeding depression in animals may be more common following matings between individuals from populations that have undergone periods of inbreeding, for example matings between immigrants and natives in an inbred population of song sparrows (*Melospiza melodia*) showed signs of reduced fitness relative to the 'pure' native sparrows (Marr, Keller and Arcese, 2002). Because inbreeding is not uncommon in endangered species, the relevance of this to conservation is apparent. In the early 20th century the Kaziranga population of Indian rhinoceroses (*Rhinoceros unicornis*) underwent a severe bottleneck that was followed by a period of substantial inbreeding and purging of deleterious alleles. A captive breeding programme mated individuals from the Kaziranga and the Chitwan populations in an attempt to reduce inbreeding, but the offspring had a higher mortality rate than those in the inbred population, suggesting genetic incompatibility and outbreeding depression (Zschokke and Baur, 2002). In conclusion, although outbreeding depression may not be a common threat to the fitness of animal populations, a number of studies have shown it to be sufficiently detrimental, at least in the short term, to make it a point of consideration in translocation programmes. It is also a potential problem that should be kept in mind when species are maintained through captive breeding, a last resort approach in conservation biology that will be the subject of our final section.

# Captive Breeding

If a species is unable to survive in the wild then the only way that it can be saved from extinction is through captive breeding. There are a number of species that would now be extinct were it not for captive breeding, including the California condor (*Gymnogyps californianus*), Père David's deer (*Elaphurus davidianus*), Arabian oryx (*Oryx leucoryx*), black-footed ferret (*Mustela nigripes*), the Franklin tree (*Franklinia alatamaha*) and the Potosi pupfish (*Cyprinodon alvarezi*). In many more cases, captive breeding programmes have been initiated for species that are rapidly dwindling in the wild in order to maintain as large a gene pool as possible and, in some cases, to provide a source of plants and animals for translocation programmes. Space, money, expertise and other resources that are needed for this costly endeavour are limited, and decisions about which species will be captively bred are often motivated by their appeal to humans, with mammals, birds and flowering plants generally receiving much more attention than invertebrates or lower plants.

## Maximizing genetic diversity

Logistical constraints often mean that captive populations are dispersed among multiple zoos, aquaria and wildlife parks. Many institutions will each house only a handful of individuals from a particular species, and this could have serious repercussions for long-term genetic diversity. Successful captive breeding therefore often requires cooperation between institutions to create what is effectively human-mediated gene flow between very small populations. This has been facilitated by a number of enterprises, including the Species Survival Program (SSP) of the American Zoo and Aquarium Association, which was started in 1981 as a cooperative conservation programme for selected species in zoos and aquaria in North America. This programme currently oversees the captive breeding of 161 different species, most of which involve many different institutes. The SSP for the golden lion tamarin, for example, is based on a captive population that is currently comprised of 445 individuals distributed around 150 different zoos.

This type of cooperation is needed if captive breeding programmes are to achieve a commonly stated goal of maintaining 90 % of genetic diversity for a period of 100 years while increasing the inbreeding levels by no more than 10 %. The effective population size needed to maintain genetic diversity for this length of time will depend to some extent on the generation time of the species in question. Assuming that the population size remains constant, and note that we are talking about $N_e$ and not $N_c$, the necessary effective population size has been derived from the the rate at which heterozygosity is expected to decline within populations of different sizes following genetic drift (Frankham, Ballou and Briscoe 2002; see also Chapter 3), and is approximately equal to:

$$N_e = 475/L \qquad (7.4)$$

where $L$ is the generation length in years. The inverse relationship between generation length and the $N_e$ necessary for the maintenance of genetic diversity is shown in Figure 7.12. The tremendous range in generation times between species means that the minimum desirable $N_e$ of captively bred populations will vary enormously; in some insects a generation will last only a few days, whereas in Caribbean flamingoes (*Phoenicopterus ruber*) it may last 26 years.

Maintaining 90 % of genetic diversity while avoiding inbreeding may be an unrealistic goal in many captively bred species, although good management can minimize the loss of diversity. We know from earlier chapters that the $N_e$ of populations, captive or otherwise, will be influenced by a number of factors, and all of these must be taken into account if maximum genetic diversity is to be maintained. A bottleneck, for example, will have a lasting effect on a population's $N_e$, and although careful management can often maintain populations at stable numbers, bottlenecks caused by founder effects are not uncommon during the establishment of captive populations. The IUCN recommends that a captive

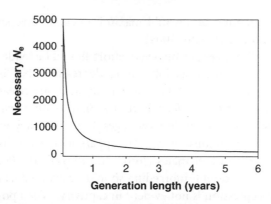

**Figure 7.12** Relationship between generation length and the minimum $N_e$ that is necessary if genetic diversity is to be maintained in captive populations. The necessary $N_e$ rapidly increases as generation times become smaller

population be established before the number of wild individuals drops below 1000, so that an adequate number (20–30) of unrelated individuals can be used as founders. Unfortunately, such forward planning is not always possible and captive populations are founded all too often by only a handful of individuals (recall the earlier example of Père David's deer). Variation in reproductive success (VRS) will also influence the $N_e$ of a captive population. Some species, including those with a socially monogamous mating system (e.g. many birds), can often be managed so as to increase the proportion of breeding adults. In other species (e.g. many mammals) breeding is most successful under a polygamous mating system, in which case the VRS of males may be high (Chapter 6).

*Captive inbreeding*

Associated with the maintenance of genetic diversity is inbreeding avoidance. This is harder to achieve in small captive populations, although careful management using pedigree information and estimates of relatedness (Chapter 6) can minimize the amount of inbreeding depression that a population may suffer. Molecular data may be particularly useful for this because pedigree information on the founders of captive populations may never be complete. Almost no pedigree information was available for the captive population of the St Vincent parrot (*Amazona guildingii*), which currently comprises around 72 individuals. This is a protected species, but the 1987 Wildlife Protection Act allowed people who had birds of their own to give them up without being prosecuted. As a result, these birds came from multiple, often anonymous, sources so there was essentially no information on their relatedness or their wild population of origin. Microsatellite data were used to estimate the relatedness between all pairs of individuals, and this formed the basis for a breeding programme that would maximize genetic diversity and minimize inbreeding (Russello and Amato, 2004).

Whenever population sizes permit, every effort should be made to avoid matings between relatives, particularly as inbreeding depression that is not evident in captive populations may manifest itself once individuals are released into the wild. Captive inbred males of the African butterfly *Bicyclus anynana* showed a small decrease in mating success when in their cages, but when they were later released into a large tropical greenhouse it became apparent that the inbred males were substantially less fit than other males (Joron and Brakefield, 2003). Cases such as this mean that, when in doubt, inbreeding should be avoided as much as possible even if inbreeding depression is not evident in captivity. When population sizes are small, the inbreeding coefficient will increase rapidly from one generation to the next (Equation 7.1; Figure 7.5). Because many zoos have only a few representatives of each species, individuals are often moved around between institutions for the purposes of breeding. However, the benefits of this in terms of reduced inbreeding must be balanced against the stress to the individuals, the potential transmission of diseases, and the reduction in $N_e$ that can follow the loss of population subdivisions.

Overall, captive breeding is costly, challenging and does not provide a long-term solution for the conservation of species. Although it may be considered the last resort for preserving a species, we shall see in the final section that there are actually even more desperate attempts under way to preserve genetic diversity for the future.

## Genetic diversity banks

Over the next 200 years, at least 2000 terrestrial vertebrate species alone will probably go extinct unless they are bred captively (Soulé *et al.*, 1986; Tudge, 1995), and it seems very unlikely that we will have sufficient resources for such a large-scale rescue operation. Even if we could preserve all of these species in captivity, their long-term survival would still depend on whether we have retained enough suitable habitat to allow for their eventual release. In the near future we will undoubtedly see a growing number of species going extinct. Because not all species that are threatened with extinction can be maintained through captive breeding, a number of organizations are setting up genetic diversity banks.

Genetic diversity banks are designed for the long-term storage of extracted DNA, seeds, tissue, sperm and other sources of genetic material. One example of this is The Millenium Seed Bank Project at the Royal Botanical Gardens in Kew, London, which aims to preserve genotypes of over 24 000 species of seed-bearing plants from around the world through the long-term storage of their seeds. Another example, albeit on a smaller scale, can be found at the Center for Plant Conservation at the Missouri Botanical Garden, which has seeds, cuttings and other material from more than 600 endangered plant species that are native to the USA.

In a similar vein, the Frozen Ark is a last ditch attempt to preserve some of the genetic diversity of animals. This project, which is a collaboration between the London Natural History Museum, the Zoological Society of London, and Nottingham University, aims to collect, store and preserve DNA and tissue from as many endangered species as possible. Priority is given to species with a high likelihood of extinction in the near future, with the first members of the ark including the yellow seahorse (*Hippocampus kuda*), Scimitar horned oryx (*Oryx dammah*), Socorro dove (*Zenaida graysoni*) and the Seychelles Fregate beetle (*Polposipus herculeanus*). If the DNA is stored appropriately it could remain intact for tens of thousands of years, possibly longer. It is too soon to know what exactly this genetic information will be used for in the future, although it is possible that cells and sperm will also be stored in the hope that future science fiction-type cloning will allow scientists to resurrect formerly extinct species.

Other genetic diversity banks have more specific short-term applications. At the Wildlife Breeding Research Centre in South Africa, for example, vets and biologists have established a sperm bank for lions and have an ongoing programme in which they travel between prides, artificially inseminating females so as to reduce inbreeding in small, isolated populations. The social structure of this species means that if a strange male is introduced to a pride then the members of that pride will either chase him away or kill him, which would do nothing to ameliorate inbreeding. Artificial insemination therefore bypasses some of the behavioural deterrents against the genetic enhancement of lion populations. Cryopreservation of sperm and oocytes may facilitate captive breeding of other taxonomic groups (e.g. Ledda *et al.*, 2001; Browne, Clulow and Mahoney, 2002), although detailed methods generally have to be worked out separately for each species, and the technology is therefore currently available for only a small proportion of animals.

## Overview

Conservation biology is very much an uphill battle. Human populations and their consumption of resources continue to grow, and more and more habitat is lost every day. The future does not look good for a growing number of species, although on a brighter note we would have lost even more species by now if we had no conservation programmes. Molecular genetics can help us to make informed decisions about the management of both wild and captive populations, and for this reason conservation genetics remains one of the most important applications of molecular ecology. There are, however, many other reasons why molecular ecology can be considered an applied science, and some of these will be considered in our next and final chapter.

# Chapter Summary

- Although in most taxonomic groups only a small proportion of species have been assessed, the numbers of threatened species have led many people to believe that we are on the brink of a sixth mass extinction.

- Conservation strategies tend to assume that species can be classified accurately, although none of the >20 species concepts currently in the literature are universally accepted. The most recent approach to species identification is DNA barcoding. Conservation may also be based on management units and evolutionary significant units, whereas the protection of hybrids is more controversial.

- Threatened populations are usually small and therefore lose genetic diversity at a relatively rapid rate following genetic drift. Because drift is more important than selection in determining the fate of alleles in small populations, deleterious alleles are more likely to reach fixation and increase the genetic load.

- Small populations are susceptible to inbreeding, and if this leads to a reduction in fitness as a result of either dominance or overdominance then the population is experiencing inbreeding depression.

- Estimates of inbreeding depression are sometimes based on correlations between one or more fitness components and individual heterozygosity values (heterozygosity fitness correlations). In some cases, purging of deleterious alleles may reduce inbreeding depression.

- Sex-biased dispersal and mate choice are two important mechanisms of inbreeding avoidance, although inbreeding may be unavoidable in populations with $N_e < 500$.

- Translocations can slow or reverse the decline of small populations through genetic rescue, although source and destination populations should be genetically compatible, and founder effects should be avoided.

- Outbreeding depression can result from the mating of genetically dissimilar individuals if locally adapted alleles are lost or if positive epistatic interactions break down. The former will be evident in the $F_1$ generation, whereas the latter will not become apparent until the $F_2$ generation.

- Small population sizes mean that captive breeding programmes must be managed carefully so as to minimize inbreeding depression and loss of genetic diversity. Pedigree information can be used to reduce inbreeding, and $N_e$ can be maximized if bottlenecks are avoided and variation in reproductive success (VRS) is kept as low as possible.

- A last resort for preserving genetic diversity is through banks that may comprise DNA, tissue, sperm, seeds and other sources of genetic material. Some of these are destined for long-term preservation, whereas others have shorter-term goals such as the avoidance of inbreeding in small populations.

## Useful Websites and Software

- Barcode of life: http://www.barcodinglife.com/

- IUCN, The World Conservation Union: http://www.iucn.org/

- BirdLife International: http://www.birdlife.net/

- American Zoo and Aquarium Association (for information on Species Survival Plans): http://www.aza.org/

- Golden lion tamarin conservation programme: http://nationalzoo.si.edu/ConservationAndScience/EndangeredSpecies/GLTProgram/ZooLife/CurrentStatus.cfm. Includes a link to a golden lion tamarin pedigree game.

- FSTAT software for calculating $F$ statistics, including inbreeding coefficients: http://www2.unil.ch/izea/softwares/fstat.html

- Kinship software program that estimates maximum likelihood ratios for testing hypotheses of relatedness: http://gsoft.smu.edu/GSoft.html

## Further Reading

### Books

Avise J.C. and Hamrick, J.L. 1996. *Conservation Genetics – Case Histories from Nature*. Chapman & Hall, London.
Frankham, R., Ballou, J.D. and Briscoe, D.A. 2002. *Introduction to Conservation Genetics*. Cambridge University Press, Cambridge.
Landweber, L.F. and Dobson, A.P. 1999. *Genetics and the Extinction of Species*. Princeton University Press, Princeton, NJ.

## Review articles

Booy, G., Hendriks, R.J.J., Smulders, M.J.M., Van Groenendael, J.M. and Vosman, B. 2000. Genetic diversity and the survival of populations. *Plant Biology* **2**: 379–395.

Cole, T.C. 2003. Genetic variation in rare and common plants. *Annual Review of Ecology, Evolution and Systematics* **34**: 213–237.

de Meeus, T., Durand, P. and Renaud, F. 2003. Species concepts: what for? *Trends in Parasitology* **19**: 425–427.

Hansson, B. and Westerberg, L. 2002. On the correlation between heterozygosity and fitness in natural populations. *Molecular Ecology* **11**: 2467–2474.

Hedrick, P.W. 2004. Recent developments in conservation genetics. *Forest Ecology and Management* **197**: 3–19.

Hedrick, P.W. and Kalinowski, S.T. 2000. Inbreeding depression in conservation biology. *Annual Review of Ecology and Systematics* **31**: 139–162.

Hufford, K.M. and Mazer, S.J. 2003. Plant ecotypes: genetic differentiation in the age of ecological restoration. *Trends in Ecology and Evolution* **18**: 147–155.

Morin, P.A., Luikart, G. and Wayne, R.K. 2004. SNPs in ecology, evolution and conservation. *Trends in Ecology and Evolution* **19**: 208–216.

## Review Questions

**7.1.** Only 0.06 % of described insect species are classified by the IUCN as threatened. Why do we nevertheless have reason to fear that we are on the brink of a sixth mass extinction?

**7.2.** Why does a reduction in population size lead to a simultaneous reduction in genetic diversity and an increase in inbreeding.

**7.3.** Add either ↑ or ↓ to each boxed variable in the following flow charts to show how different factors can influence the increment in inbreeding over time. Note that not all variables are included in each flow chart – you are asked to comment on the effects of a few key variables.

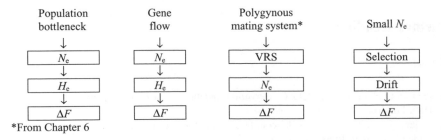

*From Chapter 6

**7.4.** Is purging more likely to reduce inbreeding depression that is caused by dominance or by overdominance?

**7.5.** Fledgling survival within a population of blue tits (*Parus major*) was found to be 80.5 % for the offspring of unrelated parents and 60 % for the offspring of related parents. What is the impact of inbreeding depression on offspring survival in this population?

**7.6.** What three aspects of population genetics are of particular relevance to successful genetic restoration programmes?

**7.7.** In the 1940s, land iguanas (*Conolophus subcristatus*) were extirpated from Isla Baltra on the Galápagos archipelago. Historical records show that some iguanas were moved from Baltra to nearby Isla Seymour Norte in the 1930s, an island that previously had lacked land iguanas. As part of a proposed translocation programme, biologists compared mitochondrial haplotypes from current populations with Isla Baltra specimens that had been collected in the 1940s to determine whether it would be appropriate to restock Isla Baltra using iguanas from Seymour Norte (Hofkin *et al.*, 2003). The six haplotypes that they found were distributed across the archipelago as follows:

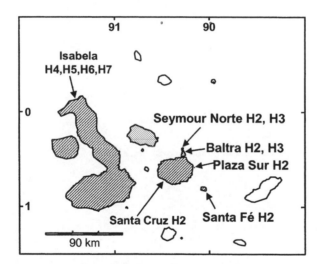

**Figure 7.13** Distribution of land iguana haplotypes (H2–H7) in past and current populations in the Galápagos archipelago. White islands have always lacked land iguana populations, grey islands represent extirpated populations, and hatched islands represent extant populations. Note that the current population on Seymour Norte is not native, and the Baltra haplotypes date from the 1940s (there is currently no population on Baltra)

On the basis of this haplotype distribution, what would be the most conservative approach to follow when translocating iguanas from Seymour Norte to Baltra?

Figure 7.12 ...

# 8

# Molecular Ecology in a Wider Context

## Applications of Molecular Ecology

By this stage in the book it should be evident that the acquisition and analysis of molecular data over the past two or three decades has provided us with considerable insight into the ecology of wild populations in virtually every major taxonomic group. Although the greatest accomplishment of molecular ecology has been to broaden the scope of ecological research, it has also provided us with a number of applications that have important social and economic consequences, and in this short, final, chapter we will take a look at molecular ecology in a wider context.

The practical applications of molecular ecology are widespread and varied. We have discussed some of these in previous chapters, one important example being conservation genetics (Chapter 6). We also saw in Chapter 5 how phylogeography can help us to determine the source, and possibly the invasion route, of introduced species, and with this knowledge we may be able to limit the ecological havoc and financial ruin that are caused each year by bioinvasions. Phylogeography and population genetics have also been applied to matters of public health, for example researchers have used phylogeography to trace the source of outbreaks of human immunodeficiency virus (HIV), dengue virus and rabies virus (Holmes, 2004), and molecular data have been used to infer patterns of dispersal and gene flow in mosquitos and other vectors of diseases such as malaria, dengue fever and West Nile disease (e.g. Tripet, Dolo and Lanzaro, 2005).

Most of this chapter will be concerned with the applications of molecular ecology to three areas that we have largely neglected so far: law enforcement, agriculture and fishing. A number of different methods that are used routinely by molecular ecologists will be the focus of these discussions, although note that

*Molecular Ecology*   Joanna Freeland
© 2005 John Wiley & Sons, Ltd.

molecular ecologists have not been solely responsible for the development of these methods. Microbiologists, for example, may routinely use species-specific geno-types for identification purposes, and there is considerable overlap between wildlife and human forensics. However, it is safe to say that without molecular ecology our ability to genetically characterize wild organisms and populations would be much poorer, and because this genetic characterization is fundamental to the applications that we will be discussing, few could question the relevance of molecular ecology.

# Wildlife Forensics

In previous chapters we have seen how genetic data can be used to identify individuals, determine their population of origin, and assign them to a particular species. In Chapter 6, for example, we looked at how individual-specific genotypes enabled researchers to link faecal samples to different coyotes. Population affinities were discussed in Chapter 4 when we looked at how assignment tests can be used to determine which population an individual originated from. We have also seen several examples in which species-specific markers were used to identify prey remains. In the following sections we will look at how these same methods are used in wildlife forensics to identify individuals and the populations and species to which they belong, thereby providing information that can help to solve cases of suspected poaching or illegal trade.

## Poaching

Poaching of wildlife species is a global problem that occurs for many reasons, for example bear paws, gall bladders and bile are used in traditional Chinese medicine, elephants are killed for their ivory, and the hunting of deer out of season is considered by some to be an enjoyable sport. When the only evidence against a suspected poacher is a carcass or a body part that was found in his or her possession, prosecution is more likely to succeed if the origin of these putatively illegal samples can be established, and this is where genetic analysis comes into play.

### Identifying individuals

A common application of individual genotyping in wildlife forensics is the linking of body parts with a recovered carcass. A typical example of this occurred in 2001 when the Colorado Division of Wildlife Officers received information about a bull elk that had been killed illegally for its antlers. A headless elk carcass was found on the Ute Mountain Indian Reservation, and a set of elk antlers was seized from the home of one of the suspected poachers. Analysis of DNA showed that the antlers had been taken from the recently killed elk, and this evidence helped the courts to convict two poachers (source: US Attorney's Office, District of Colorado). In another case, a motorist notified the California Department of Fish and Game (DFG) of a man who was parked by the highway with a dead deer in the back of his truck. By the time a DFG warden and a deputy sheriff had arrived at the scene, the man had driven away and two dead deer had been abandoned near the road. The suspect was later tracked down, and an examination of his truck yielded some blood samples. DNA analysis demonstrated that the blood had come from the dead deer that had been abandoned by the road, and again the prosecution was successful (source: DFG).

Convictions such as these are possible only if the prosecution can demonstrate that the probability of a genotype occurring in more than one individual is extremely small. One way to calculate this likelihood was shown by Equation 6.3 (see Box 6.5). Recall that the probability of identity (PI) can be calculated as:

$$\text{PI} = \sum_{i=1}^{n} (p_i^2)^2 + \sum_{i=1}^{n} \sum_{j=i+1}^{n} (2p_i p_j)^2 \qquad (6.3)$$

Forensic investigations typically employ markers from multiple polymorphic loci in order to obtain a probability of identity that will allow prosecutors to confidently exclude extraneous individuals from the case. This was illustrated by a comparison of two randomly chosen California elk (*Cervus elaphus canadensis*) that, on the basis of 11 polymorphic loci, generated an estimated PI of $1.3 \times 10^{-9}$. This means that the same genotype should be found in only one individual out of every 780 million elk (Jones, Levine and Banks, 2002). When probabilities are this low, individuals can be genetically identified with confidence (barring identical twins); however, probabilities as low as these are limited to species for which we have well-characterized hypervariable markers, and to populations for which we have robust estimates of allele frequencies.

### Identifying populations

Not all investigations of poaching are concerned with linking individuals to body parts. At times it may be more appropriate to ascertain which population an individual originated from, for example if only some populations are protected, or if there is a question over whether an animal was taken from a private collection such as a wildlife park. It may also be desirable to know which populations are targeted most commonly by poachers. This is particularly important when endangered or threatened species continue to be exploited, such as the African elephant (*Loxodonta africana*) whose total population fell from around 1.3 million to 600 000 individuals between 1979 and 1987 because of the high price of ivory.

Although a ban on ivory trade was established in 1989, the black market continues to flourish. If law enforcement officers know the origin of seized ivory goods then they may be able to identify illegal trade routes and poaching hotspots, which would benefit from additional monitoring.

With this goal in mind, researchers used 16 microsatellite markers to genotype 315 tissue samples and 84 faecal samples that were collected from elephants at 28 locations in 14 African countries. These data allowed them to map the distributions of allele frequencies across the elephants' range, and they then used assignment tests to match ivory genotypes to particular geographic origins. Preliminary analyses suggest that this method can be used to discriminate between forest and savannah elephants as the source of ivory, and in some cases to also identify specific forest blocks as the source populations (Wasser *et al.*, 2004). In general, the use of assignment tests to pinpoint the origin of a carcass has promise, although there is currently a limited number of species for which we have adequate allele frequency information for all of the candidate populations.

### Identifying species

There are also times when identifying which particular species a sample came from can be useful for investigations of suspected poaching. Species identifications are most commonly done using DNA sequences, because universal primers can be used to amplify and sequence a specific region of DNA that can then be aligned with existing sequences to determine which species the sample originated from. For example, one set of universal primers can amplify a portion of mitochondrial cytochrome *b* from at least 221 different animal species, and the sequences of these amplified fragments have so far proved to be species-specific (Verma and Singh, 2003). In one legal case, mitochondrial cytochrome *b* sequence was used to determine that traces of blood and tissue on a hunting knife had come from bushbuck (*Tragelaphus scriptus*) and not domestic cattle as the perpetrator claimed, a finding that helped prosecutors to convict the knife's owner of poaching (Pitra and Lieckfeldt, 1999). In another case, mitochondrial control region sequences that had been obtained from some hairs were shown to be from a dog and not from a wolf, thereby exonerating the suspect (Savolainen and Lundeberg, 1999).

Species can sometimes also be differentiated on the basis of microsatellite alleles. One study showed that the allele sizes of three different microsatellite loci generated genetic profiles that allowed researchers to differentiate between red deer (*Cervus elaphus*) and roe deer (*Capreolus capreolus*) (Figure 8.1), although a distinction between these two species and fallow deer (*Dama dama*) was less precise (Poetsch *et al.*, 2001). Once appropriate markers have been identified, most individuals can be assigned to a particular species with a high degree of confidence without the need first to obtain population allele frequency data. The accessibility

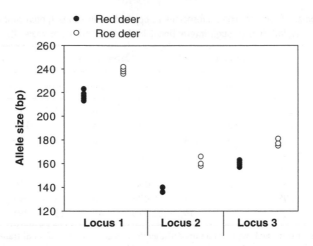

**Figure 8.1** The allele sizes at three microsatellite loci differentiate between red deer and roe deer. Adapted from Poetsch *et al*. (2001)

of species-specific genotypes means that they are used in several different areas of wildlife forensics, one of the most important being the fight against the illegal trade of endangered animals and plants.

## Illegal trade

International trade in wildlife is estimated to be worth billions of dollars each year and involves hundreds of millions of plant and animal specimens. Although some of these are traded legally, a substantial proportion involves the lucrative trade of products from endangered species. Examples of these include: caviar from Caspian Sea sturgeon; ingredients such as rhinoceros horns for traditional medicine; pets, including exotic birds and tropical fish; rainforest hardwood trees and other timber; furs and skins for items such as coats and crocodile skin bags; and tourist souvenirs such as jewelry made from coral. In an attempt to reduce the deleterious impact of illegal trade on endangered species, the Convention on International Trade in Endangered Species of Wild Fauna and Flora (CITES) came into force in 1975. Over 160 countries are now signatories to CITES and around 5000 animal species and 28 000 plant species are protected by CITES against overexploitation through international trade (Table 8.1).

Because illegally traded species are often unrecognizable by the time marketable products are seized, it can be extremely difficult to prove that illegal trafficking has occurred. For this reason, the amplification of species-specific regions of DNA has proved an invaluable method for determining which species have been included in a particular product. Traditional Chinese medicine (TCM) is an approach to healthcare that dates back at least 3000 years and treats patients using natural

**Table 8.1** Number of species (spp), subspecies (sspp) and populations (pops) that are protected by CITES against overexploitation through international trade. Data are from www.cites.org

|  | Category I[a] | | | Category II[a] | | | Category III[a] | | |
|---|---|---|---|---|---|---|---|---|---|
|  | spp | sspp | pops | spp | sspp | pops | spp | sspp | pops |
| Mammals | 228 | 21 | 13 | 369 | 34 | 14 | 57 | 11 | -- |
| Birds | 146 | 19 | 2 | 1401 | 8 | 1 | 149 | -- | -- |
| Reptiles | 67 | 3 | 4 | 508 | 3 | 4 | 25 | -- | -- |
| Amphibians | 16 | -- | -- | 90 | -- | -- | -- | -- | -- |
| Fish | 9 | -- | -- | 68 | -- | -- | -- | -- | -- |
| Invertebrates | 63 | 5 | -- | 2030 | 1 | -- | 16 | -- | -- |
| Plants | 298 | 4 | -- | 28 074 | 3 | 6 | 45 | 1 | 2 |
| Total | 827 | 52 | 19 | 32 540 | 49 | 25 | 291 | 12 | 2 |

[a]Category I: species threatened with extinction and therefore nearly all international trade is prohibited by CITES. Category II: species that may become extinct unless trade is closely controlled and therefore regulations are strict. Category III: some trading occurs but the cooperation of other countries is needed to prevent unsustainable exploitation.

plant, mineral and animal-based ingredients. Unfortunately, some of the most prized ingredients are endangered species, and TCM and the trade that surrounds it is jeopardizing the survival of some of these species. Tiger bones, for example, are used in TCM to treat joint ailments such as arthritis, but tigers (*Panthera tigris*) are an endangered species with a global population that may number as few as 5000 individuals. A tiger-specific primer pair that will amplify a region of mitochondrial cytochrome *b* sequence only from samples that contain tiger DNA (Figure 8.2) can be used to determine whether or not meat, dried skin,

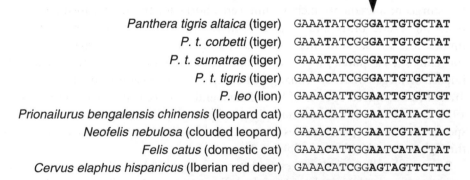

| | |
|---|---|
| *Panthera tigris altaica* (tiger) | GAAATATCGGGATTGTGCTAT |
| *P. t. corbetti* (tiger) | GAAATATCGGGATTGTGCTAT |
| *P. t. sumatrae* (tiger) | GAAATATCGGGATTGTGCTAT |
| *P. t. tigris* (tiger) | GAAACATCGGGATTGTGCTAT |
| *P. leo* (lion) | GAAACATTGGAATTGTGTTGT |
| *Prionailurus bengalensis chinensis* (leopard cat) | GAAACATTGGAATCATACTGC |
| *Neofelis nebulosa* (clouded leopard) | GAAACATTGGAATCGTATTAC |
| *Felis catus* (domestic cat) | GAAACATTGGAATCATACTAT |
| *Cervus elaphus hispanicus* (Iberian red deer) | GAAACATCGGAGTAGTTCTTC |

**Figure 8.2** Sequences from a 20 bp fragment of the mitochondrial cytochrome *b* gene from four tiger subspecies and five non-tiger species. Variable sites are in bold. The site marked with an arrow differentiates the sequences of tigers from those of other species and therefore can be used to make a tiger-specific primer that, when used in PCR reactions, will amplify a product only from tiger DNA. Adapted from Wan and Fang (2003)

hairs or other material came from a tiger (Wan and Fang, 2003). Tiger bone is sometimes added to other plant and animal derivatives to make TCM pills, and even when tiger bone is at a concentration of only 0.5 per cent it can still be detected using a highly sensitive quantitative PCR technique (Wetton *et al.*, 2002).

A thriving market for exotic food is another reason for illegal trade. Several species of sea turtles are endangered, including leatherbacks (*Dermochelys coriacea*), green turtles (*Chelonia mydas*), hawksbills (*Eretmochelys imbricata*) and flatbacks (*Natator depressus*), but they still command a high price for meat and eggs. Although trade is flourishing, prosecutions have historically been few and far between, partly because of difficulties in identifying the origin of cooked meat. In a recent study, DNA was extracted from eggs, blood, skin and muscle samples from seven species of marine turtle and two species of freshwater turtle. Analysis of these samples generated species-specific PCR-RFLP markers that could even be applied to cooked meat, and the subsequent identification of processed sea turtle meat has led to the conviction of several restaurateurs (Moore *et al.*, 2003).

Forensic techniques can also be used to characterize bushmeat, a traditional source of food in many African countries. Bushmeat is taken from a range of species, including gorillas, snakes, chimpanzees, antelopes, elephants and crocodiles. Although it can be an extremely lucrative business for the hunters, it can jeopardize endangered species; hunting helped to drive Miss Waldron's red colobus monkey (*Procolobus badius waldroni*) to extinction in 2000 after it had been a popular source of bushmeat for some years (Oates *et al.*, 2000). The market for bushmeat is global; each week Heathrow airport confiscates an average of 427 kg of animal products that often include some bushmeat. Bushmeat is processed before being transported and therefore few morphologically distinguishing characters remain at the time of seizure, but molecular identification of the component species is possible based on the PCR-RFLP analysis of cytochrome *b* (Kelly, Carter and Cole, 2003). Some other examples of how wildlife forensics can be used to identify illegally traded food are given in Table 8.2.

## Non-human perpetrators

Finally, wildlife forensics is not always used in criminal investigations against humans. In 2000, the remains of several individuals from the last population of the highly endangered hairy-nosed wombat (*Lasiorhinus krefftii*) were found in Epping Forest National Park in central Queensland. Analysis of DNA confirmed that seven individuals had been killed. The likely perpetrators were identified by the genetic analysis of nearby scats, which revealed microsatellite genotypes typical of dingoes (*Canis familiaris dingo*). When a wombat genotype that matched one of the dead individuals was amplified from one of the dingo scats, the evidence against the dingoes was compelling, although no prosecution followed! (Banks

**Table 8.2** Some species that are traded illegally as food or medicine and have been identified following PCR amplification and characterization of mitochondrial DNA

| Species | Source | Reference |
| --- | --- | --- |
| Sharks, including basking sharks (*Cetorhinus maximus*) and hammerhead sharks (*Sphyrna lewini*) | Dried fin, shark fin soup, cartilage pills | Hoelzel (2001) |
| Abalone (*Haliotis midae, H. spadicea* and *H. rubra*) | Meat confiscated from poaching syndicates | Sweijd *et al.* (1998) |
| Whales, including humpback (*Megaptera novaeangliae*), blue (*Balaenoptera musculus*) and fin whale (*Balaenoptera physalus*) | Commercial markets of Japan and Korea | Baker, Cipriano and Palumbi (1996) |
| Rhinoceroses, including white rhino (*Ceratotherium simum*) and black rhino (*Diceros bicornis*) | Powdered material containing rhino horn (used in traditional Chinese medicine) | Hsieh *et al.* (2003) |
| Sturgeon, including *Acipenser sturio, A. nudiventris* and *A. persicus* | Caviar | Birstein *et al.* (1998) |

*et al.*, 2003). In another case in which an animal was the culprit, a particular mountain lion was suspected of killing one person and injuring another in Orange County, California. As part of the investigation the lion was trapped, and DNA samples were taken from its claws and stomach contents. The recovered genotypes matched both of the victims, and investigators therefore concluded that this particular mountain lion was responsible for both attacks (source: California Department of Fish and Game).

## Agriculture

The second applied area we shall look at that has some overlap with molecular ecology is agriculture. Many of the methods we have looked at in previous chapters are used in a range of farming-related issues, and in fact the connections between molecular ecology and agriculture are so varied that they could form the basis of an entire book. In this section we shall limit ourselves to looking at just a few of the ways in which molecular ecology can be applied to agricultural concerns: retracing the spread of pests and pathogens, identifying species and varieties for the purposes of law enforcement, and assessing gene flow and hybridization between genetically modified crops and their wild relatives.

# Pests and diseases

One of the most important contributions that molecular ecology can make towards the control of pests and diseases is tracking the source of an outbreak and the subsequent dispersal patterns. The spread of a pathogen tends to occur either as the result of a rare isolated event involving long-distance dispersal, or following a pattern of gradual expansion in which the range of the pathogen increases in an ongoing and more predictable manner (Brown and Hovmller, 2002). Differentiating between the two scenarios may help us to control the spread of pathogens and can sometimes be accomplished through an analysis of population genetic structure. This is because, as we saw in Chapter 5 during our discussion of postglacial dispersal routes, founder effects should be most pronounced following rare long-distance dispersal events, and therefore populations that have expanded in this manner are expected to show relatively low levels of genetic diversity and relatively high levels of population differentiation.

Researchers applied this theory when investigating the ascomycete fungus *Mycosphaerella fijiensis*, a pathogen that causes the very serious black leaf streak disease in bananas. Believed to have originated in southeast Asia, this fungus is spreading around the world. Several PCR-RFLP markers were used to compare levels of diversity and overall genetic similarity within and among *M. fijiensis* populations from Latin America, the Caribbean and Africa. Founder effects were evident at the scales of continents and countries, suggesting that the invasion of new countries resulted from infrequent long-distance dispersal. This is compatible with either limited long-distance airborne dispersal of ascospores or the movement by humans of infected plant material over long distances. This finding suggests that improved quarantine measures may help to limit the spread of this costly pathogen (Rivas *et al.*, 2004).

Even if a pest is already widespread, it may still be important to understand dispersal patterns because of the potential for the later spread of novel strains that are resistant to pesticides. Since the 1940s pesticide resistance has evolved repeatedly, with over 500 species of insects and mites no longer susceptible to a range of chemicals (Georghiou, 1990). Pesticide resistance was an important reason for studying dispersal in the potato cyst nematode *Globodera pallida*, which parasitizes potato roots and can greatly reduce potato harvests. Larvae actively disperse through the soil over short distances, and dormant cysts containing several hundred larvae can passively disperse through water run-off or human-mediated transport of soil. Gene flow in potato cyst nematodes was for many years believed to be negligible, although this has been contradicted by a recent study in which researchers used PCR-RFLP markers to quantify gene flow at three different scales in Peru: within fields, between fields (separated by 3–35 km), and between regions (separated by 326–832 km). Although the regions were genetically distinct from one another, gene flow was extensive both within fields and between fields within the same region (Figure 8.3). This is most likely attributable to the passive

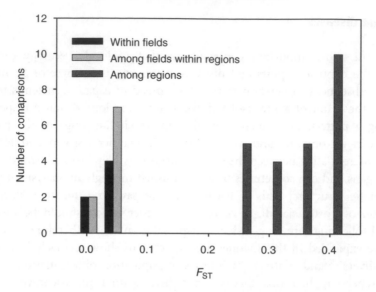

**Figure 8.3** Pairwise $F_{ST}$ values between populations of the potato cyst nematode at local and regional scales. Gene flow is extensive both within fields and between fields within the same region, although regions remain genetically distinct from one another. Adapted from Picard *et al*. (2004)

dispersal of cysts, which can remain dormant for up to 10 years. These findings suggest that potato farmers could limit the spread of G. *pallida* by disinfecting seed tubers used for planting, by regularly cleaning farm machinery, and by limiting the spread of cysts through irrigation water (Picard *et al.*, 2004).

## Law enforcement

Once again we will turn briefly to the subject of law enforcement, although this time we are specifically concerned with how molecular markers can provide evidence in legal cases that are directly relevant to agricultural concerns. One way is through the quality control of food as part of an ongoing programme to decrease the likelihood of future epidemics. The 1980s saw the start of an outbreak of bovine spongiform encephalopathy (BSE) in the UK. Also known as mad cow disease, this is a disease that affects the brain and spinal cord of cattle, and can apparently be transmitted to humans in the form of a variant of Creutzfeldt-Jakob disease (CJD). The disease can be spread between cattle if they are fed contaminated meat and bone meal, and for this reason the European Union introduced severe restrictions on the incorporation of animal protein into farm animal food. For the most part, processed animal proteins cannot be fed to animals that are destined to become food themselves, the exception to this being that fishmeal can be fed to non-ruminant animals. There is a risk, however, that fishmeal can be either contaminated or illegally supplemented with bones or tissue from other vertebrates. Fishmeal can now be efficiently checked for the presence of meat from mammals or poultry by using a set of species-specific primers that amplify a region of the 12s ribosomal RNA mitochondrial gene. The specificity of these primers means that they will amplify a product only if DNA from a particular species is present, and therefore diagnosis can be made simply on the basis of whether bands are present or absent. This technique can be used to identify mammalian or chicken DNA at concentrations as low as 0.25 per cent (Bellagamba *et al.*, 2003).

A second link between molecular markers, agriculture and law enforcement comes from the identification of different plant varieties. Registered trademarks often protect the economic rights of plant breeders who have successfully developed a novel marketable strain of a particular crop. In one case, charges were brought against some farmers in Italy who were believed to be unlawfully growing an Italian patented breed of strawberry known as Marmolada®, a variety that is sought after in some areas because of its tolerance to cold. The farmers in question denied that they were growing Marmolada® and the case could not be solved without unambiguous identification of the plants. Because the commercial production of strawberries involves micropropagation, all copies of a particular variety are genetically identical and therefore it was necessary to demonstrate that the genotypes of the known and suspected Marmolada® plants matched each other. This was accomplished with RAPD markers, which revealed Marmolada®-specific genotypes in 13 out of the 31 plants that had been genotyped. All 13 of the plants that genetically matched the Marmolada® strawberries turned out to be the plants that had come from the farm that was under investigation. The remaining 18 plants constituted a negative control that the court had included without the knowledge of the laboratory staff. On the basis of the RAPD profiles, the court concluded that the farmers had unlawfully commercialized a patented variety of strawberry (Congui *et al.*, 2000).

## Gene flow between genetically modified crops and wild relatives

One of the most controversial aspects of farming in recent years has been the actual or proposed introduction of genetically modified (GM) crops. Genetically modified crops have been altered using molecular biology techniques that allow scientists to isolate genes from one species and then incorporate these into the genome of a second species. Plants are genetically modified for a number of reasons, such as a desire to confer resistance to insect pests or herbicides such as glyphosate, to increase yield, or to make plants more resistant to harsh environmental conditions such as frost. One reason why genetically modified crops have been the subject of intensive debate is concern over the possibility that gene flow may lead to hybridization and gene introgression between domestic GM crops and

**Table 8.3**  Some examples of hybridization between crops and their wild relatives that have been detected using molecular markers. With the exception of sorghum, transgenic varieties of all of these crops have already been developed, although most are not commercially available. Adapted from Stewart, Halfhill and Warwick (2003) and references therein

| Crop | Main species | Main wild relatives | Crop-weed hybridization risk |
| --- | --- | --- | --- |
| Barley | Hordeum vulgare | H. spontaneum | Low |
| Sunflower | Helianthus annuus | H. annuus, H. petiolaris | Moderate |
| Canola | Brassica napus | B. rapa | Moderate |
| Wheat | Triticum aestivum | Aegilops cylindrica | Moderate |
| Sugar beet | Beta vulgaris vulgaris | B. v. maritime | Moderate |
| Alfalfa | Medicago sativa sativa | M. s. sativa, M. s. falcate | Moderate |
| Sorghum | S. bicolor | S. halepense, S. almum, S. propinquum | High |

their wild relatives. Some people believe that hybridization between GM crops and weeds could lead to aggressive 'superweeds' with an elevated fitness that would allow them to outcompete other wild plant species and so alter the make-up of communities and ecosystems.

Twelve of the thirteen most important crops worldwide often grow near wild plants to which they are closely related and therefore have the potential to hybridize with. By assessing levels of gene flow and hybridization, biologists have been able to identify which GM crops they believe are most likely to hybridize with wild relatives, and this has led to a classification of very low, low, moderate and high-risk crops (Stewart *et al.*, 2003). This approach may be particularly important for identifying high-risk crops for which experimental field trials could be inappropriate. Some examples of crop–weed hybrids that have been identified by molecular markers such as microsatellites, AFLPs, and RAPD are shown in Table 8.3.

Gene flow and hybridization have been particularly well studied between cultivated beets (*Beta vulgaris* ssp. *vulgaris*) and their wild relative the sea beet (*B. v. maritima*). In this case there may be a high potential for gene flow from crops to wild populations because sugar beets are outcrossing, wind pollinated, genetically compatible with sea beets, and are often grown in proximate locations. An early study that used allozyme and RFLP data to quantify patterns of gene flow among wild beet populations found that the genetic structure of wild beets was patchy and showed a pattern that could be explained by founder effects within a metapopulation structure. Under this scenario, long-distance seed dispersal may be an important mechanism of gene flow between beet populations (Raybould *et al.*, 1996). A slightly later study that investigated the possibility of gene flow between wild and cultivated beet populations found that alleles typical of the cultivated beets occurred at high frequencies in adjacent wild beet populations but were rare in geographically distant wild beets. This finding suggested that gene flow between cultivated and wild beets is relatively high when they are separated by short distances (Bartsch *et al.*, 1999).

The finding of crop–weed hybridization across small geographical distances did not preclude the possibility that rare long-distance dispersal via seeds – in a manner compatible with the metapopulation structure that was found in wild beets – was leading to occasional hybridization between cultivated and wild plants over relatively broad spatial scales. The possibility of seed-mediated gene flow from domestic to wild populations was investigated in a later study that used a chloroplast marker plus nuclear microsatellites to quantify gene flow between cultivated beets and a neighbouring coastal wild population in France (Arnoud *et al.*, 2003). Because chloroplasts are inherited maternally in beets, this combination of nuclear and chloroplast data allowed the researchers to compare pollen-mediated gene flow with seed-mediated gene flow, and they concluded that the latter was the most important mechanism for the movement of genes from beet crops into wild populations.

This was an important finding because studies on gene flow between crops and wild populations have typically concentrated on pollen-mediated gene flow. The occurrence of seed-mediated gene flow means that the potential effects of long-lived seed banks on the dispersal of transgenic organisms need to be incorporated into future studies, as does the possibility of long-distance transportation of seeds by humans and other animals (Arnaud *et al.*, 2003). Although estimates of gene flow and the likelihood of hybridization events cannot give us precise answers about the spread of transgenic crops, they do provide us with some valuable baseline data that can be added to more detailed models that may help us to predict the probability that GM crops will escape into the wild.

# Fishing

Our final example of molecular ecology in a wider context is fishing. A quick perusal through the contents of the journal *Molecular Ecology* reveals numerous studies that are important to commercial and sports fisheries. Many species of fish are in decline, and in this section we will look at a few examples that show us how population genetics can help us to understand why fish populations are dwindling so rapidly, and can assist us with the evaluation of stock enhancement programmes.

## Overfishing

Over the past few decades, overfishing has become a global concern that has serious repercussions for marine and freshwater ecosystems, and also profound and lasting economic and social impacts. Canada's northern cod fishing industry, for example, was at its peak in 1968, when 810 000 tonnes of cod were caught in a single year. A steady decline since then means that the current cod stock is now estimated at less than 2 per cent of what it was 20 years ago, and the fishery was closed indefinitely as of 2003. According to some estimates, 90 per cent of all large fish have disappeared from the world's oceans (Myers and Worm, 2003) and fishing quotas are now in effect in many countries around the world for numerous fish and shellfish species. Overfishing owes much to a complex mix of politics, economics and social concerns, but has also been influenced by a historically poor understanding of the population ecology of different species.

The census size of a population often provides a starting point for unravelling population dynamics, although obtaining accurate counts of marine populations can be logistically challenging. One commonly used method, known as virtual population analysis, is based on commercial catches, although estimates derived in this way may be inaccurate for a number of reasons, including flawed reporting. Another approach is known as the annual egg production method. This combines estimates of the number of planktonic fish eggs with various population parameters, such as average female fecundity and population sex ratio, to obtain a size estimate of the total breeding population. This technique requires accurate identification of fish eggs, which can be problematic because multiple species produce eggs of a similar size and appearance. A recent study used species-specific probes in quantitative PCR reactions to determine just what proportion of cod-like eggs in the Irish Sea had actually been laid by cod. Only 34 per cent of positively identified eggs were cod, whereas 8 per cent were haddock and 58 per cent were whiting. This study shows how breeding populations may be grossly overestimated following the inaccurate identification of morphologically similar fish eggs (Fox et al., 2005).

Even if census population size estimates are reasonably accurate, recent studies have suggested that the census population size may actually tell us very little about the effective population size ($N_e$), and hence the genetic diversity, of fish populations. Loss of genetic diversity was for many years considered to be a relatively unimportant factor in the demise of fish stocks because the numbers of dramatically depleted stocks may still be in the millions, and population genetics theory tells us that genetic diversity will not deplete substantially until population sizes become much smaller than that. However, this is true only of the *effective* population size. In one study, researchers used archived scale samples from a population of New Zealand snapper (*Pagrus auratus*) to calculate $N_e$ based on temporal changes in allele frequencies (Chapter 3), and $N_e$ was found to be approximately five orders of magnitude smaller than the census population size, $N_c$ (Hauser et al., 2002). As a result, a population of several million snappers could have genetic properties more typical of a population of several hundred fish. In this particular example the discrepancy was at least partially attributable to a high variation in individual reproductive success. It is clear from this study that $N_c$ may be a grossly inaccurate method for estimating the genetic diversity, and hence the long-term viability, of fish populationse.

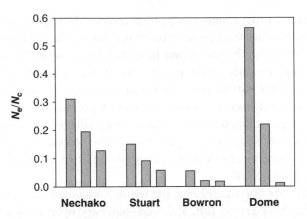

**Figure 8.4** The $N_e/N_c$ ratios of Chinook salmon in four different populations in the Upper Fraser River watershed. The $N_e$ values were calculated from temporal variations in allele frequencies across three time intervals over a 25-year period. Adapted from Shrimpton and Heath (2003)

Relatively small effective population sizes are being discovered in a growing number of fish populations. The temporal variance in allele frequencies was used to calculate the $N_e$ of red drum (*Sciaenops ocellatus*) in the northern Gulf of Mexico, and subsequent calculations put the $N_e/N_c$ ratio at around 0.001 (Turner, Wares and Gold, 2002). The same approach was used to calculate $N_e$ from scale samples that were collected over a 20-year period from five populations of Chinook salmon (*Oncorhynchus tshawytscha*) from the Upper Fraser River watershed in British Columbia, Canada. Recent increases in the census population sizes suggested that management practices have been successful, but low $N_e/N_c$ ratios showed that simple censuses of populations provide incomplete pictures of their recovery (Figure 8.4). In Chinook salmon, reduced $N_e$ values have been attributed to a combination of fluctuating $N_c$, uneven sex ratios, and variation in reproductive success (Shrimpton and Heath, 2003). Recent findings such as these are highlighting the need to manage depleted fishing stocks on the basis of their $N_e$ as opposed to their $N_c$, because the former will give us a more accurate picture of the future adaptability and persistence of fish populations.

## Stock enhancement

Remedial action aimed at halting or reversing the decline of fish populations often involves stock enhancement programmes that supplement wild populations with hatchery-reared individuals. The success of these programmes will depend in part on the survival rates of hatchery fish once they have been released into the wild, and also on the fitness of hatchery–wild hybrids. One way in which the success of stock enhancement programmes can be monitored is through genetic tags, such as a mitochondrial haplotype or an allozyme that is rare in the wild population. If the frequency of this marker increases over time then the programme is considered to be a success because this could occur only if the hatchery-reared fish were interbreeding with the natural population and producing viable offspring (Hansen *et al.*, 1995). Another way to monitor enhanced populations is through **mixed stock analysis**, a general approach for determining which of a number of possible populations are contributing to a particular mixed stock. This can be used to calculate the proportion of wild and hatchery-reared fish in an enhanced population (Debevec *et al.*, 2000).

Parentage analysis is yet another approach that was used to evaluate the fitness of wild fish, hatchery-reared fish, and wild–hatchery hybrids in a population of rainbow trout (*Oncorhynchus mykiss*). The authors of this study used six microsatellite loci that were sufficiently variable (up to 24 alleles at each locus) to exclude all but one parent-pair for each sampled offspring (Miller, Close and Kapuscinski, 2004). The survival of offspring in each class was then monitored. Survival of hatchery offspring, hybrids and back-crosses relative to the wild offspring was 0.21, 0.59 and 0.37, respectively. Survival is clearly reduced in both hatchery-produced fish and hatchery–wild hybrids, meaning that stock enhancement may actually be lowering the chances that this population will survive in the long term. Table 8.4 outlines some other ways in which molecular markers can help to assess whether or not the addition of cultured fish to wild populations has been beneficial or detrimental.

**Table 8.4** Some of the genetic effects that may follow the supplementation of a wild fish population with cultured stocks. Adapted from Utter and Epifanio (2002) and references therein

| Genetic effect | Test |
| --- | --- |
| Reduced $N_e/N_c$ following introduction of hatchery stock that has relatively low genetic diversity | Compare $N_e$ of populations with similar $N_c$ that have and have not been subjected to stock enhancement |
| Reduced $N_e$ following either an increase in competition or the introduction of pathogens | As above |
| Reduced genetic diversity following genetic swamping of the wild fish by the cultured fish | Mixed stock analysis to determine the proportion of wild, cultured and hybrid fish over several generations |
| Reduced fitness (and $N_e$) in hybrids following the loss of adaptive alleles (outbreeding depression) | Compare fitness of parents and hybrid offspring |
| Reduced fitness (and $N_e$) in hybrids following the loss of a co-adapted genome (outbreeding depression) | Compare fitness of parents with $F_2$ and subsequent generations of offspring |

# The Future of Molecular Ecology

It seems appropriate to end this book with a brief discussion on what the future may hold for molecular ecology. Although it is still a young discipline, molecular ecology has nevertheless changed dramatically over the past 20 years. Molecular data from many different genomic regions are now being acquired almost routinely from a wide range of samples, including blood, tissue, pieces of insect wings, fish scales, seeds, leaves, hairs, toenails, bone, faeces and urine, some of which can be collected without ever seeing the organism that is being genetically profiled. Increasingly refined methods of extraction mean that DNA can be obtained from extremely small samples and, although caveats such as null alleles are more of a concern for particularly small or degraded samples, non-lethal sampling is becoming an increasingly favoured option.

Over the last 10 years or so, microsatellites have emerged as one of the most commonly used type of molecular marker, but their popularity may wane in the future if more precise markers such as SNPs become available for an increasing number of species. There is also a growing movement towards the characterization of individuals and populations based on genes that have a known function, as opposed to selectively neutral markers. Although genome approaches are currently out of reach of most molecular ecologists, they are nevertheless of great interest because of their potential to identify the functions of genes. **Microarrays** can be used to simultaneously assay hundreds or even thousands of genes, and the increase or decrease in expression of these genes can be monitored under different conditions such as altered light or $CO_2$ availability. Although microarrays for some years were restricted mainly to model organisms such as *Arabidopsis*, humans and yeast, they are now being used to investigate an increasingly wide range of species. For example, DNA microarrays have been used to quantify gene expression in the digestive tracts of zebra fish (*Danoi rerio*) to see how gut microbiota affect a wide range of biological processes (Rawls, Samuel and Gordon, 2004), to study the expression of detoxifying genes in different strains of the malaria-transmitting mosquito *Anopheles gambiae* (David *et al.*, 2005), and to monitor the up-regulation of genes by the tobacco plant *Nicotiana attenuata* in response to attacks by different types of predators (Heidel and Baldwin, 2004).

Although genomics can provide tremendous benefits to ecologists, it is also true that molecular biologists can benefit from an integrative approach that embraces ecology. The Multinational Coordinating *Arabidopsis* 2010 Project aims to identify the function of all ∼25 000 *Arabidopsis* genes by the end of 2010, but the adaptive functions of an organism cannot be fully understood without exploring the interactions between its genotype, its phenotype, and the environment in which it lives (Jackson *et al.*, 2002). This is just one reason why interdisciplinary approaches to research are likely to become increasingly commonplace.

It is also probable that in the future we will see more interplay between molecular ecology and a range of traditionally non-ecological applications such as those outlined earlier in this chapter. Applied ecology may also become more prominent because of the growing number of ecological problems that continue to arise around the world following climate change and introduced species. However, although the future should see the growth of molecular ecology continue unabated, we must not lose sight of the accomplishments that have already been made. Molecular genetic data allow us to quantify the genetic diversity and infer the demographic history of a population, identify the likely roles of genetic drift versus local adaptation in the differentiation of populations, pinpoint the origin and subsequent dispersal patterns of introduced species and pathogens, identify hybrids (which can be particularly important in both conservation and evolutionary biology), quantify inbreeding and outbreeding depression in the absence of pedigrees, and ascertain levels of relatedness within both wild and captively-bred populations. This is an incomplete but nevertheless impressive list for such a young discipline and, although we cannot know what the future will bring, few could doubt that the theory, methods and applications of molecular ecology will continue to grow.

## Chapter Summary

- Many of the methods used in molecular ecology are relevant to other areas of research. As a result, interdisciplinary collaborations are becoming increasingly common between molecular ecology and a range of academic and applied fields that include epidemiology, agriculture, forensic science and fisheries.

- Molecular ecology and wildlife forensics overlap in a number of ways. If sufficiently variable markers and population allele frequency data are available, blood or tissue samples can be matched to carcasses in cases of suspected poaching. Assignment tests may enable us to identify which population a carcass or sample originated from.

- Species-specific markers, usually sequence data, can be used in poaching cases, although their use is probably more widespread in cases of illegal trafficking. By amplifying DNA from a marketable item such as food or traditional Chinese medicine, we can determine whether or not it includes material from an endangered species.

- Phylogeographic analyses can help us to identify the source and subsequent dispersal patterns of pests and pathogens. Species- or strain-specific markers can help to resolve some law enforcement issues that are relevant to agriculture, and the quantification of gene flow between genetically modified crops and their wild relatives is an important part of the risk assessments that are done on GM crops.

- High-throughput PCR analyses have shown that estimates of fish census population sizes may be inaccurate when based on eggs, because these can often be assigned to the wrong species when identifications are based solely on morphology.

- Estimates of $N_e$ have shown that, even if census population sizes are accurate, they can be extremely poor predictors of the long-term viability of fish populations; as a result, biologists now believe that a depletion in genetic diversity may be contributing to the decline of at least some fish stocks.

- Stock enhancement programmes are sometimes used to slow or reverse the decline of fish populations, and their efficacy can be evaluated through the genetic identification and survival estimates of wild versus hatchery-reared fish and their offspring.

- Although molecular ecology is a relatively young discipline, it already has an impressive list of accomplishments. These will undoubtedly grow in the near future as increasingly precise molecular markers continue to be developed, and as biologists move towards using markers that have a known genetic function.

## Useful Websites and Software

- DNA Forensics, a division of Human Genome Project Information: http://www.ornl.gov/sci/techresources/Human_Genome/elsi/forensics.shtml

- US Fish & Wildlife National Forensic Laboratory: http://www.lab.fws.gov/

- Natural Resources DNA Profiling and Forensic Centre: http://www.nrdpfc.ca/index.html

- Convention on International Trade in Endangered Species of Wild Fauna and Flora (CITES): www.cites.org

- World Wildlife Fund website on wildlife trade: http://www.worldwildlife.org/trade/index.cfm

- Functional Genomics Characterization of *Arabidopsis* Genes: http://www.arabidopsis.org/info/2010_projects/index.jsp

- A gateway to invasive species activities and programmes in the USA: http://www.invasivespecies.gov/new/new1102.shtml

- DNAMix – forensic science software to determine whether samples came from one or more individuals: http://statgen.ncsu.edu/~gwbeecha/

- SPAM – Statistics Program for Analysing Mixtures. Determines the number and proportion of different populations that are contributing to a mixed stock: http://www.genetics.cf.adfg.state.ak.us/software/download/download37.php

# Further Reading

## Books

Balding, D.J. 2005. *Weight-of-Evidence for Forensic DNA Profiles*. John Wiley, Chichester.
Pimental, D. 2002. *Biological Invasions: Economic and Environmental Costs of Alien Plant, Animal, and Microbe Species*. CRC Press, Boca Raton, FL.

## Review articles

Gibson, G. 2002. Microarrays in ecology and evolution: a preview. *Molecular Ecology* 11: 17–24.
Holmes, E.C. 2004. The phylogeography of human viruses. *Molecular Ecology* 13: 745–756.
Hutchings, J.A. 2000. Collapse and recovery of marine fishes. *Nature* 406: 882–885.
Jackson, R.B., Linder, C.R., Lynch, M., Purugganan, M., Somerville, S. and Thayer, S.S. 2002. Linking molecular insight and ecological research. *Trends in Ecology and Evolution* 17: 409–414.
Jobling, M.A. and Gill, P. 2004. Encoded evidence: DNA in forensic analysis. *Nature Reviews Genetics* 5: 739–751.
Liu, Z.J. and Cordes, J.F. 2004. DNA marker technologies and their applications in aquaculture genetics. *Aquaculture* 238: 1–37.
Macdonald, C. and Loxdale, H.D. 2004. Molecular markers to study population structure and dynamics in beneficial insects (predators and parasitoids). *International Journal of Pest Management* 50: 215–224.
Stewart, C.N., Halfhill, M.D. and Warwick, S.I. 2003. Transgene introgression from genetically modified crops to their wild relatives. *Nature Reviews Genetics* 4: 806–817.
Thomas, M.A. and Klaper, R. 2004. Genomics for the ecological toolbox. *Trends in Ecology and Evolution* 19: 439–445.

# Review Questions

**8.1.** Four microsatellite loci were used to genotype black bears from three Canadian National parks: Banff in Alberta (B), La Mauricie in Quebec (LM), and Terra Nova in Newfoundland (TN) (Paetkau and Strobeck, 1994). The results of this study are shown in Table 8.5.

**Table 8.5** Some of the results from a microsatellite-based study of black bears from three Canadian National parks

| Park | Expected heterozygosity | Total number of alleles | Probability of identity |
|------|------------------------|------------------------|------------------------|
| B | 0.801 | 32 | $1.1 \times 10^{-5}$ |
| LM | 0.783 | 35 | $2.2 \times 10^{-5}$ |
| TN | 0.360 | 9 | $4.6 \times 10^{-2}$ |

Are these microsatellite loci sufficiently variable to allow for individual genetic identification in cases of poaching? In answering this question, refer to the average number of individuals in each population that would need to be sampled before duplicate genotypes were found. Why do these markers provide greater resolution in some populations compared with others?

**8.2.** The mutilated flesh of an unknown animal was sent to a laboratory for species identification. Biologists there used universal primers to amplify and sequence part of the mitochondrial cytochrome *b* gene. Part of this sequence was:

AGGAGCAACA GTCATTACCA ACCTTCTCTC AGCAATTCCA TATATTGGTA CAAATCTAGT CGAATGGATC TGAGGGGGCT TTTCAGTAGA TAAAGCAACC

Go to the National Center for Biotechnology Information (NCBI) website (http://www.ncbi.nlm.nih.gov/) and click on the BLAST option on the horizontal bar immediately below the page title. Under the nucleotide heading, click on the Nucleotide-nucleotide BLAST (blastn) option. Enter this sequence into the box and then click on the Blast button. You will see a list of alignments, the first of which has the closest match to your sequence. What species does this mutilated carcass belong to?

**8.3.** The boll weevil *Anthonomus grandis*) has been the most destructive pest of cotton in the USA since it invaded that country around 100 years ago. RAPD markers were used to study gene flow and dispersal patterns among boll weevils in the south-central Cotton Belt of the USA and northeast Mexico (Kim and Sappington, 2004). Weevils were genotyped from 18 populations distributed among three geographical regions. Genetic diversity was highest in Mexico and lowest towards the northern limit of the boll weevil's geographical range. The *Nm* estimates (derived from $F_{ST}$ values) suggest that gene flow is relatively frequent *within* regions (populations separated by up to 500 km) but much more limited *between* regions. What do these patterns suggest to you about the colonization and subsequent dispersal of the boll weevil in the USA?

**8.4.** Why might the stocking of wild fish populations with hatchery-reared fish reduce the overall $N_e/N_c$ ratio?

# Glossary

**A** *see* allelic diversity

**Adaptive radiation** The rapid diversification of an ancestral population into several ecologically different species, each of which is adapted to a specialized environmental niche

**AFLP** *see* amplified fragment length polymorphism

**allele** A particular form of a given gene

**allele frequency** The proportion of each allele at a given locus within a designated group such as a population

**allelic diversity (A)** A measure of a population's genetic diversity, calculated as the average number of alleles per locus

**allopatric** Two or more species that occur in non-overlapping geographical areas

**allopolyploid** A polyploid that was created by hybridization between two different species. Compare with autopolyploid

**allozygous** Two alleles at a locus that have *not* recently descended from a common ancestor (note that they may still be identical). Compare with autozygous

**allozymes** Alternative forms of a protein that are encoded by different alleles at a particular locus

**amino acid** One of a group of 20 organic molecules that contain an amino group and a carboxyl group; amino acids are the usual components of proteins

**amplified fragment length polymorphism (AFLP)** A dominant molecular marker that genotypes individuals at multiple loci. Amplified DNA is digested with restriction enzymes, adapters are attached to the ends of amplified fragments, and a subset of fragments is then re-amplified to generate a multi-band pattern

*Molecular Ecology* Joanna Freeland
© 2005 John Wiley & Sons, Ltd.

**angiosperms**  Plants that produce seeds in reproductive organs that are called flowers; also known as flowering plants

**arithmetic mean**  The average of all the measurements within a data set, calculated by dividing the sum of all measurements by the total number of measurements

**assignment tests**  A method for quantifying dispersal by assigning each individual to the population in which its genotype has the highest probability of occurring, and from which it therefore most likely originated

**autopolyploid**  A polyploid that has more than two sets of homologous chromosomes, all of which came from the same species. Compare with allopolyploid

**autosomes**  Chromosomes other than sex chromosomes

**autozygous**  Two alleles that are identical by descent, i.e. they are copies of an ancestral gene (homologous). Compare with allozygous

**balancing selection**  A form of natural selection that favours the maintenance of multiple alleles within a population; includes heterozygote advantage and frequency-dependent selection

**Bayesian**  A method of statistical inference that uses prior information to test the likelihood that various parameters can explain a particular data set

**benthic**  Living in or on the substrate (e.g. mud, sand, rocks) that underlies a body of water. Compare with planktonic

**bifurcating tree**  A phylogenetic tree depicting a series of events in which one ancestral lineage splits into two descendant lineages. Contrast with network

**biological species concept (BSC)**  A concept that places individuals within the same species if they are potentially capable of interbreeding and producing viable offspring

**biparental inheritance**  Genes and genetic elements that are inherited from both parents; applies only to sexually reproducing organisms

**bottleneck**  A severe, temporary reduction in the size of a population

**bounded hybrid superiority**  A force that maintains hybrid zones when hybrids have greater fitness within a limited geographical area than non-hybrids

**BSC**  *see* biological species concept

**cDNA**  *see* complementary DNA

**census population size** $(N_c)$  The number of individuals within a population

**chloroplast**  An organelle that occurs in large numbers in the cytoplasm of plant cells and some eukaryotic algae and is responsible for photosynthesis

**chloroplast DNA (cpDNA)**  A circular DNA molecule that constitutes the chloroplast genome. This is inherited maternally in most angiosperms and paternally in most gymnosperms

**cline** A geographical gradient in one or more characteristics, such as a change in mean body size or allele frequencies. Clines are sometimes taken as evidence for natural selection along an environmental gradient

**chromosome** A structure that is located in the nucleus of a eukaryotic cell on which nuclear genes are located

**co-adapted gene complex** The group of loci that are involved in a beneficial epistatic interaction

**coalescent theory** A theory in population genetics that starts with multiple contemporary samples and works backwards through time to identify the point at which alleles coalesce (come together in a common ancestor)

**co-dominant markers** Molecular markers that allow users to distinguish between heterozygotes and homozotyes

**codon** Three consecutive nucleotides along an mRNA molecule that specify either a particular amino acid, or the end of protein synthesis

**complementary DNA (cDNA)** A segment of DNA that is generated following reverse transcription of an RNA template

**conspecific brood parasitism** (CBP) A behaviour in which a female lays an egg in the nest of another conspecific female (also known as egg dumping)

**cpDNA** *see* chloroplast DNA

**cytonuclear disequilibrium** In hybrids, the coexistence of cytoplasmic markers (mitochondria, chloroplasts) from one parental species or population, and nuclear markers from another

**cytotypes** Alternative levels of polyploidy within a single species

**deoxyribonucleic acid (DNA)** A nucleic acid macromolecule that contains the bases adenine, cytosine, guanine and thymine and carries the genetic information of all cells and many viruses

**diapause** A time of dormancy between periods of activity during which physiological activity is temporarily suspended

**dideoxyribose** A deoxyribose sugar that lacks the 3' hydroxyl group and can therefore be used to terminate the extension of a DNA sequence during a sequencing reaction

**dioecious** Plant or animal in which individuals have either male or female reproductive organs, but not both. Compare with monoecious

**diploid** An organism that has two complete sets of homologous chromosomes ($2n$)

**directional selection** Selection that favours either a higher or lower value of a particular characteristic relative to its current mean within a population

**displacement loop**    A loop that is formed during DNA replication when the origin of replication is different on the two strands of DNA. Also known as a D-loop, this is the most variable region of sequence in many animal mitochondrial genomes

**disruptive selection**    Natural selection that favours two or more extreme phenotypes that are both fitter than the intermediate phenotypes that lie between them. Also known as diversifying selection

**distance methods**    Methods of phylogenetic reconstruction that are based on the evolutionary (genetic) distances between all pairs of taxa

**diversifying selection**    *see* disruptive selection

**D-loop**    *see* displacement loop

**DNA**    *see* deoxyribonucleic acid

**DNA barcode**    *see* genetic barcode

**DNA polymerase**    An enzyme that catalyses the synthesis of new DNA strands using a template DNA sequence

**DNA replication**    Duplication of a DNA molecule using the original DNA sequence as a template for the synthesis of a new DNA sequence

**dominance**    A mechanism that promotes inbreeding depression, so-called because the favourable alleles at a locus are usually dominant whereas deleterious alleles are recessive (masked). The increased homozygosity that results from inbreeding means that deleterious recessive alleles are more likely to be expressed

**dominant markers**    Molecular markers that do not allow us to distinguish between homozygote and heterozygote individuals

**effective population size ($N_e$)**    The theoretical size of an ideal population (one with an even sex ratio, random mating and no variation in reproductive success) that will lose genetic variation through drift at the same rate as an actual population. If an actual population of 500 individuals is losing genetic variation at a rate that would be found in an ideal population of 100 individuals, then this population would have $N_c = 500$ but $N_e = 100$

**electrophoresis**    *see* gel electrophoresis

**environmental sex determination**    Occurs when certain environmental variables such as temperature play a major role in determining the sex of an organism. Compare with genetic sex determination

**enzyme**    A protein that catalyses a specific biochemical reaction

**epistasis**    The interaction of genes from multiple loci and their collective influence on particular traits

**eukaryote**    Found in higher animals and plants, fungi, protozoa and most algae, eukaryotic cells have a membrane-bound nucleus that contains chromosomes and proteins. Other cell components include membrane-bound organelles

**eusociality**  A form of social breeding that is characterized by overlapping generations, cooperative care of young, and specialized castes that do not reproduce; most common in social insects

**evolutionarily significant units**  (**ESU**) In conservation biology, these are populations that show enough genetic differentiation (monophyletic mtDNA and significantly different nuclear allele frequencies) to warrant their management as distinct units

**exon**  The part of a gene's DNA sequence that codes for amino acids and is retained in mRNA after introns are removed

**expected heterozygosity ($H_e$)**  A commonly used measure of genetic diversity that reflects the level of heterozygosity that would be expected if a population was in Hardy-Weinberg equilibrium. Compare with observed heterozygosity

**extra-pair fertilization** (**EPF**)  Achieved when an individual is fertilized by a mate that is outside its social pair bond

**$F$**  see $F_{IS}$

**$F_{IS}$**  The inbreeding coefficient (also designated $F$). In diploid species, the probability that an individual has two alleles at a particular locus that recently descended from a single common ancestor

**$F_{IT}$**  An overall inbreeding coefficient that compares the heterozygosity of an individual to that of the total population under consideration; influenced by both $F_{IS}$ and $F_{ST}$

**fitness**  The relative ability of an individual to survive and reproduce compared with other members of its population

**fitness value ($w$)**  The fitness of an individual with a particular genotype (and phenotype) relative to the genotype that has the maximum fitness within the population

**fixation**  An allele reaches fixation when it has attained a frequency of 100 per cent within a population

**founder effect**  The changes in allele frequencies, relative to the source population, that are often evident in populations that have been founded by a small number of individuals

**frameshift mutation**  A mutation that inserts or deletes one or more pairs of nucleotides within a gene and, in so doing, shifts the reading frame of all subsequent codons

**frequency-dependent selection**  If the fitness of a genotype depends on its frequency within a population, it will be selected for or against following frequency-dependent selection

**$F_{ST}$**  The probability that two random gametes, drawn from the same subpopulation, will be identical by descent, relative to gametes taken from the entire population; developed by Wright (1951), this is the most common measurement used to describe the genetic differentiation of populations

**$F$-statistics**  A series of measurements ($F_{IS}$, $F_{ST}$ and $F_{IT}$) that use inbreeding coefficients to describe the partitioning of genetic variation within and among populations

**gamete**   A mature reproductive cell such as sperm, egg or pollen; also known as a germ cell

**gel electrophoresis**   The separation of molecules (DNA or proteins) through a gel medium along an electrical field

**geminate species**   Pairs of morphologically similar, closely related species that share a recent common ancestor

**gene conversion**   A genetic process in which one sequence replaces another at a homologous locus; leads to genotypic ratios that deviate from those expected under normal Mendelian inheritance

**gene diversity** ($h$)   A measure of genetic diversity based on the probability that two alleles chosen randomly from the population will differ; often equivalent to expected heterozygosity when calculated for nuclear loci

**gene expression**   A process involving numerous steps that convert the information encoded in a gene into a protein

**gene flow**   The transfer of genetic material from one population to another following the dispersal and subsequent reproduction of individuals, propagules or gametes

**gene genealogy**   The evolutionary history of a particular gene; usually represented as a branching tree or network that shows the historical relationships of multiple lineages

**genetic barcode**   A sequence of DNA that is specific to, and hence an identifier of, a particular species. Also known as a DNA barcode

**genetic code**   A set of 64 codons, 61 of which specify one of the 20 amino acids and three of which signal the end of translation

**genetic diversity**   The amount of genetic variation that is contained within a population or species

**genetic drift**   A process that changes the allele frequencies within a population from one generation to the next because of the random sampling of gametes

**genetic load**   The decrease in the mean fitness of a population due to the presence of genotypes that confer relatively low rates of survival and reproduction

**genetic rescue**   A reduction in inbreeding depression following the introduction into a population of new alleles via immigrants

**genetic sex determination**   Occurs when sex chromosomes play a major role in determining the sex of an organism. Compare with environmental sex determination

**genome**   The full set of DNA in a cell or organism. In eukaryotes, the nuclear, mitochondrial and plastid genomes may be considered separately

**genomics**   The branch of genetics that studies the genomes of organisms and the functions of different genes

**genotype**   The set (or subset) of genes possessed by an individual

**germ cell** *see* gamete

**good genes hypothesis** A hypothesis in behavioural ecology that states that an individual will choose its reproductive partner on the basis of some measure of phenotypic superiority

$G_{ST}$ An analogue of $F_{ST}$, $G_{ST}$ is equivalent to $F_{ST}$ when there are only two alleles at a locus, and is the weighted average of $F_{ST}$ for all alleles when there are multiple alleles

**gymnosperms** Vascular plants in which the seeds develop on the scales of a cone or cone-like structure; include conifers, *Gingko* and cycads

**Haldane's rule** A rule stating that hybrids are less likely to be viable in the heterogametic sex compared with the homogametic sex

**haploid** Having only one set of chromosomes

**haploid diversity** A measure of gene diversity ($h$) that describes the numbers and frequencies of different haplotypes. Most commonly applied to mitochondrial and chloroplast data

**haplotype** A particular form of a gene or DNA sequence. In molecular ecology this is most commonly used in reference to the haploid genomes of organelles (mitochondria, chloroplasts)

**Hardy-Weinberg equilibrium (HWE)** A predictable ratio of genotype frequencies in a sexually-reproducing population of infinite size with random mating and no selection; represented as $p^2 + 2pq + q^2 = 1$

**harmonic mean** A method for calculating the mean of a set of numbers that is expressed as the reciprocal of the arithmetic mean of the reciprocals of the numbers

**hemizygous** A diploid individual who has only one allele at a particular locus; usually refers to sex-linked genes in heterogametic individuals, such as X-linked genes in male mammals

**heterogametic** Having two different sex chromosomes, e.g. male mammals (XY) or female birds (ZW). Compare with homogametic

**heteroplasmy** The state of a single individual having two or more genetic variants of mitochondria or plastids

**heterosis** The superiority of hybrids compared with either parental type with respect to one or more traits such as increased growth rate or higher fertility; also known as hybrid vigour

**heterozygote** An individual that has more than one type of allele at a particular locus

**heterozygote advantage** The condition that results when a heterozygote has a higher fitness than either homozygote at a particular locus. Also known as overdominance

**hitch-hiking effect** If an allele at one locus is linked to an allele at another locus then the frequency of that allele will deviate from the expected frequency of an unlinked allele

**homogametic**  Having only one type of sex chromosome, e.g. female mammals (XX) or male birds (ZZ). Compare with heterogametic

**homology**  Describes a characteristic present in two or more individuals or species that is similar because it was inherited from a recent common ancestor. Compare with homoplasy

**homologous chromosomes**  Two or more chromosomes within a single individual that bear the same genes and pair up during meiosis

**homoplasy**  Describes a characteristic present in two or more individuals or species that is similar for reasons other than inheritance from a recent common ancestor. Compare with homology

**homozygous**  An individual that has only one type of allele at a particular locus

**horizontal transfer**  The transfer of genes from one species to another, most commonly following hybridization

**HWE**  *see* Hardy-Weinberg equilibrium

**hybrid**  The offspring that results from mating between a male and a female from two different species or populations (hydridization)

**hybrid vigour**  *see* heterosis

**hybrid zones**  Restricted geographical areas in which the members of two different species come into contact and hybridize with each other

**hybridization**  *see* hybrid

**identical by descent**  Alleles that are copies of a single allele that existed in a recent ancestor

**inbreeding coefficient**  *see* $F_{IS}$

**inbreeding depression**  A reduction in the fitness of offspring that result from matings between close relatives

**inclusive fitness**  A fitness value that reflects the extent to which an individual's genes are transferred from one generation to the next, either through its own offspring or through the offspring of its relatives

**indel**  The collective term given to mutations that involve nucleotide insertions or deletions

**infinite alleles model**  A mutation model that assumes that the number of possible alleles at a given locus is infinite and therefore any new allele must be different from all other alleles that are already present in the species in question

**introgression**  The movement of genes from one species or population into another following hybridization

**intron** Part of a gene's DNA sequence that does not code for amino acids and is spliced out of mRNA before transcription occurs; mostly restricted to eukaryotic genes

**isolation by distance** A pattern of population differentiation in which the genetic dissimilarity of populations is correlated with the geographical distance that separates them

**karyotype** The complement of chromosomes in a cell

**kin selection** Selection that acts on an individual's inclusive fitness, i.e. it factors in both the fitness of an individual and the fitness of its kin

**lineage sorting** The process in which multiple populations or species that are recently descended from a common ancestral population or species, but are now reproductively isolated from one another, each acquires a single genetic lineage

**linkage disequilibrium** Occurs when the alleles at two or more loci co-occur more often than is expected on the basis of their frequencies. Usually involves two loci that are close together on a chromosome

**linkage equilibrium** Occurs when the alleles at two or more loci co-occur to an extent that is expected on the basis of their frequencies

**locus** The location of a particular gene or region of DNA on a chromosome (plural is loci)

**major histocompatibility complex (MHC)** A large multigene family in vertebrates that controls several aspects of the immune response

**management units** (MU) In conservation biology, these are populations that have very low levels of gene flow and can therefore be genetically differentiated from other populations

**Mantel test** A statistical procedure that uses permutations to test the null hypothesis that two variables are independent of each other. In molecular ecology this is most commonly used to compare the geographical distance and genetic differentiation among populations, i.e. to test for isolation by distance

**maximum likelihood (ML)** An analytical method that provides an explanation for a particular data set by maximizing the probability of observed data under an explicit model

**meiosis** Cell division in which a diploid parent produces four haploid daughter cells (gametes). Compare with mitosis

**messenger RNA (mRNA)** RNA that acts as a template for protein synthesis

**metapopulation** A group of populations that is prone to local extinctions and recolonizations and is interconnected by gene flow

**MHC** *see* major histocompatibility complex

**microarray**   A two-dimensional array of DNA sequences that is set up on a glass, filter or silicon wafer. Typically used as a high-throughput method for simultaneously measuring the expression of many genes under different conditions in order to deduce the functions of these genes

**microsatellite**   A stretch of DNA that consists of a short tandem sequence of up to five base pairs that is repeated multiple times, e.g. $(AG)_{10}$ represents a microsatellite in which the sequence AG is repeated ten times

**minisatellite**   Repetitive DNA sequences that are each 10-100 bases long and are dispersed throughout the genome. Also known as VNTRs (variable number of tandem repeats)

**mitochondria**   Organelles that occur in large numbers in the cytoplasm of eukaryotic cells and are central to the production of ATP

**mitochondrial DNA (mtDNA)**   A circular molecule of DNA that is located within the mitochondrion; usually inherited maternally

**mitochondrial pseudogenes**   Non-functional copies of mitochondrial genes that are located in the nuclear genome. Also known as numts

**mitosis**   The replication of genetic material in eukaryotic cells that involves the segregation of chromosomes and subsequent nuclear division. Following mitosis, a diploid parental cell will become two diploid daughter cells. Compare with meiosis

**mixed stock analysis**   A type of analysis that uses data from molecular markers to determine the relative proportions of various stocks in a mixed-stock fishery

**ML**   *see* maximum likelihood

**molecular clock**   A theory that uses a calibration date to calculate the rate of sequence divergence, and then applies this rate to pairs of sequences to determine the amount of time that has passed since they diverged from one another

**monoecious**   Plants that have unisexual flowers, i.e. a single plant is effectively both male and female. Compare with dioecious

**monogamy**   A mating system in which a single female is partnered with a single male

**monogynous colony**   A social insect colony that has a single queen

**monomorphic**   A locus that has only one allele in a population and is therefore lacking in genetic variation

**monophyletic**   A group of organisms (e.g. a population or species) that all share a common ancestor and in which all of the descendants of that ancestor can be found

**most recent common ancestor**   (MRCA) Going back in time, this refers to the most recent ancestor of two or more genetic lineages

**mRNA**   *see* messenger RNA

**mtDNA**   *see* mitochondrial DNA

**multifurcating tree**   *see* network

**mutation**   An alteration in the structure of a gene or a chromosome

**mutational meltdown**   The potentially irreversible decline in the fitness of a small population that follows the fixation of numerous mildly deleterious alleles

$N_c$   *see* census population size

$N_e$   *see* effective population size

**negative selection**   *see* purifying selection

**Network**   A phylogenetic reconstruction that depicts a series of events in which at least some of the ancestral lineages have split into three or more descendant lineages. Also known as multifurcating tree. Compare with a bifurcating tree

**non-synonymous substitution**   A nucleotide substitution that alters a codon so as to cause one amino acid to be replaced with another

**nuclear DNA**   The complement of DNA that is arranged in chromosomes and located in the nucleus of a cell

**nucleotide diversity ($\pi$)**   A measure of genetic diversity that quantifies the mean sequence divergence among several haplotypes by factoring in both the frequencies and the pairwise divergences of different sequences

**nucleotide substitution**   A mutation that results in the substitution of one nucleotide for another

**null alleles**   Alleles that fail to amplify during a PCR reaction. Microsatellite null alleles can lead to the erroneous identification of homozygotes

**numts**   *see* mitochondrial pseudogenes

**observed heterozygosity ($H_o$)**   The actual level of heterozygosity within a population, usually averaged across several loci. Compare with expected heterozygosity

**oligonucleotide primer**   A short segment of single-stranded DNA or RNA that is used as a starting point for copying template sequences, e.g. in PCR. Often referred to simply as a primer

**organelle**   A membrane-bound structure found in the cytoplasm of a cell. Organelles have specialized functions and include mitochondria and plastids (including chloroplasts)

**outbreeding depression**   A reduction in fitness following the hybridization between members of two genetically distinct populations or species

**overdominance**   *see* heterozygote advantage

$\pi$   *see* nucleotide diversity

$P$   *see* proportion of polymorphic loci

**panmixia**  Random mating; within a panmictic population there is the potential for any male to reproduce with any female

**paraphyletic**  A group of organisms (e.g. a population or species) that includes some, but not all, of the descendants of a recent common ancestor

**parthenogenesis**  The development of an individual from an egg without the contribution of a paternal genotype

**patrilineal descent**  In sexually reproducing organisms, the transmission of genes through the male line of descent, e.g. the Y chromosome in mammals

**PCR**  *see* polymerase chain reaction

**phenotype**  The physical, observable properties of an organism that result from an interaction between its genotype and its environment

**phenotypic plasticity**  The ability of a particular genotype to develop into one of several different phenotypes, depending on environmental conditions

**philopatric**  The tendency to return to, or remain at, a particular location. A philopatric animal may never leave its birthplace or, if it does disperse, will later return for breeding purposes

**phylogenetic species concept (PSC)**  A concept that defines species as groups of individuals that share at least one uniquely derived characteristic and constitute monophyletic groups

**phylogeny**  A branching diagram, usually depicted as a tree, that shows the evolutionary relationships among different genetic lineages (e.g. species or genera)

**phylogeography**  A field of study that compares the evolutionary relationships of genetic lineages with their geographical locations in an attempt to understand which factors have most influenced the current distributions of genes, populations and species

**planktonic**  Living in open water. Compare with benthic

**plastid**  A family of membrane-bound cytoplasmic organelles (including chloroplasts) that contain DNA and are found in plant cells and some protists

**polyandry**  A mating system in which one female is socially bonded with multiple males

**polygynandry**  A mating system in which multiple males are socially bonded with multiple females

**polygynous colony**  A social insect colony that has multiple queens

**polygyny**  A mating system in which one male is socially bonded with multiple females

**polymerase chain reaction (PCR)**  A procedure that uses template DNA sequences to produce numerous copies of a specific segment of DNA. The reaction typically consists of 30-40 cycles that each have three parts: denaturation of DNA, annealing of primers, and extension of sequences

**polymorphic**   A locus that has multiple alleles in a population

**polypeptide**   A linear chain of amino acids that is smaller than a protein

**polyphyletic**   A group of organisms (e.g. a population or species) that do not all share the same recent common ancestors

**polyploid**   Having three or more complete sets of homologous chromosomes

**population**   A potentially interbreeding group of individuals that belong to the same species and live within a restricted geographical area

**population bottleneck**   *see* bottleneck

**positive selection**   A form of directional selection that selects *for* a particular mutation that increases the fitness of carriers

**primary sex ratio**   The sex ratio of a brood at the time of conception. Compare with secondary sex ratio

**primer**   *see* oligonucleotide primer

**probability of identity (PI)**   The likelihood that two individuals chosen at random from a population will have the same genotype

**probe**   A length of RNA or single-stranded DNA that is labelled (usually either radio-actively or fluorescently) and then used to locate complementary sequences from a pool of DNA through hybridization with the target sequence

**prokaryotes**   Unicellular organisms that lack a nucleus

**promiscuity**   Mating in the absence of any social pair bonds

**proportion of polymorphic loci (P)**   A measure of genetic variation that reflects the proportion of loci at which two or more alleles have been identified

**PSC**   *see* phylogenetic species concept

**pseudogenes**   Non-functional copies of genes

**purging**   The elimination of deleterious alleles from an inbred population following natural selection

**purifying selection**   Selection that eliminates deleterious mutations from a population or reduces their frequencies. Also known as negative selection

**purines**   Two types of bases (adenine and guanine) that are found in DNA and RNA

**pyrimidines**   Two types of bases that are found in DNA (cytosine and thymine) and RNA (cytosine and uracil)

**qualitative traits**   Discrete traits that are controlled by one or a few loci, such as eye colour or blood type. Compare with quantitative traits

**quantitative PCR**   *see* real-time PCR

**quantitative trait locus (QTL)**   A locus that affects a quantitative trait

**quantitative traits**   Continuous traits such as height and weight that are influenced by several different genes and also tend to be influenced by environmental factors. Compare with qualitative traits

**quasi-parasitism (QP)**   Behaviour in which a female lays an egg in a nest that is shared by the male with whom she mated and his social partner

**random amplified polymorphic DNA (RAPD)**   Dominant molecular marker that generates multiple DNA fragments through the random PCR amplification of multiple regions of the genome using single arbitrary primers

**real-time PCR**   A method of PCR that allows users to monitor the amplification reaction and, in so doing, quantify the DNA (or cDNA, if the aim is to quantify RNA) that is present in the original sample. Also known as quantitative PCR

**reciprocal monophyly**   The situation that arises when two populations or species become monophyletic with respect to each other

**recombination**   The exchange of DNA segments between DNA molecules or chromosomes

**restriction enzymes**   Bacterially-derived enzymes that digest DNA sequences at specific locations; each restriction enzyme recognizes and cuts a particular sequence of 4-8 nucleotides

**restriction fragment length polymorphism (RFLP)**   Dominant molecular marker that generates multiple fragments of DNA by digesting an entire genome or a pre-selected stretch of DNA with one or more restriction enzymes. The resulting band patterns will vary depending on the underlying DNA sequences, because these will determine the number of restriction sites that are found in each individual

**reticulated evolution**   The fusing of separate branches on an evolutionary tree, e.g. following hybridization between previously distinct lineages

**RFLP**   *see* restriction fragment length polymorphism

**ribonucleic acid (RNA)**   A single-stranded nucleic acid macromolecule with ribose sugar that contains four bases: adenine, cytosine, guanine and uracil

**ribosomal RNA (rRNA)**   RNA molecules that, along with proteins, make up the structures (ribosomes) where protein synthesis occurs

**RNA**   *see* ribonucleic acid

**rRNA**   *see* ribosomal RNA

$R_{ST}$   An analogue of $F_{ST}$ that assumes a stepwise mutation model and in some situations may be more appropriate than $F_{ST}$ for quantifying population differentiation on the basis of microsatellite data

*s*   *see* selection coefficient

**scn DNA**   *see* single-copy nuclear DNA

**secondary sex ratio**   The sex ratio of a brood once the eggs have hatched. Compare with primary sex ratio

**selection coefficient (*s*)**   A measurement of the reduced probability of survival associated with an individual's fitness value (*w*); calculated as $s = 1 - w$

**self-fertilization**   Fertilization involving male and female gametes from the same individual. Also known as selfing

**selfing**   *see* self-fertilization

**sex-biased dispersal**   Occurs in a species when one sex is more likely that the other sex to disperse and subsequently reproduce at a location away from their birthplace

**sex chromosome**   A chromosome that is at least partially responsible for the determination of an individual's sex, e.g. X and Y chromosomes in mammals and Z and W chromosomes in birds

**single-copy nuclear DNA (scnDNA)**   A DNA sequence that occurs only once in each haploid nuclear genome

**single nucleotide polymorphism (SNP)**   A variation between two sequences of DNA that is caused by a single nucleotide substitution

**size homoplasy**   A phenomenon that describes microsatellite alleles that are the same size for a reason other than recent descent from a common ancestor

**slipped-strand mis-pairing**   An imperfect re-alignment of DNA strands following slippage of DNA polymerase and temporary displacement of strands during DNA replication

**SNP**   *see* single nucleotide polymorphism

**somatic cell**   Any cell other than germ cells

**stabilizing selection**   A process of natural selection that favours individuals that have the mean form of a particular character, and selects against extreme phenotypes

**standard deviation**   The square root of the variance of a data set

**statistical parsimony network**   A graphic representation of a network in which haplotypes are linked to one another through a series of evolutionary steps that require the fewest mutations possible

**statistical phylogeography**   A relatively new analytical approach in studies of phylogeography; uses models based on coalescent theory to generate statistics that provide a basis for accepting or rejecting specific hypotheses that may explain the current distributions of species and their genes

**stepwise mutation model**   A microsatellite mutation model that assumes that mutations involve either the gain or loss of a single microsatellite repeat

**sympatric**   Two or more species that occur in the same geographical area

**synonymous substitution**   A nucleotide substitution that, because of redundancy in the genetic code, does not change the amino acid that is encoded

**taxa**   *see* taxon

**taxon**   A taxonomic unit such as a subspecies, species or genus; plural = taxa

**tension zones**   Hybrid zones that are maintained by a balance between dispersal of parental genotypes into the zone and selection against hybrids within the zone

**theta**   ($\theta$) An estimate of population subdivision (analogue of $F_{ST}$) that takes into account the effects of uneven sample size and the number of sampled populations

**transcription**   The process in which the information contained within a DNA sequence is copied into an RNA molecule

**transfer RNA (tRNA)**   An RNA molecule that translates a codon into an amino acid during protein synthesis. Each tRNA molecule recognizes a specific codon on the mRNA molecule

**transgressive segregation**   The generation of extreme phenotypes in segregating hybrid populations relative to the phenotypes that are found in either parental form

**transient polymorphism**   The temporary polymorphic state of a locus that is being driven to fixation following selection or drift

**transitions**   Nucleotide substitutions in which one purine (A or G) is changed into the other purine, or one pyrimidine (T or C) is changed into the other pyrimidine

**translation**   The generation of a specific polypeptide from messenger RNA

**transposable elements**   A DNA sequence that can move from one location to another in the genome

**transversions**   Nucleotide substitutions in which a purine (A or G) is changed into a pyrimidine (T or C), or vice versa

**tRNA**   *see* transfer RNA

**uniparental inheritance**   Genes and genetic elements that are inherited from only one parent. This term is most commonly applied to the inheritance of mitochondrial and plastid genomes plus some sex chromosomes in sexually reproducing organisms

**variance**   The average squared deviation that a set of measurements show from the arithmetic mean

**variation in reproductive success (VRS)**   The variation in the number of offspring that each individual within a population manages to produce throughout its lifetime

**vicariance**   The splitting of formerly continuous populations by a geographical or ecological barrier

**VRS**   *see* variation in reproductive success

*w* *see* fitness value

**Wahlund effect**   The reduction in heterozygosity values, relative to expectations under Hardy-Weinberg equilibrium, that arises when data from two or more subpopulations with different allele frequencies are combined

# Answers to Review Questions

## Chapter 1

**1.1.** (i)  Nucleotide insertion. This is a frameshift mutation. Following this mutation, all but the first triplet encode different amino acids. The functionality is changed and therefore mutation is unlikely to be neutral.

(ii)  Nucleotide substitution in which a C was replaced with an A (transversion). The original (CCC) and derived (CCA) codons both specify proline. This is therefore a synonymous substitution and is probably neutral.

**1.2.** There are 13 transitions and 13 transversions, therefore the ratio is 1.0.

**1.3.** The primer pair is shown in bold below (in the positions that the primers would anneal to the sequences during PCR):

5′−CTCACTTTCCTCCACGAAACAGGCTCAAACAACCCAACGGGCATCCCCTCAGATTGCGAC−3′
                                                     3′−**GGGAGTCTAACGCTG**−5′

5′−**CTCACTTTCCTCCAC**−3′
3′−GAGTGAAAGGAGGTGCTTTGTCCGAGTTTGTTGGGTTGCCCGTAGGGGAGTCTAACGCTG−5′

**1.4.** The sequence is: TGTGGAAGACCTAAT.

*Molecular Ecology*   Joanna Freeland
© 2005 John Wiley & Sons, Ltd.

## Chapter 2

**2.1.** Advantages include:

- High mutation rates, therefore relatively likely to detect variation.

- Conserved arrangement of sequences, therefore many universal primers are available.

- No recombination and uniparental inheritance make it relatively easy to retrace genetic lineages.

- Small effective population size compared with most nuclear genes, therefore sensitive to demographic processes.

- Maternally inherited, therefore can be useful when studying hybridization.

Disadvantages include:

- Small effective population size compared with most nuclear genes, therefore may exaggerate the effects of past events and lead to underestimation of genetic diversity.

- Acts as a single locus and therefore there is no scope for comparing the genealogies of multiple genes.

- Maternally inherited and therefore can give an incomplete picture, e.g. if only males disperse.

- Mitochondrial pseudogenes are common in some species.

**2.2.** (i) Both chloroplasts and mitochondria are maternally inherited in angiosperms, therefore this comparison would have to compare data from one of the organelle genomes (dispersed only by seeds) to data from the nuclear genome (dispersed by both pollen and seeds).

(ii) In gymnosperms, mtDNA is maternally inherited (dispersed by seeds only), and cpDNA is paternally inherited (dispersed by pollen only) and therefore would provide a useful comparison. Alternatively, data from cpDNA (dispersed by pollen) could be compared to nuclear data (dispersed by both pollen and seeds).

(iii) Because mtDNA is inherited maternally and Y chromosomes are inherited paternally, different dispersal patterns in males and females could be deduced from a comparison of mtDNA and Y chromosome data; alternatively, a comparison between autosomal loci and mtDNA or Y chromosome markers should reveal any contradictory patterns between the sexes.

**2.3.** The total number of alleles is $2(28) = 56$. There are a total of 18 homozygotes and so there must be 10 heterozygotes, therefore the frequency of $A_1 = [2(10) + 10]/56 = 53.6$ per cent and the frequency of $A_2 = [2(8) + 10]/56 = 46.4$ per cent.

**2.4.** The recognition sites for each enzyme are shown in bold:

1: GATTATACAT**AGCT**ACTAGATACAGATACTATTTTTAGGGGCGTATGCTCGG
   ATCTATAGACCTA**GTAC**TAGATACTAGGAAAACCCGTTGTGTCGCGTGCTGA

2: GATTATACATAGTTACTAGATACAGATACTATTTTTAGGGGCGTATGCTCGG
   ATCTATAGACCTA**GTAC**TAGATACTAGGAAAACCCGTTGTGTCGCGTGCTGA

Sequence 1 will be cut at two sites and therefore will produce three bands. Sequence 2 will be cut at one site and therefore will produce two bands.

**2.5.** According to Table 2.4, the average divergence of protein-coding regions in mammalian mtDNA is 2 per cent per million years, which, if the mutation rate is constant, will equal approximately 1 per cent per 500 000 years. We would expect, therefore, to find approximate 5 bp differences in our 500 bp sequence.

**2.6.** Some of the factors are:

- Variability: are you comparing individuals, populations or species?

- Mode of inheritance: would a biparentally or uniparentally inherited marker be more appropriate?

- Dominant versus co-dominant data: do you wish to readily calculate allele frequencies?

- Will you need to infer the evolutionary histories of populations or species? If so, sequence data may be most appropriate.

- Time, money and expertise: what are your logistical constraints?

# Chapter 3

**3.1.** (i)   Total number of alleles $= 2(3969+3174+927) = 16140$

Total number of $E$ alleles $= 2(3969)+3174=11112$

Total number of $e$ alleles $= 2(927)+3174=5028$

Frequency of $E$ allele $(p)=11112/16140=0.6885$

Frequency of $e$ allele $(q)=1-p=1-0.6885=0.3115$

(ii) If the population was in HWE, we would expect the genotype frequencies to be equal to: $p^2+2pq+q^2=(0.6885)^2+2(0.6885)(0.3115)+(0.3115)^2=0.474+0.429+0.0970$.

(iii) There are a total of 8070 individuals in this population. If the population was in HWE we would expect to find $8070(0.474)=3825$ individuals with the genotype $EE$ $(p^2)$, $8070(0.429)=3462$ individuals with the genotype $Ee$ $(2pq)$ and $8070(0.097)=783$ individuals with the genotype $ee$ $(q^2)$.

(iv) $\chi^2=\Sigma(O-E)^2/E$

$\qquad = (3969-3825)^2/3825+(3174-3462)^2/3462+(927-783)^2/783$

$\qquad = 5.42+23.96+26.48$

$\qquad = 55.86$

With one degree of freedom, this $\chi^2$ is highly significant ($P<0.001$), which means that the observed distribution of genotypes in this population is significantly different from that expected if the population was in HWE.

**3.2.** $N_e=4(N_{ef})(N_{em})/(N_{ef}+N_{em})$

For population 1:

$N_e=4(68)(41)/(68+41)=11152/109$

$\qquad = 102.3$, therefore $N_e/N_c=102.3/109=0.939$

For population 2:

$N_e=4(57)(52)/(57+52)=11856/109$

$\qquad = 108.8$, therefore $N_e/N_c=108.8/109=0.998$

**3.3.** (i)   Long-term $N_e=t/[(1/N_{e1})+(1/N_{e2})+(1/N_{e3})+\cdots(1/N_{et})]$

$\qquad = 6/[(1/10^4)+(1/10^4)+(1/10^4)+(1/10^3)+(1/10^4)+(1/10^4)]$

$\qquad = 6/0.0015$

$\qquad = 4000$

(ii)   Current $N_e/N_c=4000/10000=0.4$.

**3.4.** (i)  $1/(2N_e) = 1/40 = 0.025 = 2.5\%$ lost each generation

(ii)  $1/N_{ef} = 1/10 = 0.10 = 10\%$ lost each generation

**3.5.** Possible explanations:

- Selection

- Inbreeding

- Insufficient sample size (sample size influences $H_o$ more than it influences $H_e$)

- Null alleles

- Wahlund effect

# Chapter 4

**4.1.** $F_{IT} = F_{IS} + F_{ST} - (F_{IS})(F_{ST})$

$= 0.085 + 0.136 - (0.085)(0.136)$

$= 0.085 + 0.136 - 0.01156$

$= 0.209$

**4.2.** $N_e m = (1/F_{ST} - 1)/4$

$= (1/0.136 - 1)/4$

$= 1.588$

**4.3.** Factors that may explain the discrepancy in $N_e m$ are:

- Selection of markers used to estimate $N_e m$.

- Not an island model.

- Populations are not at equilibrium.

- Mark–recapture studies targeted the wrong areas.

- Mark–recapture covered only the breeding season; adults may disperse at other times.

**4.4.** (i)   Gene flow – genetic differentiation is inversely proportional to gene flow.

(ii)   $N_e$ – in the absence of appreciable levels of gene flow at least some genetic differentiation will be attributable to genetic drift, and the rate of drift depends on $N_e$.

(iii)   Local adaptation and strength of selection pressure – if selection is strong enough then population differentiation will occur despite ongoing gene flow.

**4.5.** Yes – microsatellite divergence is generally higher than AFLP, which can be attributed to different mutation rates, but there is one outlier (AFLP locus 4) that shows particularly high divergence and may have been subjected to natural selection.

**4.6.** Assuming that $F_{ST}$ reflects genetic divergence as a result of drift, $Q_{ST}$ is lower than expected if height was a neutral QTL and therefore it appears that a fixed height has been selected for in these populations.

# Chapter 5

**5.1.** The Isthmus of Panama emerged approximately 3 million years ago. Sequence divergence $= 23/750 = 0.0307 = 3.1$ per cent per 3 million years, or approximately 1 per cent per million years.

Assumptions:

- That cytochrome $b$ evolves at the same rate in the fish species for which the clock was calibrated and the species that are the subjects of the other studies.

- That the evolutionary rate of cytochrome $b$ is constant along the entire gene (relevant if a non-homologous region of cytochrome $b$ is being compared).

- That the cytochrome $b$ clock has been constant over time (relevant if the species in the other studies had diverged substantially more or less than 3 million years ago).

**5.2.** The expected time until all gene copies coalesce is a) $4N_e = 4(200) = 800$, and b) 200. These calculations are based on a number of assumptions including constant population size, no natural selection, and random mating.

**5.3.** The most parsimonious network, i.e. the one that requires the fewest mutation steps, is:

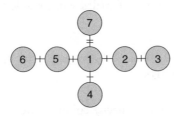

**Figure 5.18**

The bars over the connecting lines represent the number of mutations that are required between each haplotype. Haplotype 1 is most likely to be the ancestral haplotype because it is central to the network and has the largest number of connections.

**5.4.** The two processes are:

- Retention of ancestral polymorphism. If two populations or species have recently diverged, lineage sorting will not yet have resulted in monophyly and therefore alleles will be shared because they were present in the common ancestor.

- Hybridization – alleles may introgress from one species to another.

**5.5.** (i)  Haldane's rule states that these are more likely to be birds than mammals, because in birds the males are the homogametic sex.

(ii)  This is more likely to be a tension zone, because female hybrids apparently have extremely low fitness levels.

(iii)  Although nuclear DNA may introgress, mitochondrial DNA is unlikely to be exchanged between species to any extent because it is transmitted maternally and very few female hybrids survive.

## Chapter 6

**6.1.** Yes. Males can be excluded if they have no alleles that match the chick's alleles at one or more loci, and this is true of males 1, 2, 3, and 6. At locus 1, the chick's band 2 must have come from the mother, and therefore male 4 can also be excluded (he can't have provided the chick's band 1). This means that all except male 5 can be excluded from paternity.

**6.2.** Just as males tend to have a relatively high variation in reproductive success in polygynous mating systems, we may expect females to have a relatively high variation in polyandrous mating systems because, unless the population has a male-biased sex ratio, not all females will be able to defend a territory and attract multiple mates.

**6.3.** $(0.5 - 0.45) + (0.5 - 0.50)/(1 - 0.45) + (1 - 0.50) = (0.05 + 0)/(0.55 + 0.50) = 0.05/1.05 = 0.048$

**6.4.** No. In monogynous (single queen) colonies, workers have a higher relatedness with their sisters (0.75) than their brothers (0.25) and therefore their ideal sex ratio should be 3:1 compared with the queen's ideal of 1:1. In polygynous (multiple queens) colonies, full-siblings will be less frequent and therefore the ideal sex ratio for workers will be less than 3 : 1 (i.e. closer to the queen's ideal), in which case conflict should be weaker.

**6.5.** (i) Dispersal is male-biased because genetic differentiation is lower for males than for females.

  (ii) Dispersal is male-biased because genetic differentiation is higher for mtDNA (maternally inherited) loci than for nuclear (biparentally inherited) loci.

  (iii) Dispersal is male-biased because negative values of $AIc$ mean that an individual is likely to be a recent immigrant and the overall $AIc$ values were negative in males but not in females.

  (iv) Dispersal is female-biased because local relatedness values were higher for males than for females.

# Chapter 7

**7.1.** Although very few insect species have been evaluated, 72 per cent of those that have are classified as threatened. This, in conjunction with the high proportion of evaluated species that are threatened in other taxonomic groups, suggests that the demise of many species may be imminent.

**7.2.** A reduction in population size leads to an increase in the rate of genetic drift. Genetic drift reduces the diversity of populations through the loss of alleles. A reduction in alleles leads to an increase in homozygosity, which in turn is a reflection of increased inbreeding because individuals will be more likely to share two copies of an allele that is identical by descent.

**7.3.**

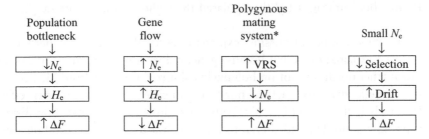

**7.4.** When inbreeding depression results from dominance, deleterious alleles are increasingly likely to be homozygous and therefore expressed, in which case they may be eliminated following natural selection, i.e. purged. In over-dominance, heterozygotes are fitter than homozygotes. Overdominance there-fore requires both relevant alleles to be maintained within the population, meaning that there is no scope for the purging of deleterious alleles.

**7.5.** $\delta = 1 - (X_I/X_0) = 1 - (0.6/0.805) = 0.255$

**7.6.** Founder effect, genetic swamping and outbreeding depression.

**7.7.** The most conservative approach would be to move only those individuals with haplotype 3 from Seymour Norte to Baltra. Because haplotype 3 has been found nowhere other than these two islands, it is reasonable to conclude that the Seymour Norte haplotype 3 individuals are direct descendants of the Baltra population that we introduced to that island in the 1930s. The individuals on Seymour Norte with haplotype 2, on the other hand, could have come from nearby Santa Cruz instead of Baltra and therefore may be less suitable for repopulating Isla Baltra.

# Chapter 8

**8.1.** The average number of individuals that would have to be screened in each population before duplicate genotypes are found is taken from the probability of identity calculations, and would be $1/1.1 \times 10^{-5} \approx 90909$ (B), $1/2.2 \times 10^{-5} \approx 45454$ (LM) and $1/4.6 \times 10^{-2} \approx 22$ (TN). These loci therefore would be suitable for individual identification in wildlife forensics when used in populations B and LM but not in TN. This difference in efficacy is due to the substantially lower levels of genetic diversity in TN (according to $H_e$ and total number of alleles) compared with the other two populations.

**8.2.** *Cervus unicolor*, the Sambar deer. This species is a prized trophy in the South Pacific, but hunting is tightly regulated throughout much of its range.

**8.3.** There are a number of possible explanations for these data, but perhaps the simplest explanation is that because genetic diversity was high in Mexico and low further north, the cotton boll likely colonized the USA from Mexico, (the low genetic diversity further north is consistent with a founder effect). Relatively high $F_{ST}$ and low $Nm$ between regions suggests that long-distance gene flow is limited and therefore the northwards spread of the boll weevil could be attributed to rare long-distance dispersal events (this is also consistent with a founder effect).

**8.4.** The $N_e/N_c$ of the composite wild-hatchery populations can be lower than that of the 'pure' wild population if:

- The genetic diversity is lower in hatchery versus wild fish.

- The overall genetic diversity is decreased following the genetic swamping of wild fish with genetically less variable hatchery fish.

- Outbreeding depression lowers the fitness of hybrids, in which case the proportion of less genetically variable hatchery fish would remain relatively high.

- There is an increase in VRS following the more intense competition that may result from an elevated $N_c$ (Chapter 3).

# References

Abbott, R. J. and Comes, H. P. 2004. Evolution in the Arctic: a phylogeographic analysis of the circumarctic plant, *Saxifraga oppositifolia* (Purple saxifrage). *New Phytologist* **161**: 211–224.

Adams, J. R., Kelly, B. T. and Waits, L. P. 2003. Using faecal DNA sampling and GIS to monitor hybridization between red wolves (*Canis rufus*) and coyotes (*Canis latrans*). *Molecular Ecology* **12**: 2175–2186.

Adams, P. B. and Ayers, W. A. 1979. Ecology of *Sclerotinia* species. *Phytopathology* **69**: 896–899.

Agapow, P. M., Bininda-Emonds, O. R. P., Crandall, K. A., Gittleman, J. L., Mace, G. M., Marshall, J. C. and A. Purvis. 2004. The impact of species concept on biodiversity studies. *Quarterly Review of Biology* **79**: 161–179.

Agusti, N., De Vicente, M. C. and Gabarra, R. 1999. Development of sequence amplified characterization region (SCAR) markers of *Helicoverpa armigera*: a new polymerase chain reaction-based technique for predator gut analysis. *Molecular Ecology* **8**: 1467–1474.

Agusti, N., Shayler, S. P., Harwood, J. D., Vaughan, I. P., Sunderland, K. D. and Symondson, W. O. C. 2003a. Collembola as alternative prey sustaining spiders in arable ecosystems: prey detection within predators using molecular markers. *Molecular Ecology* **12**: 3467–3475.

Agusti, N., Unruh, T. R. and Welter, S. C. 2003b. Detecting *Cacopsylla pyricola* (Hemiptera: Psyllidae) in predator guts using COI mitochondrial markers. *Bulletin of Entomological Research* **93**: 179–185.

Albrecht, D. J. 2000. Sex ratio manipulation within broods of house wrens, *Troglodytes aedon*. *Animal Behaviour* **59**: 1227–1234.

Allen, G. A. 2001. Hybrid speciation in *Erythronium* (Liliaceae): a new allotetraploid species from Washington State. *Systematic Botany* **26**: 263–272.

Allendorf, F. W. and Seeb, L. W. 2000. Concordance of genetic divergence among sockeye salmon populations at allozyme, nuclear DNA and mitochondrial DNA markers. *Evolution* **54**: 640–651.

Allendorf, F. W., Leary, R. F., Spruell, P. and Wenburg, J. K. 2001. The problems with hybrids: Setting conservation guidelines. *Trends in Ecology and Evolution* **16**: 613–622.

Anderson, S., Bankier, A. T., Barrell, B. G., DeBruijn, M. H. L., Coulson, A. R., Drouin, J., Eperon, I. C., Nierlich, D. P., Roe, B. A., Sanger, F., *et al.* 1981. Sequence and organization of the human mitochondrial genome. *Nature* **290**: 457–465.

*Arabidopsis* Genome Initiative. 2000. Analysis of the genome sequence of the flowering plant *Arabidopsis thaliana*. *Nature* **408**: 796–815.

Ardren, W. R. and Kapuscinski, A. R. 2003. Demographic and genetic estimates of effective population size ($N_e$) reveals genetic compensation in steelhead trout. *Molecular Ecology* **12**: 35–49.

Armbruster, P., Bradshaw, W. E. and Holzapfel, C. M. 1998. Effects of postglacial range expansion on allozyme and quantitative genetic variation of the pitcher-plant mosquito, *Wyeomyia smithii*. *Evolution* **52**: 1697–1704.

Arnaud, J. F., Viard, F., Delescluse, M. and Cuguen, J. 2003. Evidence for gene flow via seed dispersal from crop to wild relatives in *Beta vulgaris* (Chenopodiaceae): consequences for the release of genetically modified crop species with weedy lineages. *Proceedings of the Royal Society of London Series B* **270**: 1565–1571.

Arnold, M. L. 1993. *Iris nelsonii*: origin and genetic composition of a homoploid hybrid species. *American Journal of Botany* **80**: 577–583.

Aspi, J. 2000. Inbreeding and outbreeding depression in male courtship song characters in *Drosophila montana*. *Heredity* **84**: 273–282.

Austin, J. D., Davila, J. A., Lougheed, S. C. and Boag, P. T. 2000. Genetic evidence for female-biased dispersal in the bullfrog, *Rana catesbeiana* (Ranidae). *Molecular Ecology* **12**: 3165–3172.

Austin, J. D., Lougheed, S. C. and Boag, P. T. 2004. Discordant temporal and geographic patterns in maternal lineages of eastern North American frogs, *Rana catesbeiana* (Ranidae) and *Pseudacris crucifer* (Hylidae). *Molecular Phylogenetics and Evolution* **32**: 799–816.

Avise, J. C. 1992. Molecular population structure and the biogeographic history of a regional fauna: a case history with lessons for conservation. *Oikos* **63**: 62–76.

Avise, J. C. 2000. *Phylogeography. The History and Formation of Species*: Harvard University Press, Cambridge, MA.

Avise, J. C., Shapira, J. F., Daniel, S. W., Aquadro, C. F. and Lansman, R. A. 1983. Mitochondrial DNA differentiation during the speciation process in *Peromyscus*. *Molecular Biology and Evolution* **1**: 38–56.

Avise, J. C., Helfman, G. S., Saunders, N. C. and Hales, L. S. 1986. Mitochondrial DNA differentiation in North Atlantic eels: population genetic consequences of an unusual life history pattern. *Proceedings of the National Academy of Sciences USA* **83**: 4350–4354.

Avise, J. C., Arnold, J., Ball Jr., R. M., Bermingham, E., Lamb, T., Neigel, J. E., Reeb, C. A. and Saunders, N. C. 1987. Intraspecific phylogeography: the mitochondrial DNA bridge between population genetics and systematics. *Annual Review of Ecology and Systematics* **18**: 489–522.

Avise, J. C., Jones, A. G., Walker, D. and DeWoody, J. A. 2002. Genetic mating systems and reproductive natural histories of fishes: Lessons for ecology and evolution. *Annual Reviw of Genetics* **36**: 19–45.

Avise, J. C., Power, A. J. and Walker, D. 2004. Genetic sex determination, gender identification and pseudohermaphroditism in the knobbed whelk, *Busycon carica* (Mollusca: Melongenidae). *Proceedings of the Royal Society of London B* **271**: 641–646.

Ayre, D. J. and Hughes, T. P. 2000. Genotypic diversity and gene flow in brooding and spawning corals along the Great Barrier Reef, Australia. *Evolution* **54**: 1590–1605.

Ayre, D. J. and Hughes, T. P. 2004. Climate change, genotypic diversity and gene flow in reef-building corals. *Ecology Letters* **7**: 273–278.

Azad, A. A. and Deacon, N. J. 1980. The 3'-terminal primary structure of five eukaryotic 18S rRNAs determined by the direct chemical method of sequencing. *Nucleic Acids Research* **8**: 4365–4376.

Bagley, M. J. and Gall, G. A. E. 1998. Mitochondrial and nuclear DNA sequence variability among populations of rainbow trout (*Oncorhynchus mykiss*). *Molecular Ecology* **7**: 945–961.

Baker, A. J. 1992. Genetic and morphometric divergence in ancestral European and descendant New Zealand populations of chaffinches (*Fringilla coelebs*). *Evolution* **46**: 1784–1800.

Baker, C. S., Cipriano, F. and Palumbi, S. R. 1996. Molecular genetic identification of whale and dolphin products from commercial markets in Korea and Japan. *Molecular Ecology* **5**: 671–685.

Baldwin, B. G. and Robichaux, R. H. 1995. Historical biogeography and ecology of the Hawaiian silversword alliance (Asteraceae). In *Hawaiian Biogeography. Evolution on a Hot Spot Archipelago* (Wagner, W. L. & Funk, V. A., eds), pp. 259–287. Smithsonian Institution Press, Washington, DC.

Banks, S. C., Horsup, A., Wilton, A. N. and Taylor, A. C. 2003. Genetic marker investigation of the source and impact of predation on a highly endangered species. *Molecular Ecology* **12**: 1663–1667.

Barber, P. H., Palumbi, S. R., Erdmann, M. V. and Moosa, M. K. 2002. Sharp genetic breaks among populations of *Haptosquilla pulchella* (Stomatopoda) indicate limits to larval transport: patterns, causes and consequences. *Molecular Ecology* **11**: 659–674.

Barrett, S. C. H., Cole, W. W. and Herrera, C. M. 2004. Mating patterns and genetic diversity in the wild daffodil *Narcissus longispathus* (Amaryllidaceae). *Heredity* **92**: 459–465.

Barton, N. H. and Hewitt, G. M. 1985. Analysis of hybrid zones. *Annual Review of Ecology and Systematics* **16**: 113–148.

Bartsch, D., Lehnen, M., Clegg, J., Pohl-Orf, M., Schuphan, I. and Ellstrand, N. C. 1999. Impact of gene flow from cultivated beet on genetic diversity of wild sea beet populations. *Molecular Ecology* **8**: 1733–1741.

Beaumont, M. A. 2003. Estimation of population growth or decline in genetically monitored populations. *Genetics* **164**: 1139–1160.

Bekkevold, D., Hansen, M. M. and Mensberg, K. L. D. 2004. Genetic detection of sex-specific dispersal in historical and contemporary populations of anadromous brown trout *Salmo trutta*. *Molecular Ecology* **13**: 1707–1712.

Bellagamba, F., Valfrè, F., Panseri, S. and Moretti, V. M. 2003. Polymerase chain reaction-based analysis to detect terrestrial animal protein in fish meal. *Journal of Food Protection* **66**: 682–685.

Bensasson, D., Zhang, D. -X., Hartl, D. L. and Hewitt, G. M. 2001. Mitochondrial pseudogenes: evolution's misplaced witnesses. *Trends in Ecology and Evolution* **16**: 314–321.

Benton, M. J. 1993. *The Fossil Record 2*. Chapman and Hall, London.

Bermingham, E., McCafferty, S. S. and Martin, A. P. 1997. Fish biogeography and molecular clocks: perspectives from the Panamanian Isthmus. In *Molecular Systematics of Fishes* (Kocher, T. D. & Stepien, C. A., eds), pp. 113–128. Academic Press, New York.

Berry, O., Tocher, M. D. and Sarre, S. D. 2004. Can assignment tests measure dispersal? *Molecular Ecology* **13**: 551–561.

Berthier, P., Beaumont, M. A., Cornuet, J. M. and Luikart, G. 2002. Likelihood-based estimation of the effective population size using temporal changes in allele frequencies: a genealogical approach. *Genetics* **160**: 741–751.

Bettencourt, B. R., Kim, I., Hoffman, A. A. and Feder, M. E. 2002. Response to natural and laboratory selection at the *Drosophila* hsp70 genes. *Evolution* **56**: 1796–1801.

Bierne, N., David, P., Boudry, P. and Bonhomme, F. 2002. Assortative fertilization and selection at larval stage in the mussels *Mytilus edulis* and *M. galloprovincialis*. *Evolution* **56**: 292–298.

Bilton, D. T. 1992. Genetic population structure of the postglacial relict diving beetle *Hydroporus glabriusculus* Aub (Coleoptera: Dytiscidae). *Heredity* **69**: 503–511.

Birstein, V. J., Doukakis, P., Sorkin, B. and DeSalle, R. 1998. Population aggregation analysis of three caviar-producing species of sturgeons and implications for the species identification of black caviar. *Conservation Biology* **12**: 766–775.

Bishop, J. G. and Hunt, J. A. 1988. DNA divergence in and around the alcohol dehydrogenase locus in five closely related species of Hawaiian *Drosophila*. *Molecular Biology and Evolution* **5**: 415–431.

Blankenship, L. E. and Yayanos, A. A. 2005. Universal primers and PCR of gut contents to study marine invertebrate diets. *Molecular Ecology* **14**: 891–899.

Blomquist, D., Andersson, M., Küpper, C., Cuthill, I. C., Kis, J., Lanctot, R. B., Sandercock, B. K., Székely, T. Wallander, J. and Kempenaers, B. 2002. Genetic similarity between mates and extra-pair parentage in three species of shorebirds. *Nature* **419**: 613–615.

Blomster, J., Maggs, C. A. and Stanhope, M. J. 1999. Extensive intraspecific morphological variation in *Enteromorpha muscoides* (Chlorophyta) revealed by molecular analysis. *Journal of Phycology* **35**: 575–586.

Blouin, M. S., Yowell, C. A., Courtney, C. H. and Dame, J. B. 1995. Host movement and the genetic structure of populations of parasite nematodes. *Genetics* **141**: 1007–1014.

Blouin, M. S., Parsons, M., Lacaille, V. and Lotz, S. 1996. Use of microsatellite loci to classify individuals by relatedness. *Molecular Ecology* **5**: 393–401.

Blundell, G. M., Ben-David, M, Groves, P., Bowyer, R. T. and Geffen, E. 2002. Characteristics of sex-biased dispersal and gene flow in coastal river otters: implications for natural recolonization of extirpated populations. *Molecular Ecology* **11**: 289–303.

Boag, P. T. 1983. The heritability of external morphology in Darwin's ground finches (*Geospiza*) on Isla Daphne Major, Galápagos. *Evolution* **37**: 877–894.

Bohonak, A. J. 1999. Dispersal, gene flow and population structure. *The Quarterly Review of Biology* **74**: 21–45.

Boileau, M. G., Hebert, P. D. N. and Schwartz, S. S. 1992. Nonequilibrium gene frequency divergence: persistent founder effects in natural populations. *Journal of Evolutionary Biology* **5**: 25–39.

Bolch, C. J. S., Blackburn, S. I., Hallegraeff, G. M. and Vaillancourt, R. E. 1999. Genetic variation among strains of the toxic dinoflagellate *Gymnodinium catenatum* (Dinophyceae). *Journal of Phycology* **35**: 356–367.

Boulinier, T. and Lemel, J. Y. 1996. Spatial and temporal variations of factors affecting breeding habitat quality in colonial birds: Some consequences for dispersal and habitat selection. *Acta Oecologica* **17**: 531–552.

Boulton, A. M., Ramirez, M. G. and Blair, C. P. 1998. Genetic structure in a coastal dune spider (*Geolycosa pikei*) on Long Island, New York Barrier Islands. *Biological Journal of the Linnean Society* **64**: 69–82.

Bowen, B. W., Meylan, A. B., Ross, J. P., Limpus, C. J., Balazs, G. H. and Avise, J. C. 1992. Global population structure and natural history of the green sea turtle *(Chelonia mydas)* in terms of matriarchal phylogeny. *Evolution* **46**: 865–880.

Bowman, T. D., Stehn, R. A. and Scribner, K. T. 2004. Glaucous gull predation of goslings on the Yukon-Kuskokwim Delta, Alaska. *Condor* **106**: 288–298.

Briers, R. A., Gee, J. H. R., Cariss, H. M. and Geoghegan, R. 2004. Inter-population dispersal by adult stoneflies detected by stable isotope enrichment. *Freshwater Biology* **49**: 425–431.

Briskie, J. V. and Mackintosh, M. 2004. Hatching failure increases with severity of population bottlenecks in birds. *Proceedings of the National Academy of Sciences USA* **101**: 558–561.

Britten, H. B., Fleishman, E., Austin, G. T. and Murphy, D. D. 2003. Genetically effective and adult population sizes in the Apache silverspot butterfly, *Speyeria nokomis apacheana* (Lepidoptera: Nymphalidae). *Western North American Naturalist* **63**: 229–235.

Brock, M. T., Weinig, C. and Galen, C. 2005. A comparison of phenotypic plasticity in the native dandelion *Taraxacum ceratophorum* and its invasive congener *T. officinale*. *New Phytologist* **166**: 173–183.

Brower, A. V. Z. and Boyce, T. M. 1991. Mitochondrial DNA variation in Monarch butterflies. *Evolution* **45**: 1281–1286.

Brown, J. K. M. and Hovmøller, M. S. 2002. Aerial dispersal of pathogens on the global and continental scales and its impact on plant disease. *Science* **297**: 537–541.

Brown, W. M., George, M. and Wilson, A. C. 1979. Rapid evolution of animal mitochondrial DNA. *Proceedings of the National Academy of Sciences USA.* **76**: 1967–1971.

Brown, W. M., Prager, E. M., Wang, A. and Wilson, A. C. 1982. Mitochondrial DNA sequences of primates: tempo and mode of evolution. *Journal of Molecular Evolution* **18**: 225–239.

Browne, R. K., Clulow, J. and Mahony, M. 2002. The short-term storage and cryopreservation of spermatozoa from hylid and myobatrachid frogs. *Cryoletters* **23**: 129–136.

Broyles, S. B. 1998. Postglacial migration and the loss of allozyme variation in northern populations of *Asclepias exaltata* (Asclepiadaceae). *American Journal of Botany* **85**: 1091–1097.

Brumfield, R. T., Beerli, P., Nickerson, D. A. and Edwards, S. V. 2003. The utility of single nucleotide polymorphisms in inferences of population history. *Trends in Ecology and Evolution* **18**: 249–256.

Buckner, C. H. 1959. Mortality of cocoons of the Larch sawfly *Pristiphora erichsonii* (Htg.), in relation to distance from small-mammal tunnels. *Canadian Entomologist* **91**: 535–542.

Burland, T. M., Bennett, N. C., Jarvis, J. U. M. and Faulkes, C. G. 2002. Eusociality in African mole-rats: new insights from patterns of genetic relatedness in the Damaraland mole-rat (*Cryptomys damarensis*). *Proceedings of the Royal Society of London Series B* **269**: 1025–1030.

Burzynski, A., Zbawicka, M., Skibinski, D. O. F. and Wenne, R. 2003. Evidence for recombination of mtDNA in the marine mussel *Mytilus trossulus* from the Baltic. *Molecular Biology and Evolution* **20**: 388–392.

*Caenorhabditis elegans* Sequencing Consortium. 1998. Genome sequence of the nematode *C. elegans*: a platform for investigating biology. *Science* **282**: 2012–2018.

Caicedo, A. L. and Schaal, B. A. 2004. Population structure and phylogeography of *Solanum pimpinellifolium* inferred from a nuclear gene. *Molecular Ecology* **13**: 1871–1882.

Campos, W. G., Schoereder, J. H. and Sperber, C. F. 2004. Does the age of the host plant modulate migratory activity of *Plutella xylostella*? *Entomological Science* **7**: 323–329.

Capparella, A. P. 1992. Neotropical avian diversity and riverine barriers. *Acta Congressus Internationalis Ornithologici* **20**: 307–316.

Carbone, I. and Kohn, L. M. 2001. A microbial population–species interface: nested cladistic and coalescent inference with multilocus data. *Molecular Ecology* **10**: 947–964.

Carchini, G., Chiarotti, F., Di Domenico, M., Mattoccia, M. and Paganotti, G. 2001. Fluctuating asymmetry, mating success, body size and heterozygosity in *Coenagrion scitulum* (Rambur) (Odonata: Coenagrionidae). *Animal Behaviour* **61**: 661–669.

Carissan-Lloyd, F. M. M., Pipe, R. K. and Beaumont, A. R. 2004. Immunocompetence and heterozygosity in the mussel *Mytilus edulis*. *Journal of the Marine Biological Association of the United Kingdom* **84**: 377–382.

Carmichael, L. E., Nagy, J. A., Larter, N. C. and Strobeck, C. 2001. Prey specialization may influence patterns of gene flow in wolves of the Canadian Northwest. *Molecular Ecology* **10**: 2787–2798.

Carson, H. L. and Clague, D. A. 1995. Geology and biogeography of the Hawaiian Islands. In *Hawaiian Biogeography: Evolution in a Hotspot Archipelago* (Wagner, W. & Funk, V., eds), pp. 14–29. Smithsonian Institution Press, Washington, DC.

Carter, K. L., Robertson, B. C. and Kempenaers, B. 2000. A differential DNA extraction method for sperm on the perivitelline membrane of avian eggs. *Molecular Ecology* **9**: 2149–2150.

Cassens, I., Van Waerebeek, K., Best, P. B., Crespo, E. A., Reyes, J. and Milinkovitch, M. C. 2003. The phylogeography of dusky dolphins (*Lagenorhynchus obscurus*): a critical examination of network methods and rooting procedures. *Molecular Ecology* **12**: 1781–1792.

Castro, I., Mason, K. M., Armstrong, D. P. and Lambert, D. M. 2004. Effect of extra-pair paternity on effective population size in a reintroduced population of the endangered hihi and potential for behavioural management. *Conservation Genetics* **5**: 381–393.

Cegelski, C. C., Waits, L. P. and Anderson, N. J. 2003. Assessing population structure and gene flow in Montana wolverines (*Gulo gulo*) using assignment-based approaches. *Molecular Ecology* **12**: 2907–2918.

Chapman, D. D., Prodohl, P. A., Gelsleichter, J., Manire, C. A. and Shivji, M. S. 2004. Predominance of genetic monogamy by females in a hammerhead shark, *Sphyrna tiburo*: implications for shark conservation. *Molecular Ecology* **13**: 1965–1974.

Chapman, R. E., Wang, J. and Bourke, A. F. G. 2003. Genetic analysis of spatial foraging patterns and resource sharing in bumble bee pollinators. *Molecular Ecology* **12**: 2801–2808.

Chapuisat, M. and Crozier, R. 2001. Low relatedness among cooperatively breeding workers of the greenhead ant *Rhytidoponera metallica*. *Journal of Evolutionary Biology* **14**: 564–573.

Chen, Y., Giles, K. L., Payton, M. E. and Greenstone, M. H. 2000. Identifying key cereal aphid predators by molecular gut analysis. *Molecular Ecology* **9**: 1887–1898.

Chesser, R. K. and Baker, R. J. 1996. Effective sizes and dynamics of uniparentally and diparentally inherited genes. *Genetics* **144**: 1225–1235.

Cheverud, J., Routman, E., Jaquish, C., Tardif, S., Peterson, G., Belfiore, N. and Forman, L. 1994. Quantitative and molecular genetic variation in captive cotton-top tamarins. *Conservation Biology* **8**: 95–105.

Chiang, T.-Y., Schaal, B. A. and Peng, C.-I. 1998. Universal primers for amplification and sequencing a noncoding spacer between the *atpB* and *rbcL* genes of chloroplast DNA. *Botanical Bulletin of Academia Sinica* **39**: 245–250.

Christen, M. and Milinski, M. 2003. The consequences of self-fertilization and outcrossing of the cestode *Schistocephalus solidus* in its second intermediate host. *Parasitology* **126**: 369–378.

Cianchi, R., Ungaro, A., Marini, M. and Bullini, L. 2003. Differential patterns of hybridization and introgression between the swallowtails *Papilio machaon* and *P. hospiton* from Sardinia and Corsica islands (Lepidoptera, Papilionidae). *Molecular Ecology* **12**: 1461–1471.

Ciniglia, C., Yoon, H. S., Pollio, A., Pinto, G. and Bhattacharya, D. 2004. Hidden biodiversity of the extremophilic Cyanidiales red algae. *Molecular Ecology* **13**: 1827–1838.

Clavero, M. and García-Berthou, E. 2005. Invasive species are a leading cause of animal extinctions. *Trends in Ecology and Evolution* **20**: 105–149.

Clegg, S. M., Kelly, J. F., Kimura, M. and Smith, T. B. 2003. Combining genetic markers and stable isotopes to reveal population connectivity and migration patterns in a Neotropical migrant, Wilson's warbler (*Wilsonia pusilla*). *Molecular Ecology* **12**: 819–830.

Clout, M. N., Elliott, G. P. and Robertson, B. C. 2002. Effects of supplementary feeding on the offspring sex ratio of kakapo: a dilemma for the conservation of a polygynous parrot. *Biological Conservation* **107**: 13–18.

Collins, F. S., Brooks, L. D. and Chakravarti, A. 1998. A DNA polymorphism discovery resource for research on human genetic variation. *Genome Research* **8**: 1229–1231.

Colpaert, J. V., Muller, L. A. H., Lambaerts, M., Adriaensen, K. and Vangronsveld, J. 2004. Evolutionary adaptation to Zn toxicity in populations of Suilloid fungi. *New Phytologist* **162**: 549–559.

Coltman, D. W., Pilkington, J. G., Smith, J. A. and Pemberton, J. M. 1999. Parasite-mediated selection against inbred Soay sheep in a free-living, island population. *Evolution* **53**: 1259–1267.

Congui, L., Chicca, M., Cella, R., Rossi, R. and Bernacchia, G. 2000. The use of random amplified polymorphic DNA (RAPD) markers to identify strawberry varieties: a forensic application. *Molecular Ecology* **9**: 229–232.

Conrad, K. F., Clarke, M. F., Robertson, R. J. and Boag, P. T. 1998. Paternity and the relatedness of helpers in the cooperatively breeding bell miner. *Condor* **100**: 343–349.

Conrad, K. F., Willson, K. H., Harvey, I. F., Thomas, C. J. and Sherratt, T. N. 1999. Dispersal characteristics of seven odonate species in an agricultural landscape. *Ecography* **22**: 524–531.

Conrad, K. F., Johnston, P. V., Crossman, C., Kempenaers, B., Robertson, R. J., Wheelwright, N. T. and Boag, P. T. 2001. High levels of extra-pair paternity in an isolated, low-density, island population of tree swallows (*Tachycineta bicolor*). *Molecular Ecology* **10**: 1301–1308.

Conroy, C. J. and Cook, J. A. 2000. Phylogeography of a post-glacial colonizer: *Microtus longicaudus* (Rodentia: Muridae). *Molecular Ecology* **9**: 165–175.

Cook, L. M. 1965. Inheritance of shell size in the snail *Arianta arbustorum*. *Evolution* **19**: 86–94.

Coop, G. and Griffiths, R. C. 2004. Ancestral inference on gene trees under selection. *Theoretical Population Biology* **66**: 219–232.

Cooper, A., Mourer-Chauvire, C., Chambers, G. K., von Haeseler, A., Wilson, A. C. and Pääbo, S. 1992. Independent origins of New Zealand moas and kiwis. *Proceedings of the National Academy of Science USA* **89**: 8741–8744.

Cooper, S. J., Ibrahim, K. M. and Hewitt, G. M. 1995. Postglacial expansion and genome subdivision in the European grasshopper *Chorthippus parallelus*. *Molecular Ecology* **4**: 49–60.

Cornuet, J.-M., Piry, S., Luikart, G., Estoup, A. and Solignac, M. 1999. New methods employing multilocus genotypes to select or exclude populations as origins of individuals. *Genetics* **153**: 1989–2000.

Cotter, S. C. and Wilson, K. 2002. Heritability of immune function in the caterpillar *Spodoptera littoralis*. *Heredity* **88**: 229–234.

Coyer, J. A., Peters, A. F., Stam, W. T. and Olsen, J. L. 2003. Post-ice age recolonization and differentiation of *Fucus serratus* L. (Phaeophyceae; Fucaceae) populations in northern Europe. *Molecular Ecology* **12**: 1817–1829.

Cracraft, J. 1983. Species concepts and speciation analysis. In *Current Ornithology* (Johnson, R. F., ed.), pp. 159–187. Plenum Press, New York.

Crease, T. J., Lynch, M. and Spitze, K. 1990. Hierarchical analysis of population genetic variation in mitochondrial and nuclear genes of *Daphnia pulex*. *Molecular Biology and Evolution* **7**: 444–458.

Crease, T. J., Lee, S.-K., Yu, S.-L., Spitze, K., Lehman, N. and Lynch, M. 1997. Allozyme and mtDNA variation in populations of the *Daphnia pulex* complex from both sides of the Rocky Mountains. *Heredity* **79**: 242–251.

Creer, S., Thorpe, R. S., Malhotra, A., Chou, W. H. and Stenson, A. G. 2004. The utility of AFLPs for supporting mitochondrial DNA phylogeographical analyses in the Taiwanese bamboo viper, *Trimeresurus stejnegeri*. *Journal of Evolutionary Biology* **17**: 100–107.

Cristescu, M. E. A., Witt, J. D. S., Grigorovich, I. A., Hebert, P. D. N. and MacIssac, H. J. 2004. Dispersal of the Ponto-Caspian amphipod *Echinogammarus ischnus*: invasion waves from the Pleistocene to the present. *Heredity* **92**: 197–203.

Crnokrak, P. and Barrett, S. C. H. 2002. Perspective: Purging the genetic load: A review of the experimental evidence. *Evolution* **56**: 2347–2358.

Crnokrak, P. and Roff, D. A. 1999. Inbreeding depression in the wild. *Heredity* **83**: 260–270.

Cronberg, N. 2000. Genetic diversity of the epiphytic bryophyte *Leucodon sciuroides* in formerly glaciated versus nonglaciated parts of Europe. *Heredity* **84**: 710–720.

Cuervo, J. J. 2003. Parental roles and mating system in the black-winged stilt. *Canadian Journal of Zoology* **81**: 947–953.

Culley, T. M., Wallace, L. E., Gengler-Nowak, K. M. and Crawford, D. J. 2002. A comparison of two methods of calculating $G_{ST}$, a genetic measure of population differentiation. *American Journal of Botany* **89**: 460–465.

Culver, M., Johnson, W. E., Pecon-Slattery, J. and O'Brien, S. J. 2000. Genomic ancestry of the American puma (*Puma concolor*). *Journal of Heredity* **91**: 186–197.

Cunha, A. B. 1949. Genetic analysis of the polymorphism of colour pattern in *Drosophila polymorpha*. *Evolution* **3**: 239–251.

Cuthbertson, A. G. S., Fleming, C. C. and Murchie, A. K. 2003. Detection of *Rhopalosiphum insertum* (apple-grass aphid) predation by the predatory mite *Anystis baccarum* using molecular gut analysis. *Agricultural and Forest Entomology* **5**: 219–225.

Da Silva, M. N. F. and Patton, J. L. 1998. Molecular phylogeography and the evolution and conservation of Amazonian mammals. *Molecular Ecology* **7**: 475–486.

Dale, J., Montgomerie, R., Michaud, D., and Boag, P. T. 1999. Frequency and timing of extrapair fertilisation in the polyandrous red phalarope (*Phalaropus fulicarius*). *Behavioral Ecology and Sociobiology* **46**: 50–56.

Dallas, J. F. 1992. Estimation of microsatellite mutation rates in recombinant inbred strains of mouse. *Mammalian Genome* **3**: 452–456.

Dallimer, M., Blackburn, C., Jones, P. J. and Pemberton, J. M. 2002. Genetic evidence for male biased dispersal in the red-billed quelea *Quelea quelea*. *Molecular Ecology* **11**: 529–533.

David, J. P., Strode, C., Vontas, J., Nikou, D., Vaughan, A., Pignatelli, P. M., Louis, C., Hemingway, J. and Ranson, H. 2005. The *Anopheles gambiae* detoxification chip: a highly specific microarray to study metabolic-based insecticide resistance in malaria vectors. *Proceedings of the National Academy of Sciences USA* **102**: 4080–4084.

David, J. R., Alonsomoraga, A., Borai, F., Capy, P., Mercot, H., McEvey, S. F., Munozserrano, A. and Tsakas, S. 1989. Latitudinal variation of *Adh* gene frequencies in *Drosophila melanogaster*: a Mediterranean instability. *Heredity* **62**: 11–16.

Davies, C. P., Simovich, M. A. and Hathaway, S. A. 1997. Population genetic structure of a California endemic branchiopod, *Branchinecta sandiegonensis*. *Hydrobiologia* **359**: 149–158.

De Innocentiis, S., Sola, L., Cataudella, S. and Bentzen, P. 2001. Allozyme and microsatellite loci provide discordant estimates of population differentiation in the endangered dusky grouper (*Epinephelus marginatus*) within the Mediterranean Sea. *Molecular Ecology* **10**: 2163–2175.

de Meeus, T., Beati, L., Delaye, C., Aeschlimann, A. and Renaud, F. 2002. Sex-biased genetic structure in the vector of Lyme disease, *Ixodes ricinus*. *Evolution* **56**: 1802–1807.

de Meeus, T., Durand, P. and Renaud, F. 2003. Species concepts: what for? *Trends in Parasitology* **19**: 425–427.

Debevec, E. M., Gates, R. B., Masuda, M., Pella, J., Reynolds, J. and Seeb, L. W. 2000. SPAM (Version 3.2): statistics program for analyzing mixtures. *Journal of Heredity* **91**: 509–510.

Deffontaine, V., Libois, R., Kotlík, P., Sommer, R., Nieberding, C., Paradis, E., Searle, J. B. and Michaux, J. R. 2005. Beyond the Mediterranean peninsulas: evidence of central European glacial refugia for a temperate forest mammal species, the bank vole (*Clethrionomys glareolus*). *Molecular Ecology* **14**: 1727–1740.

Degnan, J. H. and Salter, L. A. 2005. Gene tree distributions under the coalescent process. *Evolution* **59**: 24–37.

Delmotte, F., Leterme, N., Gauthier, J. P., Rispe, C. and Simon, J. C. 2002. Genetic architecture of sexual and asexual populations of the aphid *Rhopalosiphum padi* based on allozyme and microsatellite markers. *Molecular Ecology* **11**: 711–723.

Demesure, B., Comps, B. and Petit, R. J. 1996. Chloroplast DNA phylogeography of the common beech (*Fagus sylvatica* L.) in Europe. *Evolution* **50**: 2515–2520.

Dever, J. A., Strauss, R. E., Rainwater, T. R., McMurry, S. and Densmore, L. D. 2002. Genetic diversity, population subdivision and gene flow in Morelet's crocodile (*Crocodylus moreletti*) from Belize, Central America. *Copeia* **4**: 1078–1091.

DeWoody, J. A., Fletcher, D. E., Wilkins, S. D., Nelson, W. S. and Avise, J. C. 2000. Genetic monogamy and biparental care in an externally fertilizing fish, the largemouth bass (*Micropterus salmoides*). *Proceedings of the Royal Society of London Series B* **267**: 2431–2437.

Dickinson, J. L. and Akre, J. J. 1998. Extrapair paternity, inclusive fitness, and within-group benefits of helping in western bluebirds. *Molecular Ecology* **7**: 95–105.

Dietz, J. M. and Baker, A. J. 1993. Polygyny and female reproductive success in golden lion tamarins (*Leontopithecus rosalia*). *Animal Behavior* **46**: 1067–1078.

Dixon, A., Ross, D., O'Malley S. L. C. and Burke, T. 1994. Paternal investment inversely related to degree of extra-pair paternity in the reed bunting. *Nature* **371**: 698–700.

Dodd, M. E. and Silvertown, J. 2000. Size-specific fecundity and the influence of lifetime size variation upon effective population size in *Abies balsamea*. *Heredity* **85**: 604–609.

Dodd, R. S. and Afzal-Rafii, Z. 2004. Selection and dispersal in a multispecies oak hybrid. *Evolution* **58**: 261–269.

Douhms, C., Cabrera, H. and Peeters, C. 2002. Population genetic structure and male-biased dispersal in the queenless ant *Diacamma cyaneiventre*. *Molecular Ecology* **11**: 2251–2264.

Dubois, S., Cheptou, P.-O., Petit, C. M., Meerts, P., Poncelet, M., Vekemans, X., Lefèbvre, C. and Escarré, J. 2003. Genetic structure and mating systems of metallicolous and nonmetallicolous populations of *Thlaspi caerulescens*. *New Phytologist* **157**: 633–641.

Dumolin-Lapègue, S., Demesure, B., Fineschi, S., LeCorre, V. and Petit, R. J. 1997. Phylogeographic structure of white oaks throughout the European continent. *Genetics* **146**: 1475–1487.

Dunn, P. O., Cockburn, A. and Mulder, R. A. 1995. Fairy-wren helpers often care for young to which they are unrelated. *Proceedings of the Royal Society of London Series B* **259**: 339–343.

Dunphy, B. K., Hamrick, J. L. and Schwagerl, J. 2004. A comparison of direct and indirect measures of gene flow in the bat-pollinated tree *Hymenaea courbaril* in the dry forest life zone of southwestern Puerto Rico. *International Journal of Plant Sciences* **165**: 427–436.

Easteal, S. and Floyd, R. B. 1986. The ecological genetics of introduced populations of the giant toad, *Bufo marinus* (Amphibia: Anura): dispersal and neighbourhood size. *Biological Journal of the Linnean Society* **27**: 17–45.

Echt, C. S., DeVerno, L. L., Anzidei, M. and Vendramin, G. G. 1998. Chloroplast microsatellites reveal population genetic diversity in red pine, *Pinus resinosa* Ait. *Molecular Ecology* **7**: 307–316.

Edmands, S. and Deimler, J. K. 2004. Local adaptation, intrinsic coadaptation and the effects of environmental stress on interpopulation hybrids in the copepod *Tigriopus californicus*. *Journal of Experimental Marine Biology and Ecology* **303**: 183–196.

El Mousadik, A. and Petit, R. J. 1996. Chloroplast DNA phylogeography of the argan tree of Morocco. *Molecular Ecology* **5**: 547–555.

Emlen, S. T. 1995. An evolutionary theory of the family. *Proceedings of the National Academy of Sciences USA* **92**: 8092–8099.

Emlen, S. T., Wrege, P. H. and Webster, M. S. 1998. Cuckoldry as a cost of polyandry in the sex-role-reversed wattled jacana, *Jacana jacana*. *Proceedings of the Royal Society of London Series B* **265**: 2359–2364.

Endler, J. A. 1977. *Geographic Variation, Speciation and Clines*. Princeton University Press, Princeton, NJ.

Endler, J. A. 1983. Natural and sexual selection on color patterns in poecilid fishes. *Environmental Biology of Fishes* **9**: 173–190.

Endler, J. A. 1986. *Natural Selection in the Wild*. Princeton University Press, Princeton, NJ.

Engel, C. R., Destombe, C. and Valero, M. 2004. Mating system and gene flow in the red seaweed *Gracilaria gracilis*: effect of haploid–diploid life history and intertidal rocky shore landscape on fine-scale genetic structure. *Heredity* **92**: 289–298.

Ennos, R. A. 1994. Estimating the relative rates of pollen and seed migration among plant populations. *Heredity* **72**: 250–259.

Epling, C., Lewis, H. and Ball, F. M. 1960. The breeding group and seed storage: a study in population dynamics. *Evolution* **14**: 238–255.

Escorza-Trevino, S. and Dizon, A. E. 2000. Phylogeography, intraspecific structure and sex-biased dispersal of Dall's porpoise, *Phocoenoides dalli*, revealed by mitochondrial and microsatellite DNA analyses. *Molecular Ecology* **9**: 1049–1060.

Estoup, A. and Cornuet, J.-M. 1999. Microsatellite evolution: inferences from population data. In *Microsatellites: Evolution and Applications* (Goldstein, D. B. & Schlötterer, C., eds), pp. 49–65. Oxford University Press, Oxford.

Evans, T. A. and Goodisman, M. A. D. 2002. Nestmate relatedness and population genetic structure of the Australian social crab spider *Diaea ergandros* (Araneae: Thomisidae). *Molecular Ecology* **11**: 2307–2316.

Ewert, M. A., Lang, J. W. and Nelson, C. E. 2005. Geographic variation in the pattern of temperature-dependent sex determination in the American snapping turtle (*Chelydra serpentina*). *Journal of Zoology* **265**: 81–95.

Favre, L., Balloux, F., Goudet, J. and Perrin, N. 1997. Female-biased dispersal in the monogamous mammal *Crocidura russula*: Evidence from field data and microsatellite patterns. *Proceedings of the Royal Society of London Series B* **264**: 127–132.

Fedriani, J. M. and Kohn, M. H. 2001. Genotyping faeces links individuals to their diet. *Ecology Letters* **4**: 477–483.

Felsenstein, J. 1981. Evolutionary trees from DNA sequences: A maximum likelihood approach. *Journal of Molecular Evolution* **17**: 368–376.

Fischer, K., Zwaan, B. J. and Brakefield, P. M. 2004. Genetic and environmental sources of egg size variation in the butterfly *Bicyclus anynana*. *Heredity* **92**: 163–169.

Fisher, M. C. and Viney, M. E. 1998. The population genetic structure of the facultatively sexual parasitic nematode *Strongyloides ratti* in wild rats. *Proceedings of the Royal Society of London Series B* **1397**: 703–709.

Fisher, R. A. 1930. *The Genetical Theory of Natural Selection*. Clarendon Press, Oxford.

Fitch, W. M. 1971. Toward defining the course of evolution: minimum charge for a specific tree topology. *Systematic Zoology* **20**: 406–416.

Fleischer, R. C., McIntosh, C. E. and Tarr, C. L. 1998. Evolution on a volcanic conveyor belt: using phylogeographic reconstructions and K-Ar-based ages of the Hawaiian Islands to estimate molecular evolutionary rates. *Molecular Ecology* **7**: 533–545.

Foitzik, S. and Herbers, J. M. 2001. Colony structure of a slavemaking ant. I. Intracolony relatedness, worker reproduction and polydomy. *Evolution* **55**: 307–315.

Fonseca, D. M., LaPointe, D. A. and Fleischer, R. C. 2000. Bottlenecks and multiple introductions: population genetics of the vector of avian malaria in Hawaii. *Molecular Ecology* **9**: 1803–1814.

Ford, M. J. 2000. Effects of natural selection on patterns of DNA sequence variation at the transferrin, somatolactin and p53 genes within and among chinook salmon (*Oncorhynchus tshawytscha*) populations. *Molecular Ecology* **9**: 843–855.

Forstmeier, W. 2003. Extra-pair paternity in the dusky warbler, *Phylloscopus fuscatus*: a test of the 'constrained female hypothesis'. *Behaviour* **140**: 1117–1134.

Fox, C. J., Taylor, M. I., Pereyra, R., Villasana, M. I. and Rico, C. 2005. TaqMan DNA technology confirms likely overestimation of cod (*Gadus morhua* L.) egg abundance in the Irish Sea: implications for the assessment of the cod stock and mapping of spawning areas using egg-based methods. *Molecular Ecology* **14**: 879–884.

Frankham, R. 1995. Effective population size/adult population size ratios in wildlife: a review. *Genetical Research* **66**: 95–107.

Frankham, R. 1996. Relationship of genetic variation to population size in wildlife. *Conservation Biology* **10**: 1500–1508.

Frankham, R., Ballou, J. D. and Briscoe, D. A. 2002. *Introduction to Conservation Genetics*. Cambridge University Press, Cambridge.

Franklin, I. R. 1980. Evolutionary change in small populations. In *Conservation Biology: an Evolutionary-Ecological Perspective* (Soulé, M. E. & Wilcox, B. A., eds), pp. 135–150. Sinauer, Sunderland, MA.

Franklin, I. R. and Frankham, R. 1998. How large must populations be to retain evolutionary potential. *Animal Conservation* **1**: 69–71.

Freeland, J. R. and Boag, P. T. 1999. The mitochondrial and nuclear genetic homogeneity of the phenotypically diverse Darwin's ground finches. *Evolution* **53**: 1553–1563.

Freeland, J. R., Hannon, S. J., Dobush, G. and Boag, P. T. 1995. Extra-pair paternity in willow ptarmigan: measuring costs of polygyny to males. *Behavioral Ecology and Sociobiology* **36**: 349–355.

Freeland, J. R., Jones, C. S., Noble, L. R. and Okamura, B. 1999. Polymorphic microsatellite loci identified in the highly clonal freshwater bryozoan Cristatella mucedo. *Molecular Ecology* **8**: 341–342.

Freeland, J. R., Noble, L. R. and Okamura, B. 2000. Genetic consequences of the metapopulation biology of a facultatively sexual freshwater invertebrate. *Journal of Evolutionary Biology* **13**: 383–395.

Freeland, J. R., Rimmer, V. and Okamura, B. 2001. Genetic changes within freshwater bryozoan populations suggest temporal gene flow from statoblast banks. *Limnology and Oceanography* **46**: 1121–1149.

Freeland, J. R., Lodge, R. J., May, M. L. and Conrad, K. F. 2003. Genetic diversity and widespread haplotypes in a migratory dragonfly, the common green darner (*Anax junius*). *Ecological Entomology* **28**: 413–421.

Freeland, J. R., Rimmer, V. K. and Okamura, B. 2004. Evidence for a residual post-glacial founder effect in a highly dispersive freshwater invertebrate. *Limnology and Oceanography* **49**: 879–883.

Friar, E. A., Ladoux, T., Roalson, E. H. and Robichaux, R. H. 2002. Microsatellite analysis of a population crash and bottleneck in the Mauna Kea silversword, *Argyroxiphium sandwicense* ssp *sandwicense* (Asteraceae) and its implications for reintroduction. *Molecular Ecology* **9**: 2027–2034.

Fu, Y.-X. and Li., W.-H. 1999. Coalescing into the 21st century: an overview and prospects of coalescent theory. *Theoretical Population Biology* **56**: 1–10.

Fumagalli, L., Taberlet, P., Stewart, D. T., Gielly, L., Hausser, J. and Vogel, P. 1999. Molecular phylogeny and evolution of *Sorex* shrews (Soricidae: Insectivora) inferred from mitochondrial DNA sequence data. *Molecular Phylogenetics and Evolution* **11**: 222–235.

Funk, D. J. and Omland, K. E. 2003. Species-level paraphyly and polyphyly: frequency, causes and consequences, with insights from animal mitochondrial DNA. *Annual Review of Ecology and Systematics* **34**: 397–423.

Furlong, J. C. and Maden, B. E. 1983. Patterns of major divergence between the internal transcribed spacers of ribosomal DNA in *Xenopus borealis* and *Xenopus laevis* and of minimal divergence within ribosomal coding regions. *EMBO Journal* **2**: 443–448.

Futuyma, D. J. 1998. *Evolutionary Biology*. (3rd. edn). Sinauer Associates, Sunderland, MA.

Gamache, I., Jaramillo-Correa, J. P., Payette, S. and Bousquet, J. 2003. Diverging patterns of mitochondrial and nuclear DNA diversity in subarctic black spruce: imprint of a founder effect associated with postglacial colonization. *Molecular Ecology* **12**: 891–901.

Gardner, M. G., Bull, C. M., Cooper, S. J. B. and Duffield, G. A. 2001. Genetic evidence for a family structure in stable social aggregations of the Australian lizard *Egernia stokesii*. *Molecular Ecology* **10**: 175–183.

Gardner, M. G., Bull, C. M. and Cooper, S. J. B. 2002. High levels of genetic monogamy in the group-living Australian lizard *Egernia stokesii*. *Molecular Ecology* **11**: 1787–1794.

Gascon, C., Malcolm, J.R., Patton, J. L., da Silva, M. N. F., Bogart, J. P., Lougheed, S. C., Peres, C. A., Neckel, S. and Boag, P. T. 2000. Riverine barriers and the geographic distribution of Amazonian species. *Proceedings of the National Academy of Sciences USA.* **97**: 13672–13677.

Gehring, J. L. and Delph, L. F. 1999. Fine-scale genetic structure and clinal variation in *Silene acaulis* despite high gene flow. *Heredity* **82**: 628–637.

Genner, M. J., Michel, E., Erpenbeck, D., de Voogd, N., Witte, F. and Pointier, J.-P. 2004. Camouflaged invasion of Lake Malawi by an Oriental gastropod. *Molecular Ecology* **13**: 2135–2141.

Georghiou, G. P. 1990. Overview of insecticide resistance. In *Managing Resistance to Agrochemicals* (Green, M. B., Le Baron, H. M. & Moberg, W. K., eds), pp. 18–41. American Chemical Society, Washington, DC.

Gerber, S., Mariette, S., Streiff, R., Bodenes, C. and Kremer, A. 2000. Comparison of micro-satellites and amplified fragment length polymorphism markers for parentage analysis. *Molecular Ecology* **9**: 1037–1048.

Gerlach, G. and Hoeck, H. N. 2001. Islands on the plains: metapopulation dynamics and female biased dispersal in hyraxes (Hyracoidea) in the Serengeti National Park. *Molecular Ecology* **10**: 2307–2317.

Gibbs, H. L., Dawson, R. J. G. and Hobson, K. A. 2000. Limited differentiation in microsatellite DNA variation among northern populations of the yellow warbler: evidence for male-biased gene flow? *Molecular Ecology* **9**: 2137–2147.

Gigord, L. D. B., Macnair, M. R. and Smithson, A. 2001. Negative frequency-dependent selection maintains a dramatic flower color polymorphism in the rewardless orchid *Dactylorhiza sambucina* (L.) Soò. *Proceedings of the National Academy of Sciences USA* **98**: 6253–6255.

Gilk, S. E., Wang, I. A., Hoover, C. L., Smoker, W. W., Taylor, S. G., Gray, A. K. and Gharrett, A. J. 2004. Outbreeding depression in hybrids between spatially separated pink salmon, *Oncorhynchus gorbuscha*, populations: marine survival, homing ability and variability in family size. *Environmental Biology of Fishes* **69**: 287–297.

Glaubitz, J. C., Rhodes, O. E. and DeWoody, J. A. 2003. Prospects for inferring pairwise relationships with single nucleotide polymorphisms. *Molecular Ecology* **12**: 1039–1047.

Godoy, J. A. and Jordano, P. 2001. Seed dispersal by animals: exact identification of source trees with endocarp DNA microsatellites. *Molecular Ecology* **10**: 2275–2283.

Goffeau, A., Barrell, B. G., Bussey, H., Davis, R. W., Dujon, B., Feldmann, H., Galibert, F., Hoheisel, J. D., Jacq, C., Johnston, M. *et al.* 1996. Life with 6000 genes. *Science* **274**: 563–567.

Goodisman, M. A. D. and Hahn, D. A. 2004. Colony genetic structure of the ant *Componotus ocreatus* (Hymenoptera: Formicidae). *Sociology* **44**: 21–33.

Goodman, S. J., Barton, N. H., Swanson, G., Abernethy, K. and P. J.M. 1999. Introgression through rare hybridization: A genetic study of a hybrid zone between red and sika deer (genus *Cervus*) in Argyll, Scotland. *Genetics* **152**: 355–371.

Grant, P. R. and Grant, B. R. 1992a. Demography and the genetically effective size of two populations of Darwin's finches. *Ecology* **73**: 766–784.

Grant, P. R. and Grant, B. R. 1992b. Hybridization of bird species. *Science* **256**: 193–197.

Grant, V. 1981. *Plant Speciation*. Columbia University Press, New York.

Graur, D. and Martin, W. 2004. Reading the entrails of chickens: molecular timescales of evolution and the illusion of precision. *Trends in Genetics* **20**: 80–86.

Greene, E. 1996. Effect of light quality and larval diet on morph induction in the polymorphic caterpillar *Nemoria arizonaria* (Lepidoptera: Geometridae). *Biological Journal of the Linnean Society* **58**: 277–285.

Greenwood, P. J. 1980. Mating systems, philopatry and dispersal in birds and mammals. *Animal Behavior* **28**: 1140–1162.

Griffith, S. C., Owens, I. P. F. and Thuman, K. A. 2002. Extra pair paternity in birds: a review of interspecific variation and adaptive function. *Molecular Ecology* **11**: 2195–2212.

Griffiths, R., Double, M. C., Orr, K. and Dawson, R. J. G. 1998. A DNA test to sex most birds. *Molecular Ecology* **7**: 1071–1075.

Gu, R. S., Fonseca, S., Puskas, L. G., Hackler, L., Zvara, A., Dudits, D. and Pais, M. S. 2004. Transcript identification and profiling during salt stress and recovery of *Populus euphratica*. *Tree Physiology* **24**: 265–276.

Gubernick, D. J. and Teferi, T. 2000. Adaptive significance of male parental care in a monogamous mammal. *Proceedings of the Royal Society of London Series B* **267**: 147–150.

Guidot, A., Johannesson, H., Dahlberg, A. and Stenlid, J. 2003. Parental tracking in the postfire wood decay ascomycete *Daldinia loculata* using highly variable nuclear gene loci. *Molecular Ecology* **12**: 1717–1730.

Gyllensten, U. and Wilson, A. C. 1987. Interspecific mitochondrial DNA transfer and the colonization of Scandinavia by mice. *Genetical Research Cambridge* **49**: 25–29.

Gyllensten, U., Wharton, D., Josefsson, A. and Wilson, A. C. 1991. Paternal inheritance of mitochondrial DNA in mice. *Nature* **352**: 255–257.

Hadly, E. A., Ramakrishnan, U., Chan, Y. L., van Tuinen, M., O'Keefe, K., Spaeth, P. A. and Conroy, C. J. 2004. Genetic response to climate change: insights from ancient DNA and phylochronology. *PLoS Biology* **2**: 1–10.

Haig, S. M. and Avise, J. C. 1996. Avian conservation genetics. In *Conservation Genetics: Case Histories from Nature* (Avise, J. C. & Hamrick, J. R., eds), pp. 160–189. Chapman and Hall, New York

Hairston, N. G. J., Van Brunt, R. A. and Kearns, C. M. 1995. Age and survivorship of diapausing eggs in a sediment egg bank. *Ecology* **76**: 1706–1711.

Haldane, J. B. S. 1922. Sex-ratio and unidirectional sterility in hybrid animals. *Journal of Genetics* **12**: 101–109.

Haldane, J. B. S. 1924. A mathematical theory of natural and artificial selection. Part I. *Transactions of the Cambridge Philosophical Society* **23**: 10–41.

Hamilton, M. B. and Miller, J. R. 2002. Comparing relative rates of pollen and seed gene flow in the island model using nuclear and organelle measures of population structure. *Genetics* **162**: 1897–1909.

Hamilton, W. D. 1964. The genetical evolution of social behavior. *Journal of Theoretical Biology* **7**: 1–52.

Hammond, R. L., Bruford, M. W. and Bourke, A. F. G. 2002. Ant workers selfishly bias sex ratios by manipulating female development. *Proceedings of the Royal Society of London Series B* **269**: 173–178.

Hamrick, J. L. and Godt, M. J. W. 1990. Allozyme diversity in plant species. In *Plant population genetics, breeding and genetic resources* (Brown, A. H. D., Clegg, M. T., Kahler, A. L. B. S. & Weir, B. S., eds), pp. 43–63. Sinauer, Sunderland, MA.

Hancock, J. M. 1999. Microsatellites and other simple sequences: genomic context and mutational mechanisms. In *Microsatellites: Evolution and Applications* (Goldstein, D. B. & Schlötterer, eds), pp. 1–9. Oxford University Press, Oxford.

Hansen, M. M., Hynes, R. A., Loeschcke, V. and Rasmusen, G. 1995. Assessment of the stocked or wild origin of anadromous brown trout (*Salmo trutta* L.) in a Danish river system, using mitochondrial DNA RFLP analysis. *Molecular Ecology* **4**: 189–198.

Hansson, B., Bensch, S. and Hasselquist, D. 2003. A new approach to study dispersal: immigration of novel alleles reveals female-biased dispersal in great reed warblers. *Molecular Ecology* **12**: 631–637.

Hare, M. P. 2001. Prospects for nuclear gene phylogeography. *Trends in Ecology and Evolution* **16**: 700–706.

Harper, G. L., King, R. A., Dodd, C. S., Harwood, J. D., Glen, D. M., Bruford, M. W. and Symondson, W. O. C. 2005. Rapid screening of invertebrate predators for multiple prey DNA targets. *Molecular Ecology* **14**: 819–827.

Harris, H. 1966. Enzyme polymorphisms in man. *Proceedings of the Royal Society of London Series B* **164**: 298–310.

Harrison, R. G. 1986. Pattern and process in a narrow hybrid zone. *Heredity* **56**: 337–349.

Harrison, R. G. 1990. Hybrid zones: windows on evolutionary process. *Oxford Surveys in Evolutionary Biology* **7**: 69–128.

Hartl, D. L. and Clark, A. G. 1989. *Principles of Population Genetics* (2nd edn). Sinauer Associates, Sunderland, MA.

Hasegawa, E. 1994. Sex allocation in the ant *Colobopsis nipponicus* (Wheeler). I. Population sex ratio. *Evolution* **48**: 1121–1129.

Hasselquist, D. and Kempenaers, B. 2002. Parental care and adaptive brood sex ratio manipulation in birds. *Philosophical Transactions of the Royal Society of London B* **357**: 363–372.

Hasselquist, D., Bensch, S. and von Schantz, T. 1996. Correlation between male song repertoire, extra-pair paternity and offspring survival in the great reed warbler. *Nature* **381**: 229–232.

Hauser, L., Adcock, G. J., Smith, P. J., Ramirez, J. H. B. and Carvalho, G. R. 2002. Loss of microsatellite diversity and low effective population size in an overexploited population of New Zealand snapper (*Pagrus auratus*). *Proceedings of the National Academy of Sciences USA* **99**: 11742–11747.

Hebert, P. D. N., Penton, E. H., Burns, J. M., Janzen, D. H. and Hallwachs, W. 2004a. Ten species in one: DNA barcoding reveals cryptic species in the neotropical skipper butterfly *Astraptes fulgerator*. *Proceedings of the National Academy of Science USA* **101**: 14812–14817.

Hebert, P. D. N., Stoeckle, M. E., Zemlak, T. S. and Francis, C. M. 2004b. Identification of birds through DNA barcodes. *PLoS Biology* **2**: 1657–1663.

Hedgecock, D., Chow, V. and Waples, R. S. 1992. Effective population numbers of shellfish broodstocks estimated from temporal variance in allelic frequencies. *Aquaculture* **108**: 215–232.

Hedrick, P. W. and Kim, T. J. 2000. Genetics of complex polymorphisms: parasites and maintenance of MHC variation. In *Evolutionary Genetics from Molecules to Morphology* (Singh, R. S. & Krimbas, C. K., eds). pp. 204–234. Cambridge University Press, Cambridge.

Hedrick, P. W., Parker, K. M., Gutierrez-Espeleta, G. A., Rattink, A. and Lievers, K. 2000. Major histocompatibility complex variation in the Arabian oryx. *Evolution* **54**: 2145–2151.

Heidel, A. J. and Baldwin, I. T. 2004. Microarray analysis of salicylic acid- and jasmonic acid-signalling in responses of *Nicotiana attenuata* to attack by insects from multiple feeding guilds. *Plant Cell and Environment* **27**: 1362–1373.

Hellberg, M. E. and Vacquier, V. D. 1999. Rapid evolution of fertilization selectivity and lysin cDNA sequences in teguline gastropods. *Molecular Biology and Evolution* **16**: 839–848.

Henter, H. J. 2003. Inbreeding depression and haplodiploidy: experimental measures in a parasitoid and comparisons across diploid and haplodiploid insect taxa. *Evolution* **57**: 1793–1803.

Hewitt, G. M. 1989. The subdivision of species by hybrid zones. In *Speciation and Its Consequences* (Otte, D. & Endler, J. A., eds). pp. 85–110. Sinauer Associates, Inc., Sunderland, MA.

Hewitt, G. M. 1999. Post-glacial re-colonization of European biota. *Biological Journal of the Linnean Society* **68**: 87–112.

Hewitt, G. M. 2004. Genetic consequences of climatic oscillations in the Quaternary. *Philosophical Transactions of the Royal Society of London Series B* **359**: 183–195.

Hey, J., Waples, R. S., Arnold, M. L., Butlin, R. K. and Harrison, R. G. 2003. Understanding and confronting species uncertainty in biology and conservation. *Trends in Ecology and Evolution* **18**: 597–603.

Hildner, K. K. and Soulé, M. E. 2004. Relationship between the energetic cost of burrowing and genetic variability among populations of the pocket gopher, *T. bottae*: does physiological fitness correlate with genetic variability? *Journal of Experimental Biology* **207**: 2221–2227.

Hill, G. E., Inouye, C. Y. and Montgomerie, R. 2002. Dietary carotenoids predict plumage coloration in wild house finches. *Proceedings of the Royal Society of London Series B* **269**: 1119–1124.

Hill, W. G. 1981. Estimation of effective population size from data on linkage disequilibrium. *Genetical Research* **38**: 209–216.

Hitchings, S. P. and Beebee, T. J. C. 1998. Loss of genetic diversity and fitness in Common Toad (*Bufo bufo*) populations isolated by inimical habitat. *Journal of Evolutionary Biology* **11**: 269–283.

Hodkinson, T. R., Chase, M. W., Takahashi, C., Leitch, I. J., Bennett, M. D. and Renvoize, S. A. 2002. The use of DNA sequencing (ITS and *TRNL-F*), AFLP and fluorescent *in situ* hybridization to study allopolyploid *Miscanthus* (Poaceae). *American Journal of Botany* **89**: 279–286.

Hoelzel, A. R. 2001. Shark fishing in fin soup. *Conservation Genetics* **2**: 69–72.

Hoelzel, A. R., Hancock, J. M. and Dover, G. A. 1991. Evolution of the cetacean mitochondrial D-loop region. *Molecular Biology and Evolution* **8**: 475–493.

Hofkin, B. V., Wright, A., Altenback, J., Rassmann, K., Snell, H. M., Miller, R. D., Stone, A. C. and Snell, H. L. 2003. Ancient DNA gives light to Galápagos Land Iguana repatriation. *Conservation Genetics* **4**: 105–108.

Holmes, E. C. 2004. The phylogeography of human viruses. *Molecular Ecology* **13**: 745–756.

Holmes, E. C., Worobey, M. and Rambaut, A. 1999. Phylogenetic evidence for recombination in dengue virus. *Molecular Biology and Evolution* **16**: 405–409.

Hoogendoorn, M. and Heimpel, G. E. 2001. PCR-based gut content analysis of insect predators: using ribosomal ITS-1 fragments from prey to estimate predation frequency. *Molecular Ecology* **10**: 2059–2067.

Houlden, B. A., England, P. R., Taylor, A. C., Greville, W. D. and Sherwin, W. B. 1996. Low genetic variability of the koala *Phascolarctos cinereus* in south-eastern Australia following a severe population bottleneck. *Molecular Ecology* **5**: 269–281.

Howarth, F. G., Nishida, G. and Asquith, A. 1995. Insects of Hawaii. In *Our Living Resources: a Report to the Nation on the Distribution, Abundance and Health of U.S. Plants, Animals and Ecosystems* (LaRoe, E. T., Farris, G. S., Puckett, C. E., Doran, P. D. & Mac, M. J., eds). pp. 365–368. US Department of the Interior, Washington, DC.

Hsieh, H. M., Huang, L. H., Tsai, L. C., Kuo, Y. C., Meng, H. H., Linacre, A. and Lee, J. C. I. 2003. Species identification of rhinoceros horns using the cytochrome b gene. *Forensic Science International* **136**: 1–11.

Hudson, R. R. 1990. Gene genealogies and the coalescent process. In *Oxford Surveys in Evolutionary Biology* (Futuyma, D. J. & Antonovics, J., eds). pp. 1–44. Oxford University Press, New York.

Huelsenbeck, J. P., Ronquist, F., Nielsen, R. and Bollback, J. P. 2001. Bayesian inference of phylogeny and its impact on evolutionary biology. *Science* **294**: 2310–2314.

Hufford, K. M. and Mazer, S. J. 2003. Plant ecotypes: genetic differentiation in the age of ecological restoration. *Trends in Ecology and Evolution* **18**: 147–155.

Hughes, J. M., Mather, P. B., Toon, A., Ma, J., Rowley, I. and Russell, E. 2003. High levels of extra-group paternity in a population of Australian magpies *Gymnorhina tibicen*: evidence from microsatellite analysis. *Molecular Ecology* **12**: 3441–3450.

Hunt, J. and Simmons, L. W. 2002. Confidence of paternity and paternal care: covariation revealed through the experimental manipulation of the mating system in the beetle *Onthophagus taurus*. *Journal of Evolutionary Biology* **15**: 784–795.

Hunt, W. G. and Selander, R. K. 1973. Biochemical genetics of hybridisation in European house mice. *Heredity* **31**: 11–33.

Hurtado, L. A., Erez, T., Castrezana, S. and Markow, T. A. 2004. Contrasting population genetic patterns and evolutionary histories among sympatric Sonoran Desert cactophilic *Drosophila*. *Molecular Ecology* **13**: 1365–1375.

Husmeier, D. and Wright, F. 2001. Probabilistic divergence measures for detecting interspecies recombination. *Bioinformatics* **17** (Suppl. 1): S123–S131

Huyse, T., Van Houdt, J. and Volckaert, F. A. M. 2004. Paleoclimatic history and vicariant speciation in the "sand goby" group (Gobiidae, Teleostei). *Molecular Phylogenetics and Evolution* **32**: 324–336.

Ibarguchi, G., Gissing, G. J., Gaston, A. J., Boag, P. T. and Friesen, V. L. 2004. Male-biased mutation rates and the overestimation of extrapair paternity: problem, solution, and illustration using thick-billed murres (*Uria lomvia*, Alcidae). *Journal of Heredity* **95**: 209–216.

Ibrahim, K., Nichols, R. A. and Hewitt, G. M. 1996. Spatial patterns of genetic variation generated by different forms of dispersal during range expansion. *Heredity* **77**: 282–291.

Ingvarsson, P. K. and Olsson, K. 1997. Hierarchical genetic structure and effective population sizes in *Phalacrus substriatus*. *Heredity* **79**: 153–161.

International Human Genome Mapping Consortium. 2001. A physical map of the human genome. *Nature* **409**: 934–941.

International Human Genome Sequencing Consortium. 2004. Finishing the euchromatic sequence of the human genome. *Nature* **431**: 931–945.

Jackson, R. B., Linder, C. R., Lynch, M., Purugganan, M., Somerville, S. and Thayer, S. S. 2002. Linking molecular insight and ecological research. *Trends in Ecology and Evolution* **17**: 409–414.

Jehle, R., Arntzen, W., Burke, T., Krupa, A. P. and Hodl, W. 2001. The annual number of breeding adults and the effective population size of syntopic newts (*Triturus cristatus, T. marmoratus*). *Molecular Ecology* **10**: 839–850.

Jehle, R., Wilson, G. A., Arntzen, J. W. and Burke, T. 2005. Contemporary gene flow and the spatio-temporal genetic structure of subdivided newt populations (*Triturus cristatus, T. marmoratus*). *Journal of Evolutionary Biology* **18**: 619–628.

Jenkins, D. G. 1995. Dispersal-limited zooplankton distribution and community composition in new ponds. *Hydrobiologia* **313**: 15–20.

Ji, Y.-J., Zhang, D.-X. and He, L.-J. 2003. Evolutionary conservation and versatility of a new set of primers for amplifying the ribosomal internal transcribed spacer regions in insects and other invertebrates. *Molecular Ecology Notes* **3**: 581–585.

Jiang, C. S., Gibson, J. B. and Chen, H. Z. 1989. Genetic differentiation in populations of *Drosophila melanogaster* from the Peoples Republic of China: comparison with patterns on other continents. *Heredity* **62**: 193–198.

John, B. and Miklos, G. L. G. 1988. *The Eukaryote Genome in Development and Evolution*. Allen and Unwin, London.

Johnson, N. K. and Cicero, C. 2004. New mitochondrial DNA data affirm the importance of Pleistocene speciation in North American birds. *Evolution* **58**: 1122–1130.

Jones, K. C., Levine, K. F. and Banks, J. D. 2002. Characterization of 11 polymorphic tetra-nucleotide microsatellites for forensic applications in California elk (*Cervus elaphus canadensis*). *Molecular Ecology Notes* **2**: 425–427.

Jordan, M. A., Snell, H. M. and Jordan, W. C. 2005. Phenotypic divergence despite high levels of gene flow in Galápagos lava lizards (*Microlophus albemarlensis*). *Molecular Ecology* **14**: 859–867.

Jordan, W. C. and Bruford, M. W. 1998. New perspectives on mate choice and the MHC. *Heredity* **81**: 239–245.

Joron, M. and Brakefield, P. M. 2003. Captivity masks inbreeding effects on male mating success in butterflies. *Nature* **424**: 191–194.

Joseph, L., Moritz, C. and Hugall, A. 1995. Molecular support for vicariance as a source of diversity in rainforests. *Proceedings of the Royal Society of London B* **260**: 177–182.

Kaeuffer, R., Pontier, D., D. S. and Perrin, N. 2004. Effective size of two feral domestic cat populations (*Felis catus* L.): effect of the mating system. *Molecular Ecology* **13**: 483–490.

Karhu, A., Hurme, P., Karjalainen, M., Karvonen, P., Karkkainen, K., Neale, D. and Savolainen, O. 1996. Do molecular markers reflect patterns of differentiation in adaptive traits of conifers? *Theoretical and Applied Genetics* **93**: 215–221.

Karl, S. A. and Avise, J. C. 1992. Balancing selection at allozyme loci in oysters: implications from nuclear RFLPs. *Science* **256**: 100–102.

Karl, S. A., Bowen, B. W. and Avise, J. C. 1992. Global population genetic structure and male-mediated gene flow in the green turtle (*Chelonia mydas*): RFLP analysis of anonymous nuclear loci. *Genetics* **131**: 163–173.

Kasper, M. L., Reeson, A. F., Cooper, S. J. B., Perry, K. D. and Austin, A. D. 2004. Assessment of prey overlap between a native (*Polistes humilis*) and an introduced (*Vespula germanica*) social wasp using morphology and phylogenetic analyses of 16S rDNA. *Molecular Ecology* **13**: 2037–2048.

Keller, M., Kollmann, J. and Edwards, P. J. 2000. Genetic introgression from distant provenances reduces fitness in local weed populations. *Journal of Applied Ecology* **37**: 647–659.

Kelly, J., Carter, T. and Cole, M. D. 2003. Forensic identification of non-human mammals involved in the bushmeat trade using restriction digests of PCR amplified mitochondrial DNA. *Forensic Science International* **136**(Suppl. 1): 379–380.

Kempenaers, B., Everding, S., Bishop, C., Boag, P. T. and Robertson, R. J. 2001. Extra-pair paternity and the reproductive role of male floaters in the tree swallow (*Tachycineta bicolor*). *Behavioral Ecology and Sociobiology* **49**: 251–259.

Kent, A. D. and Triplett, E. W. 2002. Microbial communities and their interactions in soil and rhizosphere ecosystems. *Annual Review of Microbiology* **56**: 211–236.

Kerth, G., Mayer, F. and Petit, E. 2002a. Extreme sex-biased dispersal in the communally breeding, nonmigratory Bechstein's bat (*Myotis bechsteinii*). *Molecular Ecology* **11**: 1491–1498.

Kerth, G., Safi, K. and Konig, B. 2002b. Mean colony relatedness is a poor predictor of colony structure and female philopatry in the communally breeding Bechstein's bat (*Myotis bechsteinii*). *Behavioral Ecology and Sociobiology* **52**: 203–210.

Kim, K. S. and Sappington, T. W. 2004. Genetic structuring of boll weevil populations in the US based on RAPD markers. *Insect Molecular Biology* **13**: 293–303.

Kimura, M. 1968. Evolutionary rate at the molecular level. *Nature* **217**: 624–626.

Kimura, M. 1980. A simple method for estimating evolutionary rate of base substitutions through comparative studies of nucleotide sequences. *Journal of Molecular Evolution* **16**: 111–120.

Kimura, M. and Crow, J. F. 1964. The number of alleles that can be maintained in a finite population. *Genetics* **49**: 725–738.

Kimura, M. and Ohta, T. 1978. Stepwise mutation model and distribution of alellic frequencies in a finite population. *Proceedings of the National Academy of Sciences of the USA* **75**: 2868–2872.

Kingman, J. F. C. 1982. The coalescent. *Stochastic Processes and Their Applications* **13**: 235–248.

Klebanov, S., Flurkey, K., Roderick, T. H., Archer, J. R., Astle, M. C., Chen, J. and Harrison, D. E. 2000. Heritability of life span in mice and its implication for direct and indirect selection for longevity. *Genetica* **110**: 209–218.

Knapp, M., Stöckler, K., Havell, D., Delsuc, F., Sebastiani, F. and Lockhart, P. J. 2005. Relaxed molecular clock provides evidence for long-distance dispersal of *Nothofagus* (Southern Beech). *PLoS Biology* **3**: 38–43.

Knight, M. E., Van Oppen, M. J. H., Smith, H. L., Rico, C., Hewitt, G. M. and Turner, G. F. 1999. Evidence for male-biased dispersal in Lake Malawi cichlids from microsatellites. *Molecular Ecology* **8**: 1521–1527.

Knoblach, I. W. 1972. Intergeneric hybridization in flowering plants. *Taxon* **21**: 97–103.

Knowles, L. L. 2001. Did the Pleistocene glaciations promote divergence? Tests of explicit refugial models in montane grasshoppers. *Molecular Ecology* **10**: 691–701.

Knowles, L. L. 2004. The burgeoning field of statistical phylogeography. *Journal of Evolutionary Biology* **17**: 1–10.

Knowles, L. L. and Maddison, W. P. 2002. Statistical Phylogeography. *Molecular Ecology* **11**: 2623–2635.

Kocher, T. D., Thomas, W. K., Meyer, A., Edwards, S. V., Pääbo, S., Villablanca, F. X. and Wilson, A. C. 1989. Dynamics of mitochondrial DNA evolution in animals: amplification and sequencing with conserved primers. *Proceedings of the National Academy of Sciences of the USA* **86**: 6196–6200.

Koenig, W. D., van Vuren, D. and Hooge, P. N. 1996. Detectability, philopatry and the distribution of dispersal distances in vertebrates. *Trends in Ecology and Evolution* **11**: 514–518.

Kohlmann, K., Gross, R., Murakaeva, A. and Kersten, P. 2003. Genetic variability and structure of common carp (*Cyprinus carpio*) populations throughout the distribution range inferred from allozyme, microsatellite and mitochondrial DNA markers. *Aquatic Living Resources* **16**: 421–431.

Komdeur, J. and Pen, I. 2002. Adaptive sex allocation in birds: the complexities of linking theory and practice. *Philosophical Transactions of the Royal Society of London Series B* **357**: 373–380.

Kooistra, W. H. C. F. and Medlin, L. K. 1996. Evolution of the diatoms (Bacillariophyta). 4. Reconstruction of their age from small subunit rRNA coding regions and the fossil record. *Molecular Phylogenetics and Evolution* **6**: 391–407.

Kornfield, I. and Parker, A. 1997. Molecular systematics of a rapidly evolving species flock: the mbuna of Lake Malawi and the search for phylogenetic signals. In *Molecular Systematics of Fishes* (Kocher, T. & Stepien, C., eds), pp. 25–37. Academic Press, New York.

Kornfield, I. and Smith, P. F. 2000. African cichlid fishes: model systems for evolutionary biology. *Annual Review of Ecology and Systematics* **31**: 163–196.

Kraaijeveld-Smit, F. J. L., Ward, S. J., Temple-Smith, P. D. and Paetkau, D. 2002. Factors influencing paternity success in *Antechinus agilis*: last-male sperm precedence, timing of mating and genetic compatibility. *Journal of Evolutionary Biology* **15**: 100–107.

Kretzmann, M. B., Capote, N., Gautschi, B., Godoy, J. A., Donazar, J. A. and Negro, J. J. 2003. Genetically distinct island populations of the Egyptian vulture (*Neophron percnopterus*). *Conservation Genetics* **4**: 697–706.

Krieger, M. J. B. and Keller, L. 2000. Mating frequency and genetic structure of the Argentine ant *Linepithema humile*. *Molecular Ecology* **9**: 119–126.

Krings, M., Stone, A., Schmitz, R. W., Krainitzki, H., Stoneking, M. and Pääbo, S. 1997. Neanderthal DNA sequence and the origin of modern humans. *Cell* **90**: 19–30.

Krützen, M., Barre, L. M., Connor, R. C., Mann, J. and Sherwin, W. B. 2004. 'O father: where art thou?' – paternity assessment in an open fission-fusion society of wild bottlenose dolphins (*Tursiops* sp.) in Shark Bay, Western Australia. *Molecular Ecology* **13**: 1975–1990.

Kumar, S. and Hedges, S. B. 1998. A molecular timescale for vertebrate evolution. *Nature* **392**: 917–920.

Kvist, L., Martens, J., Nazarenko, A. A. and Orell, M. 2003. Paternal leakage of mitochondrial DNA in the great tit (*Parus major*). *Molecular Biology and Evolution* **20**: 243–247.

Lacey, E. A., Wieczorek, J. R. and Tucker, P. K. 1997. Male mating behaviour and patterns of sperm precedence in Arctic ground squirrels. *Animal Behaviour* **53**: 767–779.

Lacy, R. C. and Ballou, J. D. 1998. Effectiveness of selection in reducing the genetic load in populations of *Peromyscus polionotus* during generations of inbreeding. *Evolution* **52**: 900–909.

Ladoukakis, E. D. and Zouros, E. 2001. Direct evidence for homologous recombination in mussel (*Mytilus galloprovincialis*) mitochondrial DNA. *Molecular Biology and Evolution* **18**: 1168–1175.

Ladoukakis, E. D., Saavedra, C., Magoulas, A. and Zouros, E. 2002. Mitochondrial DNA variation in a species with two mitochondrial genomes: the case of *Mytilus galloprovincialis* from the Atlantic, the Mediterranean and the Black Sea. *Molecular Ecology* **11**: 755–769.

Laidlaw, H. H. and Page, R. E. 1984. Polyandry in honey bees (*Apis mellifera* L.): sperm utilization and intracolony genetic relationships. *Genetics* **108**: 985–997.

Lampert, K. P., Rand, A. S., Mueller, U. G. and Ryan, M. J. 2003. Fine-scale genetic pattern and evidence for sex-biased dispersal in the tungara frog, *Physalaemus pustulosus*. *Molecular Ecology* **12**: 3325–3334.

Land, E. D. and Lacy, R. C. 2000. Introgression level achieved through Florida panther genetic restoration. *Endangered Species Update* **17**: 100–105.

Landry, C. and Bernatchez, L. 2001. Comparative analysis of population structure across environments and geographical scales at major histocompatibility complex and microsatellite loci in Atlantic salmon (*Salmo salar*). *Molecular Ecology* **10**: 2525–2539.

Landweber, L. F. 1999. Something old for something new: The future of ancient DNA in conservation biology. In *Genetics and the Extinction of Species* (Landweber, L. F. & Dobson, A. P., eds), pp. 163–186. Princeton University Press, Princeton, NJ.

Lascoux, M., Palme, A. E., Cheddadi, R. and Latta, R. G. 2004. Impact of Ice Ages on the genetic structure of trees and shrubs. *Philosophical Transactions of the Royal Society of London Series B* **359**: 197–207.

Latta, R. G. and Mitton, J. B. 1997. A comparison of population differentiation across four classes of gene marker in limber pine (*Pinus flexilis* James). *Genetics* **146**: 1153–1163.

Lawler, R. R., Richard, A. F. and Riley, M. A. 2003. Genetic population structure of the white sifaka (*Propithecus verreauxi verreauxi*) at Beza Mahafaly Special Reserve, southwest Madagascar (1992–2001). *Molecular Ecology* **12**: 2307–2317.

Le Clerc, V., Briard, M., Granger, J. and Delettre, J. 2003. Genebank biodiversity assessments regarding optimal sample size and seed harvesting techniques for the regeneration of carrot accessions. *Biodiversity and Conservation* **12**: 2227–2236.

Leakey, R. and Lewin, R. 1995. *The Sixth Extinction: Patterns of Life and the Future of Humankind*. Doubleday, New York.

Ledda, S., Leoni, G., Bogliolo, L. and Naitana, S. 2001. Oocyte cryopreservation and ovarian tissue banking. *Theriogenology* **55**: 1359–1371.

Lehman, N., Eisenhawer, A., Hansen, K., Mech, L. D., Peterson, R. O., Gogan, P. J. P. and Wayne, R. K. 1991. Introgression of coyote mitochondrial DNA into sympatric North American gray wolf populations. *Evolution* **45**: 104–119.

Lericolais, G., Auffret, J. P. and Bourillet, J. F. 2003. The quaternary channel river: seismic stratigraphy of its palaeo-valleys and deeps. *Journal of Quaternary Science* **18**: 245–260.

Lessios, H. A., Kane, J. and Robertson, D. R. 2003. Phylogeography of the pantropical sea urchin *Tripneustes*: Contrasting patterns of population structure between oceans. *Evolution* **57**: 2026–2036.

Levins, R. 1969. Some demographic and genetic consequences of environmental heterogeneity for biological control. *Bulletin of the Entomological Society* **15**: 237–240.

Lewontin, R. C. and Hubby, J. 1966. A molecular approach to the study of genic heterozygosity in natural populations. II. Amounts of variation and degree of heterozygosity in natural populations of *Drosophila pseudoobscura*. *Genetics* **54**: 595–609.

Li, W.-H. 1997. *Molecular Evolution*. Sinauer Associates, Sunderland, MA.

Li, W.-H. and Graur, D. 1991. *Fundamentals of Molecular Evolution*. Sinauer Associates, Sunderland, MA.

Li, W.-H. and Sadler, L. A. 1991. Low nucleotide diversity in man. *Genetics* **129**: 513–523.

Liebherr, J. K. 1988. Gene flow in ground beetles (Coleoptera: Carabidae) of differing habitat preference and flight-wing development. *Evolution* **42**: 129–137.

Lippe, C., Dumont, P. and Bernatchez, L. 2004. Isolation and identification of 21 microsatellite loci in the Copper redhorse (*Moxostoma hubbsi*; Catostomidae) and their variability in other catostomids. *Molecular Ecology Notes* **4**: 638–641.

Llewellyn, K. S., Loxdale, H. D., Harrington, R., Brookes, C. P., Clark, S. J. and Sunnucks, P. 2003. Migration and genetic structure of the grain aphid (*Sitobion avenae*) in Britain related to climate and clonal fluctuation as revealed using microsatellites. *Molecular Ecology* **12**: 21–34.

Lodge, R. J. and Freeland, J. R. 2003. The use of odonata museum specimens in questions of molecular evolution. *Odonatologica* **32**: 375–380.

Lohm, J., Grahn, M., Langefors, A., Andersen, O., Storset, A. and T., von, Schantz. 2002. Experimental evidence for major histocompatibility complex-allele-specific resistance to a bacterial infection. *Proceedings of the Royal Society of London Series B* **269**: 2029–2033.

Lopez, J. V., Culver, M., Stephens, J. C., Johnson, W. E. and O'Brien., S. J. 1997. Rates of nuclear and cytoplasmic mitochondrial DNA sequence divergence in mammals. *Molecular Biology and Evolution* **14**: 277–286.

Lougheed, S. C., Gascon, C., Jones, D. A., Bogart, J. P. and Boag, P. T. 1999. Ridges and rivers: a test of competing hypotheses of Amazonian diversification using a dart-poison frog (*Epipedobates femoralis*). *Proceedings of the Royal Society of London B* **266**: 1829–1835.

Lougheed, S. C., Freeland, J. R., Handford, P. and Boag, P. T. 2000. A molecular phylogeny of warbling-finches (*Poospiza*): paraphyly in a neotropical emberizid genus. *Molecular Phylogenetics and Evolution* **17**: 367–378.

Lovette, I. J., Clegg, S. M. and Smith, T. B. 2004. Limited utility of mtDNA markers for determining connectivity among breeding and overwintering locations in three neotropical migrant birds. *Conservation Biology* **18**: 156–166.

Lucchini, V., Fabbri, E., Marucco, F., Ricci, S., Boitani, L. and Randi, E. 2002. Noninvasive molecular tracking of colonizing wolf (*Canis lupus*) packs in the western Italian Alps. *Molecular Ecology* **11**: 857–868.

Luikart, G. and Cornuet, J.-M. 1998. Empirical evaluation of a test for identifying recently bottlenecked populations from allele frequency data. *Conservation Biology* **13**: 523–530.

Luikart, G., Sherwin, W. B., Steele, B. M. and Allendorf, F. W. 1998. Usefulness of molecular markers for detecting population bottlenecks via monitoring genetic change. *Molecular Ecology* **7**: 963–974.

Luke, K., Horrocks, J. A., LeRoux, R. A. and Dutton, P. H. 2004. Origins of green turtle (*Chelonia mydas*) feeding aggregations around Barbados, West Indies. *Marine Biology* **144**: 799–805.

Luppi, T. A., Spivak, E. D. and Bas, C. C. 2003. The effects of temperature and salinity on larval development of *Armases rubripes* Rathbun, 1897 (Brachyura, Grapsoidea, Sesarmidae) and the southern limit of its geographical distribution. *Estuarine Coastal and Shelf Science* **58**: 575–585.

Ma, L. J., Rogers, S. O., Catranis, C. M. and Starmer, W. T. 2000. Detection and characterization of ancient fungi entrapped in glacial ice. *Mycologia* **92**: 286–295.

MacIsaac, H. J., Grigorovich, I. A., Hoyle, J. A., Yan, N. D. and Panov, V. E. 1999. Invasion of Lake Ontario by the Ponto-Caspian predatory cladoceran *Cercopagis Pengoi*. *Canadian Journal of Fisheries and Aquatic Sciences* **56**: 1–5.

Madsen, T., Shine, R., Loman, J. and Hakansson, T. 1992. Why do female adders copulate so frequently? *Nature* **355**: 440–441.

Madsen, T., Shine, R., Olsson, M. and Wittzell, H. 1999. Conservation biology – restoration of an inbred adder population. *Nature* **402**: 34–35.

Madsen, T., Ujvari, B. and Olsson, M. 2004. Novel genes continue to enhance population growth in adders (*Vipera berus*). *Biological Conservation* **120**: 145–147.

Magrath, R. D. and Whittingham, L. A. 1997. Subordinate males are more likely to help if unrelated to the breeding female in cooperatively breeding white-browed scrubwrens. *Behavioral Ecology and Sociobiology* **41**: 185–192.

Maguire, T. L., Peakall, R. and Saenger, P. 2002. Comparative analysis of genetic diversity in the mangrove species *Avicennia marina* (Forsk.) Vierh. (Avicenniaceae) detected by AFLPs and SSRs. *Theoretical and Applied Genetics* **104**: 388–398.

Manel, S., Berthier, P. and Luikart, G. 2002. Detecting wildlife poaching: Identifying the origin of individuals with Bayesian assignment tests and multilocus genotypes. *Conservation Biology* **16**: 650–659.

Markert, J. A., Danley, P. D. and Arnegard, M. E. 2001. New markers for new species: microsatellite loci and the East African cichlids. *Trends in Ecology and Evolution* **16**: 100–107.

Marr, A. B., Keller, L. F. and Arcese, P. 2002. Heterosis and outbreeding depression in descendants of natural immigrants to an inbred population of song sparrows (*Melospiza melodia*). *Evolution* **56**: 131–142.

Marshall, T. C., Slate, J., Kruuk, L. E. B. and Pemberton, J. M. 1998. Statistical confidence for likelihood-based paternity inference in natural populations. *Molecular Ecology* **7**: 639–655.

Martin, A. P. and Palumbi, S. R. 1993. Body size, metabolic rate, generation time and the molecular clock. *Proceedings of the National Academy of Sciences USA* **90**: 4087–4091.

Martin, A. P., Naylor, G. J. P. and Palumbi, S. R. 1992. Rates of mitochondrial DNA evolution in sharks are slow compared with mammals. *Nature* **357**: 153–155.

Martinez, J. G., Soler, J. J., Soler, M., Møller, A. P. and Burke, T. 1999. Comparative population structure and gene flow of a brood parasite, the great spotted cuckoo (*Clamator glandarius*) and its primary host, the magpie (*Pica pica*). *Evolution* **53**: 269–278.

Martins, T. L. F., Blakey, J. K. and Wright, J. 2002. Low incidence of extra-pair paternity in the colonially nesting common swift *Apus apus*. *Journal of Avian Biology* **33**: 441–446.

Maynard Smith, J. 1978. *The Evolution of Sex*. Cambridge University Press, Cambridge, MA.

Maynard Smith, J. and Haigh, J. 1974. The hitch-hiking effect of a favourable gene. *Genetical Research* **23**: 23–35.

Mayr, E. 1942. *Systematics and the Origin of Species*. Columbia University Press, New York.

McCauley, D. E. 1997. The relative contributions of seed and pollen movement to the local genetic structure of *Silene alba*. *Journal of Heredity* **88**: 257–263.

McFadden, C. S. and Hutchinson, M. B. 2004. Molecular evidence for the hybrid origin of species in the soft coral genus *Alcyonium* (Cnidaria: Anthozoa: Octocorallia). *Molecular Ecology* **13**: 1495–1505.

McKay, J. K. and Latta, R. G. 2002. Adaptive population divergence: markers, QTL and traits. *Trends in Ecology and Evolution* **17**: 285–291.

McMillan, W. O., Raff, R. A. and Palumbi, S. R. 1992. Population genetic consequences of developmental evolution in sea urchins (genus *Heliocidaris*). *Evolution* **46**: 1299–1312.

Merilä, J. and Crnokrak, P. 2001. Comparison of genetic differentiation at marker loci and quantitative traits. *Journal of Evolutionary Biology* **14**: 892–903.

Merilä, J., Björklund, M. and Baker, A. J. 1996. The successful founder: Genetics of introduced *Carduelis chloris* (greenfinch) populations in New Zealand. *Heredity* **77**: 410–422.

Metzlaff, M., Borner, T. and Hagemann, R. 1981. Variations of chloroplast DNAs in the genus *Pelargonium* and their biparental inheritance. *Theoretical and Applied Genetics* **60**: 37–41.

Michaux, J. R., Magnanou, E., Paradis, E., Nieberding, C. and Libois, R. 2003. Mitochondrial phylogeography of the Woodmouse (*Apodemus sylvaticus*) in the Western Palearctic region. *Molecular Ecology* **12**: 685–697.

Michot, B., Bachellerie, J. P., Raynal, F. and Renalier, M. H. 1982. Sequence of the 3'-terminal domain of mouse 18S rRNA. Conservation of structural features with other pro- and eukaryotic homologs. *FEBS Letters* **142**: 260–266.

Miller, C. R. and Waits, L. P. 2003. The history of effective population size and genetic diversity in the Yellowstone grizzly (*Ursus arctos*). *Proceedings of the National Academy of Sciences USA* **100**: 4334–4339.

Miller, L. M., Close, T. and Kapuscinski, A. R. 2004. Lower fitness of hatchery and hybrid rainbow trout compared with naturalized populations in Lake Superior tributaries. *Molecular Ecology* **13**: 3379–3388.

Milot, E., Gibbs, H. L. and Hobson, K. A. 2000. Phylogeography and genetic structure of northern populations of the yellow warbler (*Dendroica petechia*). *Molecular Ecology* **9**: 667–681.

Mindell, D. P. and Honeycutt, R. L. 1990. Ribosomal RNA in vertebrates and phylogenetic applications. *Annual Review of Ecology and Systematics* **21**: 541–566.

Miner, B. G. 2005. Evolution of feeding structure plasticity in marine invertebrate larvae: a possible trade-off between arm length and stomach size. *Journal of Experimental Marine Biology and Ecology* **315**: 117–125.

Miura, I., Ohtani, H., Nakamura, M., Ichikawa, Y. and Saitoh, K. 1998. The origin and differentiation of the heteromorphic sex chromosomes Z, W, X and Y in the frog *Rana rugosa*, inferred from the sequences of a sex-linked gene, ADP/ATP translocase. *Molecular Biology and Evolution* **15**: 1612–1619.

Moore, M. K., Bemiss, J. A., Rice, S. M., Quattro, J. M. and Woodley, C. 2003. Use of restriction fragment length polymorphisms to identify sea turtle eggs and cooked meats to species. *Conservation Genetics* **4**: 95–103.

Moore, W. S. 1977. An evaluation of narrow hybrid zones in vertebrates. *Quarterly Review of Biology* **52**: 263–277.

Morand, M. E., Brachet, S., Rossignol, P., Dufour, J. and Frascaria-Lacoste, N. 2002. A generalized heterozygote deficiency assessed with microsatellites in French common ash populations. *Molecular Ecology* **11**: 377–385.

Morin, P. A. and Woodruff, D. S. 1996. Noninvasive genotyping for vertebrate conservation. In *Molecular Genetic Approaches in Conservation* (Smith, T. B. & Wayne, R. K., eds), pp. 298–313. Oxford University Press, New York.

Morin, P. A., Luikart, G. and Wayne, R. K. 2004. SNPs in ecology, evolution and conservation. *Trends in Ecology and Evolution* **19**: 208–216.

Moritz, C. C. 1994. Defining 'evolutionarily significant units' for conservation. *Trends in Ecology and Evolution* **9**: 373–375.

Moriuchi, K. S. and Winn, A. A. 2005. Relationships among growth, development and plastic response to environment quality in a perennial plant. *New Phytologist* **166**: 149–158.

Morjan, C. L. and Rieseberg, L. H. 2004. How species evolve collectively: implications of gene flow and selection for the spread of advantageous alleles. *Molecular Ecology* **13**: 1341–1356.

Mullis, K. B. and Faloona, F. A. 1987. Specific synthesis of DNA *in vitro* via a polymerase-catalyzed chain reaction. *Methods in Enzymology* **155**: 335–350.

Mulvey, M., Aho, J. M., Lydeard, C., Leberg, P. L. and Smith, M. H. 1991. Comparative population genetic structure of a parasite (*Fascioloides magna*) and its definitive host. *Evolution* **45**: 1628–1640.

Murata, K., Satou, M., Matsushima, K., Satake, S. and Yamamoto, Y. 2004. Retrospective estimation of genetic diversity of an extinct Oriental white stork (*Ciconia boyciana*) population in Japan using mounted specimens and implications for reintroduction programs. *Conservation Genetics* **5**: 553–560.

Myers, R. A. and Worm, B. 2003. Rapid worldwide depletion of predatory fish communities. *Nature* **423**: 280–283.

Nebel, S., Cloutier, A. and Thompson, G. J. 2004. Molecular sexing of prey remains permits a test of sex-biased predation in a wintering population of western sandpipers. *Proceedings of the Royal Society of London B* **271** (Suppl.): S321–S323.

Neff, B. D. 2001. Genetic paternity analysis and breeding success in bluegill sunfish (*Lepomis macrochirus*). *Journal of Heredity* **92**: 111–119.

Neff, B. D. 2003. Decisions about parental care in response to perceived paternity. *Nature* **422**: 716–719.

Nei, M. 1972. Genetic distance between populations. *American Naturalist* **106**: 283–292.

Nei, M. 1973. Analysis of gene diversity in subdivided populations. *Proceedings of the National Academy of Sciences USA* **70**: 3321–3323.

Nei, M. 1987. *Molecular Evolutionary Genetics*. Columbia University Press, New York.

Nei, M. and Tajima, F. 1981. Genetic drift and estimation of effective population size. *Genetics* **98**: 625–640.

Nuismer, S. L. and Thompson, J. N. 2001. Plant polyploidy and non-uniform effects on insect herbivores. *Proceedings of the Royal Society of London B* **268**: 1937–1940.

Nunez-Farfan, J., Dominguez, C. A., Eguiarte, L. E., Cornejo, A., Quijano, M., Vargas, J. and Dirzo, R. 2002. Genetic divergence among Mexican populations of red mangrove (*Rhizophora mangle*): geographic and historic effects. *Evolutionary Ecology Research* **4**: 1049–1064.

Nürnberger, B. and Harrison, R. G. 1995. Spatial population structure in the whirligig beetle *Dineutus assimilis*. Evolutionary inferences based on mitochondrial DNA and field data. *Evolution* **49**: 266–275.

Nybom, H. 2004. Comparison of different nuclear DNA markers for estimating intraspecific genetic diversity in plants. *Molecular Ecology* **13**: 1143–1155.

Oates, J. F., Abedi-Lartey, M., McGraw, W. S., Struhsaker, T. T. and Whitesides, G. H. 2000. Extinction of a West African red colobus monkey. *Conservation Biology* **14**: 1526–1532.

Oddou-Muratorio, S., Petit, R. J., Le Guerroue, B., Guesnet, D. and Demesure, B. 2001. Pollen- versus seed-mediated gene flow in a scattered forest tree species. *Evolution* **55**: 1123–1135.

Ohyama, K., Fukuzawa, H., Kohchi, T., Shirai, H., Sano, T., Sano, S., Umesono, K., Shiki, M., Takeuchi, Y., Chang, Z., Aota, S., Inokuchi, H. and Ozeki, H. 1986. Chloroplast gene organization deduced from complete sequence of liverwort *Marchantia polymorpha* chloroplast DNA. *Nature* **322**: 572–574.

Olsson, M., Shine, R., Madsen, T., Gullberg, A. and Tegelström., H. 1996. Sperm selection by females. *Nature* **383**: 585.

Olsson, M., Madsen, T., Nordby, J., Wapstra, E., Ujvari, B. and Wittsell, H. 2003. Major histocompatibility complex and mate choice in sand lizards. *Proceedings of the Royal Society of London Series B* **270** (Suppl. 2): S254–S256.

Paetkau, D. and Strobeck, C. 1994. Microsatellite analysis of genetic variation in black bear populations. *Molecular Ecology* **3**: 489–495.

Paetkau, D., Calvert, W., Stirling, I. and Strobeck, C. 1995. Microsatellite analysis of population structure in Canadian polar bears. *Molecular Ecology* **4**: 347–354.

Paetkau, D., Waits, L. P., Clarkson, P. L., Craighead, L., Vyse, E., Ward, R. and Strobeck, C. 1998. Variation in genetic diversity across the range of North American brown bears. *Conservation Biology* **12**: 418–429.

Pajunen, V. I. and Pajunen, I. 2003. Long-term dynamics in rock pool *Daphnia* metapopulations. *Ecography* **26**: 731–738.

Pakkasmaa, S., Merila, J. and O'Hara, R. B. 2003. Genetic and maternal effect influences on viability of common frog tadpoles under different environmental conditions. *Heredity* **91**: 117–124.

Palmer, J. D. 1991. Plastid chromosomes: structure and evolution. In *The Molecular Biology of Plastids*. (Bogorad, L. & Vasil, I. K., eds), pp. 5–53. Academic Press, San Diego.

Palmer, J. D. and Herbon, L. A. 1988. Plant mitochondrial DNA evolves rapidly in structure, but slowly in sequence. *Journal of Molecular Evolution* **28**: 87–97.

Palo, J. U., O'Hara, R. B., Laugen, A. T., Laurila, A., Primmer, C. R. and Merila, J. 2003. Latitudinal divergence of common frog (*Rana temporaria*) life history traits by natural selection: evidence from a comparison of molecular and quantitative genetic data. *Molecular Ecology* **12**: 1963–1978.

Palumbi, S. R. and Baker, C. S. 1994. Contrasting population structure from nuclear intron sequences and mtDNA of humpback whales. *Molecular Biology and Evolution* **11**: 426–435.

Pamilo, P. and Nei, M. 1988. Relationships between gene trees and species trees. *Molecular Biology and Evolution* **5**: 568–583.

Papura, D., Simon, J. C., Halkett, F., Delmotte, F., Le Gallic, J. F. and Dedryver, C. A. 2003. Predominance of sexual reproduction in Romanian populations of the aphid *Sitobion avenae* inferred from phenotypic and genetic structure. *Heredity* **90**: 397–404.

Parkash, R., Shamin, A. and Vashist, M. 1992. Latitudinal *Adh* allozymic variation and ethanol tolerance in Indian populations of *Drosophila melanogaster*. *Zeitschrift fur Zoologische Systematik und Evolutionsforschung* **34**: 64–72.

Parker, P. G. and Waite, T. A. 1997. Mating systems, effective population size and conservation of natural populations. In *Behavioural Approaches to Conservation in the Wild* (Clemmons, J. R. & Buchhotz, R., eds), pp. 244–262. Cambridge University Press, London.

Passera, L., Aron, S., Vargo, E. L. and Keller, L. 2001. Queen control of sex ratio in fire ants. *Science* **293**: 1308–1310.

Paxinos, E. E., James, H. F., Olson, S. L., Ballou, J. D., Leonard, J. A. and Fleischer, R. C. 2002. Prehistoric decline of genetic diversity in the néné. *Science* **296**: 1827–1827.

Pearson, B., Raybould, A. F. and Clarke, R. T. 1995. Breeding behavior, relatedness and sex-investment ratios in *Leptothorax tuberum* Fabricius. *Entomologia Experimentalis et Applicata* **75**: 165–174.

Pemberton, J. 2004. Measuring inbreeding depression in the wild: the old ways are the best. *Trends in Ecology and Evolution* **19**: 613–615.

Penalva, M. A., Maya, A., Dopazo, J. and Ramon, D. 1990. Sequences of isopenicillin N synthetase genes suggest horizontal gene transfer from prokaryotes to eukaryotes. *Proceedings of the Royal Society of London Series B* **241**: 164–169.

Peres, C. A., Patton, J. L. and Da Silva, M. N. F. 1996. Riverine barriers and gene flow in Amazonian saddle-back tamarins. *Folia Primatologica* **67**: 113–124.

Perrin, N. and Mazalov, V. 1999. Local competition, inbreeding and the evolution of sex-biased dispersal. *American Naturalist* **154**: 116–127.

Peterson, A. T. and Navarro-Siguenza, A. G. 1999. Alternate species concepts as bases for determining priority conservation areas. *Conservation Biology* **13**: 427–431.

Peterson, M. A. and Denno, R. F. 1998a. The influence of dispersal and diet breadth on patterns of genetic isolation by distance in phytophagous insects. *American Naturalist* **152**: 428–446.

Peterson, M. A. and Denno, R. F. 1998b. Life-history strategies and the genetic structure of phytophagous insect populations. In *Genetic Structure and Local Adaptation in Natural Insect Populations* (Mopper, S. & Strauss, S. Y., eds), pp. 263–322. Chapman and Hall, New York.

Petit, C., Fréville, H., Mignot, A., Colas, B., Riba, M., Imbert, E., Hurtrez-Boussés, S., Virevaire, M. and Olivieri, I. 2001. Gene flow and local adaptation in two endemic plant species. *Biological Conservation* **100**: 21–34.

Petit, R. J., Duminil, J., Fineschi, S., Hampe, A., Salvini, D. and Vendramin, G. G. 2005. Comparative organization of chloroplast, mitochondrial and nuclear diversity in plant populations. *Molecular Ecology* **14**: 689–702.

Philipp, D. P. and Gross, M. R. 1994. Genetic evidence for cuckoldry in bluegill *Lepomis macrochirus*. *Molecular Ecology* **3**: 563–569.

Picard, D., Plantard, O., Scurrah, M. and Mugniéry, D. 2004. Inbreeding and population structure of the potato cyst nematode (*Globodera pallida*) in its native area (Peru). *Molecular Ecology* **13**: 2899–2908.

Piertney, S. B., MacColl, A. D. C., Bacon, P. J., Racey, P. A., Lambin, X. and Dallas, J. F. 2000. Matrilineal genetic structure and female-mediated gene flow in red grouse (*Lagopus lagopus scoticus*): An analysis using mitochondrial DNA. *Evolution* **54**: 279–289.

Pike, T. M. and Petrie, M. 2003. Potential mechanisms of avian sex manipulation. *Biological Review* **78**: 553–574.

Pilastro, A., Griggio, M., Biddau, L. and Mingozzi, T. 2002. Extrapair paternity as a cost of polygyny in the rock sparrow: behavioural and genetic evidence of the 'trade-off' hypothesis. *Animal Behaviour* **63**: 967–974.

Pillmann, A., Woolcott, G. W., Olsen, J. L., Stam, W. T. and King, R. J. 1997. Inter- and intraspecific genetic variation in *Caulerpa* (Chlorophyta) based on nuclear rDNA ITS sequences. *European Journal of Phycology* **32**: 379–386.

Pires, A. E. and Fernandes, M. L. 2003. Last lynxes in Portugal? Molecular approaches in a pre-extinction scenario. *Conservation Genetics* **4**: 525–532.

Pitra, C. and Lieckfeldt, D. 1999. Molecular-forensic contribution to the conviction of an alleged poacher: A case report. *Zeitschrift fur Jagdwissenschaft* **45**: 270–275.

Plantard, O. and Porte, C. 2004. Population genetic structure of the sugar beet cyst nematode *Heterodera schachtii*: a gonochoristic and amphimictic species with highly inbred but weakly differentiated populations. *Molecular Ecology* **13**: 33–41.

Pluess, A. R. and Stocklin, J. 2004. Genetic diversity and fitness in *Scabiosa columbaria* in the Swiss Jura in relation to population size. *Conservation Genetics* **5**: 145–156.

Poetsch, M., Seefeldt, S., Maschke, M. and Lignitz, E. 2001. Analysis of microsatellite polymorphism in red deer, roe deer and fallow deer – possible employment in forensic applications. *Forensic Science International* **116**: 108.

Poirier, N. E., Whittingham, L. A., and Dunn, P. O. 2004. Males achieve greater reproductive success through multiple broods, than through extrapair mating in house wrens. *Animal Behaviour* **67**: 1109–1116.

Ponniah, M. and Hughes, J. M. 2004. The evolution of Queensland spiny mountain crayfish of the genus *Euastacus*. I. Testing vicariance and dispersal with interspecific mitochondrial DNA. *Evolution* **58**: 1073–1085.

Posada, D. and Crandall, K. A. 2001. Intraspecific gene genealogies: trees grafting into networks. *Trends in Ecology and Evolution* **16**: 37–45.

Powell, J. A. 2001. Longest insect dormancy: Yucca moth larvae (Lepidoptera: Prodoxidae) metamorphose after 20, 25 and 30 years in diapause. *Annals of the Entomological Society of America* **94**: 677–680.

Powell, W., Morgante, M., Andre, C., Hanafey, M., Vogel, J., Tingey, S. and Rafalski, A. 1996. The comparison of RFLP, RAPD, AFLP and SSR (microsatellite) markers for germplasm analysis. *Molecular Breeding* **2**: 225–238.

Primack, R. B. 1998. *Essentials of Conservation Biology*. Sinauer Associates, Sunderland, MA.

Primmer, C. R., Borge, T., Lindell, J. and Saetre, G.-P. 2002. Single nucleotide polymorphism characterization in species with limited available sequence information: high nucleotide diversity revealed in the avian genome. *Molecular Ecology* **11**: 603–612.

Pritchard, J. K., Stephens, M. and Donnelly, P. 2000. Inference of population structure using multilocus genotype data. *Genetics* **155**: 945–959.

Provan, J., Biss, P. M., McMeel, D. and Mathews, S. 2004. Universal primers for the amplification of chloroplast microsatellites in grasses (Poaceae). *Molecular Ecology Notes* **4**: 262–264.

Provan, J., Wattier, R. A. and Maggs, C. A. 2005. Phylogeographic analysis of the red seaweed *Palmaria palmata* reveals a Pleistocene marine glacial refugium in the English Channel. *Molecular Ecology* **14**: 793–803.

Pudovkin, A. I., Zaykin, D. V. and Hedgecock, D. 1996. On the potential for estimating the effective number of breeders from heterozygote excess in progeny. *Genetics* **144**: 383–387.

Pusey, A. E. 1987. Sex-biased dispersal and inbreeding avoidance in birds and mammals. *Trends in Ecology and Evolution* **2**: 295–299.

Queller, D. C. and Goodnight, K. F. 1989. Estimating relatedness using genetic markers. *Evolution* **43**: 258–275.

Quilichini, A., Debussche, M. and Thompson, J. D. 2001. Evidence for local outbreeding depression in the Mediterranean island endemic *Anchusa crispa* Viv. (Boraginaceae). *Heredity* **87**: 190–197.

Rand, D. M. 1994. Thermal habit, metabolic rate and the evolution of mitochondrial DNA. *Trends in Ecology and Evolution* **9**: 238.

Rand, D. M. and Harrison, R. G. 1989. Ecological genetics of a mosaic hybrid zone: mitochondrial, nuclear and reproductive differentiation of crickets by soil type. *Evolution* **43**: 432–449.

Randi, E., Lucchini, V., Christensen, M. F., Mucci, N., Funk, S. M., Dolf, G. and Loeschcke, V. 2000. Mitochondrial DNA variability in Italian and East European wolves: detecting the consequences of small population size and hybridization. *Conservation Biology* **14**: 464–473.

Rassmann, K., Tautz, D., Trillmich, F. and Gliddon, C. 1997. The microevolution of the Galápagos marine iguana *Amblyrhynchus cristatus* assessed by nuclear and mitochondrial genetic analyses. *Molecular Ecology* **6**: 437–452.

Raup, D. M. 1994. The role of extinction in evolution. *Proceedings of the National Academy of Sciences USA* **91**: 6758–6763.

Rawls, J. F., Samuel, B. S. and Gordon, J. I. 2004. Gnotobiotic zebrafish reveal evolutionarily conserved responses to the gut microbiota. *Proceedings of the National Academy of Sciences USA* **101**: 4596–4601.

Raybould, A. F., Goudet, J., Mogg, R. J., Gliddon, C. J. and Gray, A. J. 1996. Genetic structure of a linear population of *Beta vulgaris* ssp *maritima* (sea beet) revealed by isozyme and RFLP analysis. *Heredity* **76**: 111–117.

Reed, D. H. and Frankham, R. 2003. Correlation between fitness and genetic diversity. *Conservation Biology* **17**: 230–237.

Reineke, A., Schmidt, O. and Zebitz, C. P. W. 2003. Differential gene expression in two strains of the endoparasitic wasp *Venturia canescens* identified by cDNA-amplified fragment length polymorphism analysis. *Molecular Ecology* **12**: 3485–3492.

Rendell, S. and Ennos, R. A. 2002. Chloroplast DNA diversity in *Calluna vulgaris* (heather) populations in Europe. *Molecular Ecology* **11**: 69–78.

Rhode, J. M. and Duffy, J. E. 2004. Seed production from the mixed mating system of Chesapeake Bay (USA) eelgrass (*Zostera marina*; Zosteraceae). *American Journal of Botany* **91**: 192–197.

Rhymer, J. M., Williams, M. J. and Braun, M. J. 1994. Mitochondrial analysis of gene flow between New Zealand mallards (*Anax platyrhynchos*) and grey ducks (*A. superciliosa*). *Auk* **111**: 970–978.

Richards, C. and Leberg, P. L. 1996. Temporal changes in allele frequencies and a population's history of severe bottlenecks. *Conservation Biology* **10**: 832–839.

Rieseberg, L. H., Choi, H. C., Chan, R. and Spore, C. 1993. Genomic map of a diploid hybrid species. *Heredity* **70**: 285–293.

Rivas, G.-G., Zapater, M.-F., Abadie, C. and Carlier, J. 2004. Founder effects and stochastic dispersal at the continental scale of the fungal pathogen of bananas *Mycosphaerella fijiensis*. *Molecular Ecology* **13**: 471–482.

Roelke, M. E., Martenson, J. S. and O'Brien, S. J. 1993. The consequences of demographic reduction and genetic depletion in the endangered Florida panther. *Current Biology* **3**: 340–350.

Roemer, G. W., Smith, D. A., Garcelon, D. K. and Wayne, R. K. 2001. The behavioural ecology of the island fox (*Urocyon littoralis*). *Journal of Zoology* **255**: 1–14.

Rose, O. C., Brookes, M. I. and Mallet, J. L. B. 1994. A quick and simple non-lethal method for extracting DNA from butterfly wings. *Molecular Ecology* **3**: 275

Rosenberg, N. A. and Nordborg, M. 2002. Genealogical trees, coalescent theory and the analysis of genetic polymorphisms. *Nature Reviews Genetics* **3**: 380–390.

Roskam, J. C. and Brakefield, P. M. 1999. Seasonal polyphenism in *Bicyclus* (Lepidoptera: Satyridae) butterflies: different climates need different cues. *Biological Journal of the Linnean Society* **66**: 315 356.

Ross, K. G. 2001. Molecular ecology of social behaviour: analyses of breeding systems and genetic structure. *Molecular Ecology* **10**: 265–284.

Rothstein, S. I. 1990. A model system for coevolution: Avian brood parasitism. *Annual Review of Ecology and Systematics* **21**: 481–508.

Rülicke, T., Chapuisat, M., Homberger, F. R., Macas, E. and Wedekind, C. 1998. MHC-genotype of progeny influenced by parental infection. *Proceedings of the Royal Society of London Series B* **265**: 711–716.

Russello, M. A. and Amato, G. 2004. *Ex situ* population management in the absence of pedigree information. *Molecular Ecology* **13**: 2829–2840.

Saccheri, I., Kuussaari, M., Kankare, M., Vikman, P., Fortelius, W. and Hanski, I. 1998. Inbreeding and extinction in a butterfly metapopulation. *Nature* **392**: 491–494.

Saitou, N. and Nei, M. 1987. The neighbour-joining method: A new method for reconstructing phylogenetic trees. *Molecular Biology and Evolution* **4**: 406–425.

Sakisaka, Y., Yahara, T., Miura, I. and Kasuya, E. 2000. Maternal control of sex ratio in *Rana rugosa*: evidence from DNA sexing. *Molecular Ecology* **9**: 1711–1715.

Saltonstall, K. 2002. Cryptic invasion by a non-native genotype of the common reed, *Phragmites australis*, into North America. *Proceedings of the National Academy of Sciences USA* **99**: 2445–2449.

Sandell, M. I. and Diemer, M. 1999. Intraspecific brood parasitism: a strategy for floating females in the European starling. *Animal Behavior* **57**: 197–202.

Saux, C., Simon, C. and Spicer, G. S. 2003. Phylogeny of the dragonfly and damselfly order odonata as inferred by mitochondrial 12S ribosomal RNA sequences. *Annals of the Entomological Society of America* **96**: 693–699.

Savolainen, P. and Lundeberg, J. 1999. Forensic evidence based on mtDNA from dog and wolf hairs. *Journal of Forensic Sciences* **44**: 77–81.

Schmitt, T. and Seitz, A. 2002. Postglacial distribution area expansion of *Polyommatus coridon* (Lepidoptera: Lycaenidae) from its Ponto-Mediterranean glacial refugium. *Heredity* **89**: 20–26.

Schneider, C. J., Cunningham, M. and Moritz, C. 1998. Comparative phylogeography and the history of endemic vertebrates in the Wet Tropics rainforests of Australia. *Molecular Ecology* **7**: 487–498.

Schultheis, A. S., Weigt, L. A. and Hendricks, A. C. 2002. Gene flow, dispersal and nested clade analysis among populations of the stonefly *Peltoperla tarteri* in the southern Appalachians. *Molecular Ecology* **11**: 317–327.

Schwartz, F. J. 1972. World literature to fish hybrids, with an analysis by family, species and hybrid. *Publication of Gulf Coast Research Laboratory Museum* **3**: 1–328.

Schwartz, F. J. 1981. World literature to fish hybrids, with an analysis by family, species and hybrid: Supplement 1. *NOAA Technical Report NMFS SSRF-750. US Dept. of Commerce*, Washington, DC.

Schwartz, M. and Vissing, J. 2002. Paternal inheritance of mitochondrial DNA. *New England Journal of Medicine* **347**: 576–580.

Scofield, V. L., Schlumpberger, J. M., West, L. A. and Weissman, I. L. 1982. Protochordate allorecognition is controlled by a MHC-like gene system. *Nature* **295**: 499–502.

Scribner, K. T. and Bowman, T. D. 1998. Microsatellites identify depredated waterfowl remains from glaucous gull stomachs. *Molecular Ecology* **7**: 1401–1405.

Seddon, J. M., Parker, H. G., Ostrander, E. A. and Ellegren, H. 2005. SNPs in ecological and conservation studies: a test in the Scandinavian wolf population. *Molecular Ecology* **14**: 503–511.

Seehausen, O. 2004. Hybridization and adaptive radiation. *Trends in Ecology and Evolution* **19**: 198–207.

Seymour, A. M., Montgomery, M. E., Costello, B. H., Ihle, S., Johnsson, G., St John, B., Taggart, D. and Houlden, B. A. 2001. High effective inbreeding coefficients correlate with morphological abnormalities in populations of South Australian koalas (*Phascolarctos cinereus*). *Animal Conservation* **4**: 211–219.

Sheldon, B. C., Andersson, S., Griffith, S. C., Ornborg, J. and Sendecka, J. 1999. Ultraviolet colour variation influences blue tit sex ratios. *Nature* **402**: 874–877.

Shen, P., Wang, F., Underhill, P. A., Franco, C., Yang, W. H., Roxas, A., Sung, R., Lin, A. A., Hyman, R. W., Vollrath, D. *et al.* 2000. Population genetic implications from sequence variation in four Y chromosome genes. *Proceedings of the National Academy of Sciences USA* **97**: 7354–7359.

Shen-Miller, J., Mudgett, M. B., Schopf, J. W., Clarke, S. and Berger, R. 1995. Exceptional seed longevity and robust growth: ancient sacred lotus from China. *American Journal of Botany* **82**: 1367–1380.

Sheppard, S. K., Henneman, M. L., Memmott, J. and Symondson, W. O. C. 2004. Infiltration by alien predators into invertebrate food webs in Hawaii: a molecular approach. *Molecular Ecology* **13**: 2077–2088.

Shinozaki, K., Ohme, M., Tanaka, M., Wakasugi, N., Hayashida, T., Matsubayashi, N., Zaita, J., Chunwongse, J., Obokata, K., Yamaguchi-Shinozaki, C. *et al.* 1986. The complete nucleotide sequence of tobacco chloroplast genome: Its gene organization and expression. *EMBO Journal* **5**: 2043–2049.

Shrimpton, J. M. and Heath, D. D. 2003. Census vs. effective population size in chinook salmon: large- and small-scale environmental perturbation effects. *Molecular Ecology* **12**: 2571–2583.

Siegel, D. A., Kinlan, B. P., Gaylord, B. and Gaines, S. D. 2003. Lagrangian descriptions of marine larval dispersion. *Marine Ecology Progress Series* **260**: 83–96.

Silva-Montellano, A. and Eguiarte, L. E. 2003a. Geographic patterns in the reproductive ecology of *Agave lechuguilla* in the Chihuahuan Desert. I. Floral characteristics, visitors and fecundity. *American Journal of Botany* **90**: 377–387.

Silva-Montellano, A. and Eguiarte, L. E. 2003b. Geographic patterns in the reproductive ecology of *Agave lechuguilla* (Agavaceae) in the Chihuahuan Desert. II. Genetic variation, differentiation and inbreeding estimates. *American Journal of Botany* **90**: 700–706.

Simon, J.-C., Baumann, S., Sunnucks, P., Hebert, P. D. N., Pierre, J.-S., Le Gallic, J.-F. and Dedryver, C.-A. 1998. Reproductive mode and population genetic structure of the cereal aphid *Sitobion avenue* studied using phenotypic and microsatellite markers. *Molecular Ecology* **8**: 531–545.

Slate, J., Kruuk, L. E. B., Marshall, T. C., Pemberton, J. M. and Clutton-Brock, T. H. 2000. Inbreeding depression influences lifetime breeding success in a wild population of red deer (*Cervus elaphus*). *Proceedings of the Royal Society of London Series B* **267**: 1657–1662.

Slatkin, M. 1995. A measure of population subdivision based on microsatellite allele frequencies. *Genetics* **139**: 457–462.

Sloane, M. A., Sunnucks, P., Alpers, D., Beheregaray, L. B. and Taylor, A. C. 2000. Highly reliable genetic identification of individual northern hairy-nosed wombats from single remotely collected hairs: a feasible censusing method. *Molecular Ecology* **9**: 1233–1240.

Smith, H. G. 1993. Heritability of tarsus length in cross-fostered broods of the European starling (*Sturnus vulgaris*). *Heredity* **71**: 318–322.

Smith, R. L. and Sytsma, K. J. 1990. Evolution of *Populus nigra* (sect. *Aigeiros*): introgressive hybridization and the chloroplast contribution of *Populus alba* L. (sect. *Populus*). *American Journal of Botany* **77**: 1176–1187.

Smith, S., Hughes, J. and Wardell-Johnson, G. 2003. High population differentiation and extensive clonality in a rare mallee eucalypt: *Eucalyptus curtisii* – Conservation genetics of a rare mallee eucalypt. *Conservation Genetics* **4**: 289–300.

Solé, M., Durka, W., Eber, S. and Brandl, R. 2004. Genotypic and genetic diversity of the common weed *Cirsium arvense* (Asteraceae). *International Journal of Plant Sciences* **165**: 437–444.

Sorci, G. and Clobert, J. 1995. Effects of maternal parasite load on offspring life-history traits in the common lizard (*Lacerta vivipara*). *Journal of Evolutionary Biology* **8**: 711–723.

Soulé, M., Gilpin, M., Conway, W. and Foose, T. 1986. The millennium ark: how long a voyage, how many staterooms, how many passengers? *Zoo Biology* **5**: 101–113.

Spencer, C. C., Neigel, J. E. and Leberg, P. L. 2000. Experimental evaluation of the usefulness of microsatellite DNA for detecting demographic bottlenecks. *Molecular Ecology* **9**: 1517–1528.

Squirrell, J., Hollingsworth, P. M., Bateman, R. M., Dickson, J. H., Light, M. H. S., MacConaill, M. and Tebbitt, M. C. 2001. Partitioning and diversity of nuclear and organelle markers in native and introduced populations of *Epipactis helleborine* (Orchidaceae). *American Journal of Botany* **88**: 1409–1418.

Stam, W. T., Olsen, J. L. and Coyer, J. A. 2001. Post-glacial recolonization and biogeographic patterns in the North Atlantic. *Phycologia* **40s**: 46.

Steiner, C. and Catzeflis, F. M. 2004. Genetic variation and geographical structure of five mouse-sized opossums (Marsupialia, Didelphidae) throughout the Guiana Region. *Journal of Biogeography* **31**: 959–973.

Stenson, A. G., Halhotra, A. and Thorpe, R. S. 2002. Population differentiation and nuclear gene flow in the Dominican anole (*Anolis oculatis*). *Molecular Ecology* **11**: 1679–1688.

Stewart Jr., C. N., Halfhill, M. D. and Warwick, S. I. 2003. Transgene introgression from genetically modified crops to their wild relatives. *Nature Reviews Genetics* **4**: 806–817.

Stilwell, K. L., Wilbur, H. M., Werth, C. R. and Taylor, D. R. 2003. Heterozygote advantage in the American chestnut, *Castanea dentata* (Fagaceae). *American Journal of Botany* **90**: 207–213.

Strand, M., Prolla, T. A., Liskay, R. M. and Petes, T. D. 1993. Destabilization of tracts of simple repetitive DNA in yeast by mutations affecting DNA mismatch repair. *Nature* **365**: 274–276.

Sugiura, M. 1992. The chloroplast genome. *Plant Molecular Biology* **19**: 149–168.

Sültmann, H. and Mayer, E. W. 1997. Reconstruction of cichlid fish phylogeny using nuclear DNA markers. In *Molecular Systematics of Fishes* (Kocher, T. D. & Stepien, C. A., eds), pp. 39–51. Academic Press, New York.

Sun, G. L., Diaz, O., Salomon, B. and Von Bothmer., R. 1998. Microsatellite variation and its comparison with allozyme and RAPD variation in *Elymus fibrosus* (Schrenk) Tzvel. (Poaceae). *Hereditas* **129**: 275–282.

Sundström, L., Keller, L. and Chapuisat, M. 2003. Inbreeding and sex-biased gene flow in the ant *Formica exsecta*. *Evolution* **57**: 1552–1561.

Sutherland, G. D., Harestad, A. S., Price, K. and Lertzman, K. P. 2000. Scaling of natal dispersal distances in terrestrial birds and mammals. *Conservation Ecology* **4**: 16.

Sweigart, A. L. and Willis, J. H. 2003. Patterns of nucleotide diversity in two species of *Mimulus* are affected by mating system and asymmetric introgression. *Evolution* **57**: 2490–2506.

Sweijd, N. A., Bowie, R. C. K., Lopata, A. L., Marinaki, A. M., Harley, E. H. and Cook, P. A. 1998. A PCR technique for forensic, species-level identification of abalone tissue. *Journal of Shellfish Research* **17**: 889–895.

Taberlet, P. 1998. Biodiversity at the intraspecific level: The comparative phylogeographic approach. *Journal of Biotechnology* **64**: 91–100.

Taberlet, P. and Bouvet, J. 1994. Mitochondrial DNA polymorphism, phylogeography and conservation genetics of the brown bear *Ursus arctos* in Europe. *Proceedings of the Royal Society of London Series B* **255**: 195–200.

Taberlet, P., Fumagalli, L., Wust-Saucy, A.-G. and Cosson, J.-F. 1998. Comparative phylogeography and postglacial colonization routes in Europe. *Molecular Ecology* **7**: 453–464.

Takahashi, J., Akimoto, S., Martin, S. J., Tamukae, M. and Hasegawa, E. 2004. Mating structure and male production in the giant hornet *Vespa mandarinia* (Hymenoptera: Vespidae). *Applied Entomology and Zoology* **39**: 343–349.

Takahata, N. and Nei, M. 1990. Allelic genealogy under overdominant and frequency-dependent selection and polymorphism of major histocompatibility complex loci. *Genetics* **124**: 967–978.

Tan, A. M. and Wake, D. B. 1995. MtDNA phylogeography of the California newt, *Taricha torosa* (Caudata, salamandridae). *Molecular Phylogenetics and Evolution* 4: 383–394.

Tanguy, A., Guo, X. M. and Ford, S. E. 2004. Discovery of genes expressed in response to *Perkinsus marinus* challenge in Eastern (*Crassostrea virginica*) and Pacific (*C. gigas*) oysters. *Gene* 338: 121–131.

Taylor, D. J., Finston, T. L. and Hebert, P. D. N. 1998. Biogeography of a widespread freshwater crustatcean: pseudocongruence and cryptic endemism in the North American *Daphnia laevis* complex. *Evolution* 52: 1648–1670.

Taylor, M. I., Morley, J. I., Rico, C. and Balshine, S. 2003. Evidence for genetic monogamy and female-biased dispersal in the biparental mouthbrooding cichlid *Eretmodus cyanostictus* from Lake Tanganyika. *Molecular Ecology* 12: 3173–3177.

Tegelström, H. 1987. Transfer of mitochondrial DNA from the northern red-backed vole (*Clethrionomys rutilus*) to the bank vole (*C. glareolus*). *Journal of Molecular Evolution* 24: 218–227.

Templeton, A. R. 2004. Statistical phylogeography: methods of evaluating and minimizing inference errors. *Molecular Ecology* 13: 789–809.

Templeton, A. R., Crandall, K. A. and Sing, C. F. 1992. A cladistic analysis of phenotypic associations with haplotypes inferred from restriction endonuclease mapping and DNA sequence data. III. Cladogram estimation. *Genetics* 132: 619–633.

Templeton, A. R., Routman, E. and Phillips, C. A. 1995. Separating population structure from population history: a cladistic analysis of the geographical distribution of mitochondrial DNA haplotypes in the tiger salamander, *Ambystoma tigrinum*. *Genetics* 140: 767–782.

Tero, N., Aspi, J., Siikamaki, P., Jakalaniemi, A. and Tuomi, J. 2003. Genetic structure and gene flow in a metapopulation of an endangered plant species, *Silene tatarica*. *Molecular Ecology* 12: 2073–2085.

Thalman, O., Hebler, J., Poinar, H. N., Pääbo, S. and Vigilant, L. 2004. Unreliable mtDNA data due to nuclear insertions: a cautionary tale from analysis of humans and other great apes. *Molecular Ecology* 13: 321–335.

Thomas, B. R., Macdonald, S. E., Hicks, M., Adams, D. L. and Hodgetts, R. B. 1999. Effects of reforestation methods on genetic diversity of lodgepole pine: an assessment using microsatellite and randomly amplified polymorphic DNA markers. *Theoretical and Applied Genetics* 98: 793–801.

Thompson, J. D., Gibson, T. J., Plewniak, F., Jeanmougin, F. and Higgins, D. G. 1997. The ClustalX windows interface: flexible strategies for multiple sequence alignment aided by quality analysis tools. *Nucleic Acids Research* 24: 4876–4882.

Thrall, P. H., Richards, C. M., McCauley, D. E. and Antonovics, J. 1998. Metapopulation collapse: the consequences of limited gene flow in spatially structured populations. In *Modeling Spatiotemporal Dynamics in Ecology* (Bascompte, J. & Solé, R. V., eds), pp. 83–104. Springer-Verlag, Berlin.

Tiedemann, R., Paulus, K. B., Scheer, M., Von Kistowski, K. G., Skirnisson, K., Bloch, D. and Dam, M. 2004. Mitochondrial DNA and microsatellite variation in the eider duck (*Somateria mollissima*) indicate stepwise postglacial colonization of Europe and limited current long-distance dispersal. *Molecular Ecology* 13: 1481–1494.

Torres, R. and Velando, A. 2005. Male preference for female foot colour in the socially monogamous blue-footed booby, *Sula nebouxii*. *Animal Behaviour* 69: 59–65.

Travis, S. E., Slobodchikoff, C. N. and Keim, P. 1995. Social assemblages and mating relationships in prairie dogs: a DNA fingerprint analysis. *Behavioral Ecology* 7: 95–100.

Tregenza, T. and Wedell, N. 1998. Benefits of multiple mates in the cricket *Gryllus bimaculatus*. *Evolution* 52: 1726–1730.

Tripet, F., Dolo, G. and Lanzaro, G. C. 2005. Multilevel analyses of genetic differentiation in *Anopheles gambiae* ss reveal patterns of gene flow important for malaria-fighting mosquito projects. *Genetics* **169**: 313–324.

Trivers, R. L. and Willard, D. E. 1973. Natural selection of parental ability to vary the sex ratio of offspring. *Science* **179**: 90–92.

Trottier, R. 1966. The emergence and sex ratio of *Anax junius* (Odonata: Aeshnidae) in Canada. *Canadian Entomologist* **98**: 794–798.

Tudge, C. 1995. Captive audiences for future conservation. *New Scientist* **145**: 51–52.

Turner, T. F., Wares, J. P. and Gold, J. R. 2002. Genetic effective size is three orders of magnitude smaller than adult census size in an abundant, estuarine-dependent marine fish (*Sciaenops ocellatus*). *Genetics* **162**: 1329–1339.

Turpeinen, T., Vanhala, T., Nevo, E. and Nissilä, E. 2003. AFLP genetic polymorphism in wild barley (*Hordeum spontaneum*) populations in Israel. *Theoretical and Applied Genetics* **106**: 1333–1339.

Tuttle, E. M. 2003. Alternative reproductive strategies in the white-throated sparrow: behavioral and genetic evidence. *Behavioral Ecology* **14**: 425–432.

Ulstrup, K. E. and Van Oppen, M. J. H. 2003. Geographic and habitat partitioning of genetically distinct zooanthellae (*Symbiodinium*) in *Acropora* corals in the Great Barrier Reef. *Molecular Ecology* **12**: 3477–3484.

Unwin, R. and Maiden, M. C. J. 2003. Multi-locus sequence typing: a tool for global epidemiology. *Trends in Microbiology* **11**: 479–487.

Utter, F. and Epifanio, J. 2002. Marine aquaculture: Genetic potentialities and pitfalls. *Reviews in Fish Biology and Fisheries* **12**: 59–77.

Valiere, N. and Taberlet, P. 2000. Urine collected in the field as a source of DNA for species and individual identification. *Molecular Ecology* **9**: 2150–2152.

Van der Velde, M. and Bijlsma, R. 2003. Phylogeography of five *Polytrichum* species within Europe. *Biological Journal of the Linnean Society* **78**: 203–213.

van der Wurff, A. W. G., Isaaks, J. A., Ernsting, G. and Van Straalen, N. M. 2003. Population substructures in the soil invertebrate *Orchesella cincta*, as revealed by microsatellite and TE-AFLP markers. *Molecular Ecology* **12**: 1349–1359.

van Oppen, M. J. H., Turner, G. F., Rico, C., Robinson, R. L., Deutsch, J. C., Genner, M. J. and Hewitt, G. M. 1998. Assortative mating among rock-dwelling cichlid fishes supports high estimates of species richness from Lake Malawi. *Molecular Ecology* **7**: 991–1001.

Velazquez, R. R. E., Zurdo-Pineiro, J. L., Mateos, P. F. and Molina, E. M. 2004. Identification of microorganisms by PCR amplification and sequencing of a universal amplified ribosomal region present in both prokaryotes and eukaryotes. *Journal of Microbial Methods* **56**: 413–426.

Vences, M., Vieites, D. R., Glaw, F., Brinkmann, H., Kosuch, J., Veith, M. and Meyer, A. 2003. Multiple overseas dispersal in amphibians. *Proceedings of the Royal Society of London Series B* **270**: 2435–2442.

Verma, S. K. and Singh, L. 2003. Novel universal primers establish identity of an enormous number of animal species for forensic identification. *Molecular Ecology Notes* **3**: 28–31.

Vermeij, G. J. 1978. *Biogeography and Adaptation*. Harvard University Press, Cambridge, Massachusetts.

Vilà, C., Sundqvist, A.-K., Flagstad, Ø., Seddon, J., Björnerfeldt, S., Kojola, I., Casulli, A., Sand, H., Wabakken, P. and Ellegren, H. 2003a. Rescue of a severely bottlenecked wolf (*Canis lupus*) population by a single immigrant. *Proceedings of the Royal Society of London Series B* **270**: 91–97.

Vilà, C., Walker, C., Sundqvist, A. K., Flagstad, O., Andersone, Z., Casulli, A., Kojola, I., Valdmann, H., Halverson, J. and Ellegren, H. 2003b. Combined use of maternal, paternal and bi-parental genetic markers for the identification of wolf-dog hybrids. *Heredity* **90**: 17–24.

Vogel, J. C., Rumsey, F. J., Schneller, J. J., Barrett, J. A. and Gibby, M. 1999. Where are the glacial refugia in Europe? Evidence from pteridophytes. *Biological Journal of the Linnean Society* **66**: 23–37.

Vos, P., Hogers, R., Bleeker, M., Reijans, M., Van de Lee, T., Hornes, M., Frijters, A., Pot, J., Peleman, J., Kuiper, M. and Zabeau, M. 1995. AFLP: a new technique for DNA fingerprinting. *Nucleic Acids Research* **23**: 4407–4414.

Waits, L., Taberlet, P., Swenson, J. E., Sandegren, F. and Franzen, R. 2000. Nuclear DNA microsatellite analysis of genetic diversity and gene flow in the Scandinavian brown bear (*Ursus arctos*). *Molecular Ecology* **9**: 421–431.

Waldman, B., Rice, J. E. and Honeycutt, R. L. 1992. Kin recognition and incest avoidance in toads. *American Zoologist* **32**: 18–30.

Wallace, A. R. 1876. *The Geographical Distribution of Animals*, Vol. 1. Macmillan, London.

Wan, Q.-H. and Fang, S.-G. 2003. Application of species-specific polymerase chain reaction in the forensic identification of tiger species. *Forensic Science International* **131**: 75–78.

Wang, H.-Y., Tsai, M.-P., Tu, M.-C. and Lee, S.-C. 2000. Universal primers for amplification of the complete mitochondrial 12S rRNA gene in vertebrates. *Zoological Studies* **39**: 61–66.

Wang, J. L. and Whitlock, M. C. 2003. Estimating effective population size and migration rates from genetic samples over space and time. *Genetics* **163**: 429–446.

Wang, M. and Schreiber, A. 2001. The impact of habitat fragmentation and social structure on the population genetics of roe deer (*Capreolus capreolus* L.) in Central Europe. *Heredity* **86**: 703–715.

Waples, R. S. 1989. A generalised approach for estimating effective population size from temporal changes in allele frequency. *Genetics* **121**: 379–391.

Ward, P. I. 1998. A possible explanation for cryptic female choice in the yellow dung fly, *Scathophaga stercoraria*. *Ethology* **104**: 97–110.

Ward, R. D., Appleyard, S. A., Daley, R. K. and Reilly, A. 2001. Population structure of pink ling (*Genypterus blacodes*) from south-eastern Australian water, inferred from allozyme and microsatellite analyses. *Marine and Freshwater Research* **52**: 965–973.

Waser, N. M. and Price, M. V. 1994. Crossing distance effects in *Delphinium nelsonii*: outbreeding and inbreeding depression in progeny fitness. *Evolution* **48**: 842–852.

Waser, N. M., Price, M. V. and Shaw, R. G. 2000. Outbreeding depression varies among cohorts of *Ipomopsis aggregata* planted in nature. *Evolution* **54**: 485–491.

Wasser, S. K., Shedlock, A. M., Comstock, K., Ostrander, E. A., Mutayoba, B. and Stephens, M. 2004. Assigning African elephant populations to geographic region of origin: Applications to the ivory trade. *Proceedings of the National Academy of Sciences USA* **101**: 14847–14852.

Waterston, R. H., Lindblad-Toh, K., Birney, E., Rogers, J., Abril, J. F., Agarwal, P., Agarwala, R., Ainscough, R., Alexandersson, M. and An, P. 2002. Initial sequencing and comparative analysis of the mouse genome. *Nature* **420**: 520–562.

Watt, W. B., Wheat, C. W., Meyer, E. H. and Martin, J. F. 2003. Adaptation at specific loci. VII. Natural selection, dispersal and the diversity of molecular-functional variation patterns among butterfly species complexes (*Colias* : Lepidoptera, Pieridae). *Molecular Ecology* **12**: 1265–1275.

Weir, B. S. and Cockerham, C. C. 1984. Estimating *F*-statistics for the analysis of population structure. *Evolution* **38**: 1358–1370.

Welsh, J. and McClelland, M. 1990. Fingerprinting genomes using PCR with arbitrary primers. *Nucleic Acids Research* **18**: 7213–7218.

Wennerberg, L. 2001. Breeding origin and migration pattern of dunlin (*Calidris alpina*) revealed by mitochondrial DNA analysis. *Molecular Ecology* **10**: 1111–1120.

Westerdahl, H., Hansson, B., Bensch, S. and Hasselquist, D. 2004. Between-year variation of MHC allele frequencies in great reed warblers: selection or drift? *Journal of Evolutionary Biology* **17**: 485–492.

Wetton, J. H., Tsang, C. S. F., Roney, C. A. and Spriggs, A. C. 2002. An extremely sensitive species-specific ARMS PCR test for the presence of tiger bone DNA. *Forensic Science International* **126**: 137–144.

Wheelwright, N. T. and Mauck, R. A. 1998. Philopatry, natal dispersal and inbreeding avoidance in an island population of Savannah Sparrows. *Ecology* **79**: 755–767.

Whitlock, M. C. and McCauley, D. E. 1999. Indirect measures of gene flow and migration: $F_{ST} \neq 1/(4Nm+1)$. *Heredity* **82**: 117–125.

Wiley, R. H. 1973. Territoriality and non-random mating in the sage grouse *Centrocercus urophasianus*. *Animal Behaviour Monographs* **6**: 87–169.

Wilkinson-Herbots, H. M. and Ettridge, R. 2004. The effect of unequal migration rates on $F_{ST}$. *Theoretical Population Biology* **66**: 185–197.

Williams, B. L. 2002. Conservation genetics, extinction, and taxonomic status: a case history of the regal fritillary. *Conservation Biology* **16**: 148–157.

Williams, B. L., Brawn, J. D. and Paige, K. N. 2003. Landscape scale genetic effects of habitat fragmentation on a high gene flow species: *Speyeria idalia* (Nymphalidae). *Molecular Ecology* **12**: 11–20.

Williams, J. G. K., Kubelik, A. R., Livak, K. J., Rafalski, J. A. and Tingey, S. V. 1990. DNA polymorphisms amplified by arbitrary primers are useful as genetic markers. *Nucleic Acids Research* **18**: 6531–6535.

Williams, S. L. 2001. Reduced genetic diversity in eelgrass transplantations affects both population growth and individual fitness. *Ecological Applications* **11**: 1472–1488.

Willis, K. J., Rudner, E. and Sumegi, P. 2000. The full-glacial forests of central and southeastern Europe. *Quaternary Research* **53**: 203–213.

Wilson, A. C., Cann, R. L., Carr, S. M., George, M., Gyllensten, U. B., Helm-Bychowski, K. M., Higuchi, R. G., Palumbi, S. R., Prager, E. M., Sage, R. D. and Stoneking, M. 1985. Mitochondrial DNA and two perspectives on evolutionary genetics. *Biological Journal of the Linnean Society* **26**: 375–400.

Wilson, G. A. and Rannala, B. 2003. Bayesian inference of recent migration rates using multilocus genotypes. *Genetics* **163**: 1177–1191.

Wilson, G. A. and Strobeck, C. 1999. Genetic variation within and relatedness among wood and plains bison populations. *Genome* **42**: 483–496.

Wilson, N., Tubman, S. C., Eady, P. E. and Robertson, G. W. 1997. Female genotype affects male success in sperm competition. *Proceedings of the Royal Society of London Series B* **264**: 1491–1495.

Wirth, T. and Bernatchez, L. 2001. Genetic evidence against panmixia in the European eel. *Nature* **409**: 1037–1040.

Witzell, W. N., Bass, A. L., Bresette, M. J., Singewald, D. A. and Gorham, J. C. 2002. Origin of immature loggerhead sea turtles (*Caretta caretta*) at Hutchinson Island, Florida: evidence from mtDNA markers. *Fishery Bulletin* **100**: 624–631.

Wolfe, K. H., Li, W.-H. and Sharp, P. M. 1987. Rates of nucleotide substitution vary greatly among plant mitochondrial, chloroplast and nuclear DNAs. *Proceedings of the National Academy of Sciences USA* **84**: 9054–9058.

Wolfe, K. H., Sharp, P. M. and Li, W.-H. 1989. Rates of synonymous substitutions in plant nuclear genes. *Journal of Molecular Evolution* **29**: 208–211.

Wolff, J. O. and Macdonald, D. W. 2004. Promiscuous females protect their offspring. *Trends in Ecology and Evolution* **19**: 127–134.

Woodward, S. R., Weyand, N. J. and Bunnell, M. 1994. DNA sequence from cretaceous period bone fragments. *Science* **266**: 1229–1232.

Wright, S. 1931. Evolution in Mendelian populations. *Genetics* **16**: 97–159.

Wright, S. 1943. Isolation by distance. *Genetics* **28**: 139–156.

Wright, S. 1951. The genetical structure of populations. *Annals of Eugenics* **15**: 323–354.

Wright, S. 1969. *Evolution and the Genetics of Populations, Vol. 2: The Theory of Gene Frequencies*. University of Chicago Press, Chicago.

Yalden, D. 1999. *The History of British Mammals*. Academic Press, London

Yu, J., Hu, S., Wang, J., Wong, G. K., Li, S., Liu, B., Deng, Y., Dai, L., Zhou, Y., Zhang, X., *et al.* 2002. A draft sequence of the rice genome (*Oryza sativa* L. ssp. *indica*). *Science* **296**: 79–92.

Yue, G. H. and Orban, L. 2001. Rapid isolation of DNA from fresh and preserved fish scales for polymerase chain reaction. *Marine Biotechnology* **3**: 199–204.

Zaidi, R. H., Jaal, Z., Hawkes, N. J., Hemingway, J. and Symondson, W. O. C. 1999. Can multiple-copy sequences of prey DNA be detected amongst the gut contents of invertebrate predators? *Molecular Ecology* **8**: 2081–2087.

Zane, L., Ostellari, L., Maccatrozzo, L., Bargelloni, L., Cuzin-Roudy, J., Buchholz, F. and Patarnello, T. 2000. Genetic differentiation in a pelagic crustacean (*Meganyctiphanes norvegica*: Euphausiacea) from the North East Atlantic and the Mediterranean Sea. *Marine Biology* **136**: 191–199.

Zenger, K. R., Eldridge, M. D. B. and Cooper, D. W. 2003. Intraspecific variation, sex-biased dispersal and phylogeography of the eastern grey kangaroo (*Macropus giganteus*). *Heredity* **91**: 153–162.

Zerjal, T., Xue, Y., Bertorelle, G., Wells, R. S., Bao, W., Zhu, S., Qamar, R., Ayub, Q., Mohyuddin, A., Fu, S., Li, P., Yuldasheva, A., *et al.* 2003. The genetic legacy of the Mongols. *American Journal of Human Genetics* **72**: 717–721.

Zhang, X., Leung, F. C., Chang, D. K. O., Chen, Y. and Wu, C. 2002. Comparative analysis of allozyme, random amplified polymorphic DNA and microsatellite polymorphism on Chinese native chickens. *Poultry Science* **81**: 1093–1098.

Zigler, K. S. and Lessios, H. A. 2004. Speciation on the coasts of the new world: phylogeography and the evolution of bindin in the sea urchin genus *Lytechinus*. *Evolution* **58**: 1225–1241.

Zink, R. M. 2004. The role of subspecies in obscuring avian biological diversity and misleading conservation policy. *Proceedings of the Royal Society of London Series B* **271**: 561–564.

Zischler, H., Höss, M., von Haeseler, A., van der Kuyl, A. C., Goudsmit, J. and Pääbo, S. 1995. Detecting dinosaur DNA. *Science* **268**: 1192–1193.

Zschokke, S. and Baur, B. 2002. Inbreeding, outbreeding, infant growth and size dimorphism in captive Indian rhinoceros (*Rhinoceros unicornis*). *Canadian Journal of Zoology* **80**: 2014–2023.

Zuckerkandl, E. and Pauling, L. 1965. Evolutionary divergence and convergence in proteins. In *Evolving Genes and Proteins* (Bryson, V. & Vogel, H. J., eds), pp. 97–166. Academic Press, New York.

REFERENCES

# Index

*Molecular Ecology*   Joanna Freeland
© 2005 John Wiley & Sons, Ltd.